Microbial Biofuel

Microbial Biofuel: A Sustainable Source of Renewable Energy explores microbial biofuel production from a technical standpoint addressing a wide range of topics including bio alcohol, biodiesel, biohydrogen, biomethane, biohythane, jet fuel, drop-in fuel, bioelectricity, bio-oil, biomass to biofuel, carbon capture, and more.

Each chapter provides an in-depth examination of a specific biofuel type, discussing the underlying science, production processes, challenges, and its potential applications. The title draws examples from the latest research and advancements in the field, including cutting-edge technologies, methodologies, and case studies. It covers advances in fermentation strategies and commercial-scale implementation of microbial technology for biofuel production along with comprehensive information on bio alcohol, biodiesel, biohydrogen, biomethane, etc. The book explores practical applications of microbial biofuels and uses real-life examples and case studies. Moreover, the book discusses sustainability and environmental benefits of using renewable energy.

The title is an ideal read for graduate students and researchers specialising in bioenergy and chemical engineering.

Microbial Biofuel

A Sustainable Source of Renewable Energy

Edited by
Shashi Kant Bhatia,
Parmjit Singh Panesar, and
Ranjit Gurav

CRC Press
Taylor & Francis Group
Boca Raton London New York

CRC Press is an imprint of the
Taylor & Francis Group, an **informa** business

Designed cover image: Toa55 [ShutterStock ID: 2374858505]

First edition published 2025
by CRC Press
2385 NW Executive Center Drive, Suite 320, Boca Raton FL 33431

and by CRC Press
4 Park Square, Milton Park, Abingdon, Oxon, OX14 4RN

CRC Press is an imprint of Taylor & Francis Group, LLC

ISBN: 978-1-032-66203-9 (hbk)
ISBN: 978-1-032-95543-8 (pbk)
ISBN: 978-1-003-58539-8 (ebk)

DOI: 10.1201/9781003585398

Typeset in Times
by SPi Technologies India Pvt Ltd (Straive)

Contents

Editors

Shashi Kant Bhatia is Associate Professor in the Department of Biological Engineering, Konkuk University, Seoul, South Korea, and has more than ten years' experience in biowaste valorization into bioenergy, biochemicals, and biomaterials. He holds an MSc and a PhD in biotechnology from Himachal Pradesh University (India). Dr. Bhatia has worked as a Brain Pool Post Doc Fellow at Konkuk University (2014–2016) and has contributed extensively to the industrial press and served as an editorial board member of the *Sustainability* and *Energies* journals and an associate editor of *Frontier of Microbiology*, *Microbial Cell Factories*, *Biomass Conversion and Biorefinery*, *Bioprocess and Biosystems Engineering*, *Carbohydrate Technology and Applications*, *3 Biotech*, and *PLOS One* journals. He is Editor in Chief of the *Biotechnology for Sustainable Materials* journal. He has published more than 200 research and review articles on industrial biotechnology, bioenergy production, biomaterial, biotransformation, microbial fermentation, and enzyme technology in international scientific peer-reviewed journals and holds 15 international patents. He has also edited four books on microbial biotechnology. Dr. Bhatia has successfully supervised and completed three projects funded by NRF Korea ($600,000).

Parmjit S. Panesar is working as Professor at Sant Longowal Institute of Engineering and Technology, Longowal, Punjab, India. Panesar has also worked as Dean (Research and Consultancy) and Head of the Department of Food Engineering & Technology, SLIET Longowal, India. His research is focused on the area of Food Biotechnology, especially bioprocessing of food industry by-products, food enzymes, prebiotics (lactulose, GOS), biopigments, etc. In 2005, he was awarded a BOYSCAST (Better Opportunities for Young Scientists in Chosen Areas of Science and Technology) fellowship by the Department of Science & Technology (DST), Government of India, to carry out advanced research at Chembiotech labs, University of Birmingham Research Park, UK. Professor Panesar has successfully completed seven research projects funded by DBT, CSIR, MHRD, AICTE, and New Delhi and 20 projects are in progress. He has published more than 200 international/national scientific papers, 50 book reviews in peer-reviewed journals, and 32 chapters and has authored/edited 16 books. He has guided 19 PhD students and eight students are under progress. He is a member of the editorial advisory boards of national/international journals including the *International Journal of Biological Macromolecules*, *Journal of Food Science & Technology*, and *Carbohydrate Polymer Technologies & Applications*. In recognition of his work, Panesar was awarded the 2018 Fellow Award by the Biotech Research Society of India (BRSI) and the 2019 Fellow Award by the National Academy of Dairy Science India (NADSI). He was also selected for the award of prestigious INSA Teachers Award by the Indian National Science Academy (INSA) in 2020. He has served as the National Vice President of the Association of

Food Scientists & Technologists of India (AFSTI) in 2019. He has also visited several countries like the UK, the USA, Canada, Switzerland, New Zealand, Australia, Germany, France, Singapore, Malaysia, China, Iran, and Thailand.

Ranjit Gurav is Associate Professor at UPES Dehradun, India. He received his PhD in Biotechnology in 2013 and has since conducted extensive research for a decade in various countries, including the USA, South Korea, India, and China. Throughout his career, he has held roles such as associate professor, research professor, assistant professor, postdoctoral research associate, and young scientist, contributing significantly to advancements in his field. His research primarily focuses on harnessing agriculture and food industry waste biomass, and by-products to produce bioelectricity, polyhydroxyalkanoates, biodiesel, exopolysaccharides, and biohydrogen. Dr. Gurav is highly skilled in implementing microbial, bioelectrochemical, hydrogel, biochar, and self-activated carbon-based technologies for the remediation of organic contaminants and heavy metals in water and soil. Using a microbial system, he has established a unique technology for converting poultry feather biomass into a growth stimulator, and chitin biomass into an antifungal agent. In addition, he developed green infrastructure to manage urban stormwater runoff through biologically active pervious structures. Gurav has secured research funding totalling $100,000 from organizations such as the NSF I-Corps of the United States, FIAP of Texas State University, and DST-SERB of India. Gurav is deeply involved in the academic community, serving as an associate editor, academic editor, topical editor, and editor for ten international journals and reviewer for over 70 international journals. He holds life memberships with the Biotech Research Society of India and the Association of Microbiologists of India.

Contributors

Siddhika Ajmera
Department of Chemical Engineering
Manipal Institute of Technology
Manipal Academy of Higher Education
Manipal, India

Eunice O. Babatunde
Ingram School of Engineering
Texas State University
San Marcos, Texas

Ravi Kant Bhatia
Department of Biotechnology
Himachal Pradesh University
Shimla, India

Shashi Kant Bhatia
Department of Biological Engineering
Konkuk University
Seoul, South Korea

Paramjeet Dhull
Department of Environmental Science & Engineering
Guru Jambheshwar University of Science & Technology
Hisar, India

Ranjit Gurav
Sustainability Cluster, School of Advanced Engineering
UPES Dehradun
Dehradun, India

Hayat Ali Haddad
Department of Chemical Engineering
King Fahd University of Petroleum & Minerals
Dhahran, Saudi Arabia

Sangchul Hwang
Ingram School of Engineering
Texas State University
San Marcos, Texas

Jyoti Jadhav
Department of Biotechnology
Shivaji University
Kolhapur, India
and
Department of Biotechnology
Willingdon College
Sangli, India

Rahul Jadhav
Department of Biotechnology
Shivaji University
Kolhapur, India
and
Department of Biotechnology
Willingdon College
Sangli, India

Yura Kim
Department of Chemical Engineering
Pukyong National University
Busan, South Korea

D. Jaya Prasanna Kumar
Department of Chemical Engineering
M.S. Ramaiah Institute of Technology
Bangalore, India

Pradeep Kumar
Department of Forensic Science
Himachal Pradesh University
Shimla, India

Sachin Kumar
Biochemical Conversion Division
Sardar Swaran Singh National Institute of Bio-Energy
Kapurthala, India

Priyanka Magadum
Department of Environment Science
College of Non-Conventional Vocational Courses for Women
Kolhapur, India

Sujata Mandal
Ingram School of Engineering
Texas State University
San Marcos, Texas

Ranjeet Kumar Mishra
Department of Chemical Engineering
Manipal Institute of Technology, Manipal Academy of Higher Education
Manipal, India

Yash Misra
Department of Chemical Engineering
M.S. Ramaiah Institute of Technology
Bangalore, India

Akansha Mohanty
Department of Chemical Engineering
Manipal Institute of Technology
Manipal Academy of Higher Education
Manipal, India

Ahmad Nawaz
Interdisciplinary Research Center for Refining and Advanced Chemicals
King Fahd University of Petroleum & Minerals
Dhahran, Saudi Arabia

Rajee Olaganathan
Aviation Department
2000 University Ave
Dubuque, Iowa

Parmjit S. Panesar
Department of Food Engineering and Technology
Sant Longowal Institute of Engineering and Technology
Longowal, India

Anmol Patil
Department of Biotechnology
Shivaji University, Vidyanagar
Kolhapur, India
and
Department of Biotechnology
Smt. Kasturbai Walchand College
Sangli, India

Shaikh Abdur Razzak
Interdisciplinary Research Center for Refining and Advanced Chemicals
King Fahd University of Petroleum & Minerals
and
Department of Chemical Engineering
King Fahd University of Petroleum & Minerals
Dhahran, Saudi Arabia

Neha Saini
Department of Environmental Science & Engineering
Guru Jambheshwar University of Science & Technology
Hisar, India

Ravi Sankannavar
School of Chemical and Materials Sciences
Indian Institute of Technology Goa
Ponda, India

Falak Shaheen
National Institute of Oceanography
Panaji, India

Naimi Sirjohn
Faculty of Applied Sciences and Biotechnology
Shoolini University of Biotechnology and Management Sciences
Solan, India

Ayah Stif
Department of Chemical Engineering
OnDokuz Mayis University
Atakum/Samsun, Turkiye

Sugandhi
Dr. S.S. Bhatnagar University Institute of Chemical Engineering and Technology
Panjab University
Chandigarh, India

Rohit Topkar
Department of Biotechnology
Shivaji University
Kolhapur, India
and
Department of Biotechnology
Willingdon College
Sangli, India

Sarita Wadmare
Department of Biotechnology
Willingdon College
and
Department of Microbiology
Willingdon College
Sangli, India

Ying Xu
State Key Laboratory of Pollution Control and ResourceReuse
School of Environmental Science and Engineering
Tongji University
and
Shanghai Institute of Pollution Control and Ecological Security
Shanghai, China

Yung-Hun Yang
Department of Biological Engineering
College of Engineering, Konkuk University
Seoul, South Korea

Chen Zhang
Shanghai Institute of Pollution Control and Ecological Security
Shanghai, China

Preface

Welcome to *Microbial Biofuel: A Sustainable Source of Renewable Energy*. As the global energy demand continues to escalate and concerns over climate change mount, the imperative to find sustainable solutions has never been more pressing. Fossil fuels, the cornerstone of our energy infrastructure, are finite resources, with projections suggesting their depletion within decades. It is evident that alternative, renewable energy sources are essential for meeting future demands while mitigating environmental impact. In this book, we embark on a comprehensive exploration of microbial biofuels, showcasing their potential as a sustainable pathway to address the world's energy challenges. This book includes 10 chapters, and each chapter explores a distinct facet of microbial biofuel technology, offering insights into its production, applications, and potential impact on energy sustainability. Chapter 1, Introduction to microbial biofuel, sets the stage by providing a comprehensive overview of the subject matter, highlighting the urgent need for sustainable energy solutions in the face of depleting fossil fuel reserves. Chapter 2, Bioalcohol production, explores the fermentation processes that transform nature's bounty into bioalcohols, offering renewable alternatives for transportation and industrial applications. Chapter 3, Biodiesel production, discusses about conversion of organic oils into biodiesel, offering cleaner alternatives for diesel engines while reducing greenhouse gas emissions. Chapter 4, Biohydrogen production, harnesses microbial activity to generate clean energy, presenting a promising avenue for hydrogen production from renewable sources. Chapter 5, Biomethane production, showcases the microbial conversion of organic waste into green gas, offering renewable alternatives for natural gas and reducing reliance on fossil fuels. Chapter 6, Biohythane production, merges biohydrogen and biomethane production processes, offering a sustainable solution with enhanced energy content and combustion characteristics. Chapter 7, Jet fuel production explores the microbial innovation driving the development of sustainable aviation fuels, essential for reducing emissions in the aviation industry. Chapter 8, Drop-in fuel production, focuses on the transformation of biomass into next-generation fuels compatible with existing infrastructure, facilitating a smoother transition to renewable energy sources. Chapter 9, Bioelectricity generation, showcases microbial fuel cells to produce electricity from organic feedstocks, offering sustainable alternatives for power generation and environmental remediation. Chapter 10, Bio-oil production, unveils the potential of converting biomass into liquid bio-oil, offering renewable substitutes for petroleum-based fuels and chemicals.

Through meticulous research, technological innovation, and collaborative efforts, scientists and engineers worldwide are unlocking the latent potential of microbial bioconversion, paving the way for a more sustainable and resilient energy future. This book serves as a comprehensive resource for students, academics, industry professionals, and policymakers alike, inspiring curiosity, fostering innovation, and catalysing meaningful contributions towards a greener and more sustainable world.

Shashi Kant Bhatia
Parmjit Singh Panesar
Ranjit Gurav
Editors

1 Introduction to Microbial Biofuels

Nature's Solution to Energy Sustainability

Shashi Kant Bhatia, Parmjit S. Panesar, and Ranjit Gurav

Global energy demand is continually increasing due to ever-growing populations and industrialization which ultimately enforces over-exploitation of various fossil-based energy resources (Bhatia et al., 2021c). The traditional fossil-based resources (oil, coal, and natural gas) fulfill almost 80% of energy demand and are available in limited amounts. It is expected that all these resources will be extinct in the next few decades, i.e., oil in 40 years, coal in 250 years, and natural gas in 60 years (Bhatia et al., 2023). There is an urgency to find sustainable and alternative energy resources to meet future energy demands. While renewable energy sources such as solar, wind, and hydroelectric power have gained attention, their intermittent nature poses challenges to achieving a consistent energy supply (Ahuja et al., 2024). In this context, the potential of harnessing microbial activity as a viable energy solution emerges as a compelling option. Microorganisms possess remarkable metabolic capabilities that enable them to convert organic matter into energy-rich compounds (bioalcohols, biodiesel, biohydrogen, and methane) through processes like fermentation and anaerobic digestion (Figure 1.1) (Cherwoo et al., 2023).

This inherent ability makes microbes a promising candidate for generating biofuels and biogas from a wide range of feedstocks, including agricultural waste, organic residues, and wastewater (Bhatia et al., 2018; Raj et al., 2022). Moreover, microbial-based energy production offers the advantage of scalability, adaptability to diverse environmental conditions, and the potential for decentralized deployment, thereby presenting a sustainable and resilient pathway toward meeting the world's growing energy needs. Through meticulous research and technological innovation, scientists and engineers have been able to unlock the latent potential of microbial bioconversion, harnessing nature's own biochemical machinery to address the pressing challenges of energy sustainability. According to the Scopus database, around 4200 research articles were published in 2023. The top five countries involved in the microbial biofuel research area include China (961), India (799), the United States (495), Brazil (265), and Saudi Arabia (207) (Figure 1.2).

DOI: 10.1201/9781003585398-1

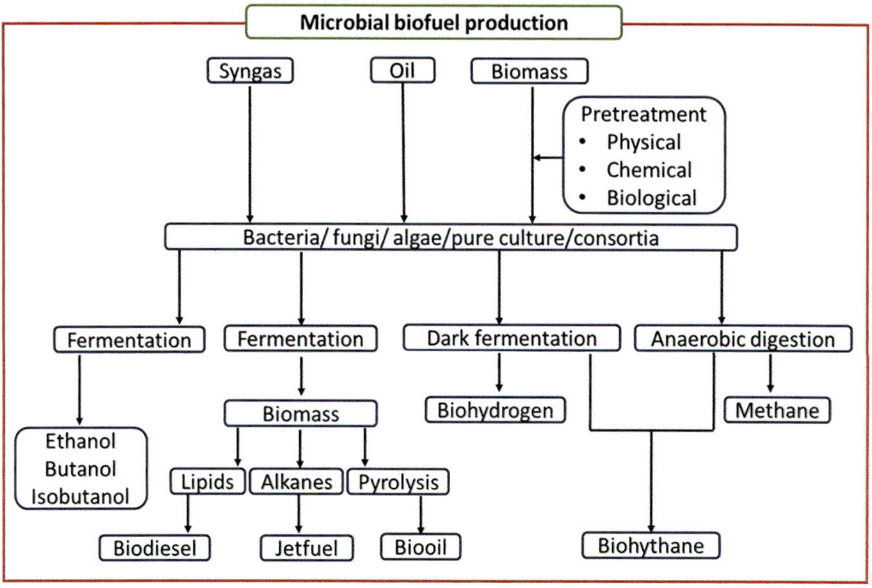

FIGURE 1.1 Microbial fuel production from different feedstocks.

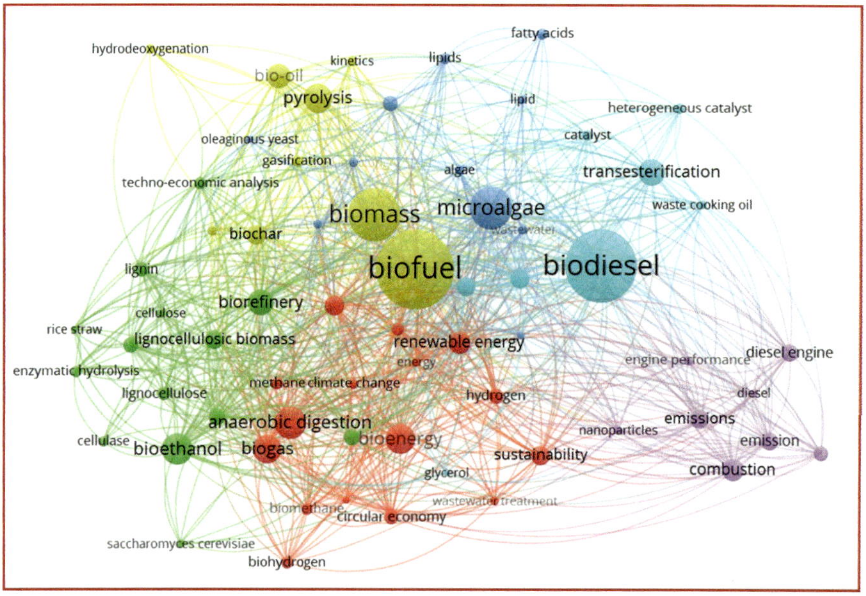

FIGURE 1.2 Co-occurrence mapping of keywords from publications related to microbial biofuel (min. number of occurrence 20).

Unlike traditional fossil fuels, bioalcohols (ethanol and butanol) offer a renewable and cleaner alternative that contributes to mitigating greenhouse gas emissions. A majority of ethanol at the industrial scale is produced using the yeast *Saccharomyces cerevisiae* due to its high tolerance to ethanol (da Silva Fernandes et al., 2022). Moreover, bioalcohols possess higher octane ratings and oxygen content, which can enhance the efficiency of combustion engines and reduce harmful emissions such as carbon monoxide and particulate matter (Khan et al., 2022). In addition to their stand-alone use as fuels, bioalcohols can be blended with conventional gasoline or diesel, offering a pathway to reduce the carbon intensity of transportation fuels. Ethanol, for instance, can be blended with gasoline to create ethanol-gasoline blends such as E10 (10% ethanol, 90% gasoline) or E85 (85% ethanol, 15% gasoline) (Zhao & Wang, 2020). Similarly, butanol exhibits compatibility with existing infrastructure and can be blended with gasoline at higher concentrations due to its lower hygroscopicity and higher energy content compared to ethanol (Xue et al., 2013). This versatility in blending options allows for seamless integration into existing transportation systems, offering consumers a greener choice without requiring significant modifications to vehicles or refueling infrastructure.

A variety of microbes such as *Yarrowia lipolytica*, *Rhodococcus opacus*, *Chlorella pyrenoidosa*, and *Botryococcus braunii* are able to accumulate lipids which can be extracted using various methods and further transformed into biodiesel through transesterification (Bhatia et al., 2017). Transesterification can be carried out using enzymes (lipase) or chemical based (acid and alkali) as a catalyst. The properties of biodiesel depend on the fatty acids compositions, chain length, and degree of unsaturation. Burning of biodiesel does not contribute to net atmospheric CO_2 levels, because it originates from renewable carbon sources and has high oxygen content (Bhatia et al., 2021b). The compatibility of biodiesel with existing diesel engines and infrastructure, alongside its reduced emissions profile, underscores its potential for widespread adoption in the transportation sector. Furthermore, biodiesel exhibits superior lubricating properties compared to petroleum diesel, which can enhance engine longevity and reduce maintenance costs.

Hydrogen is another form of energy source and can be produced through various technologies including photolysis, thermochemical reactions like pyrolysis and gasification of biomass, photo fermentation, and dark fermentation (Brar et al., 2022; Hwang et al., 2024). Biohydrogen production using fermentation (*Enterobacter*, *Clostridium*, *Shewanella*, and *Bacillus*) has garnered attention due to their ability to utilize a wide range of organic waste materials (Bhatia et al., 2023; Hwang et al., 2024; Kim et al., 2023). However, the amount of biohydrogen produced depends on factors such as the type of strain used and the substrate's disintegration rate, influenced by competing microorganisms like methanogens and acetogens (Kumar et al., 2022). Although pure cultures yield higher hydrogen, mixed cultures provide versatility and easier control. Lignocellulosic biomass, a plentiful and inexpensive raw material, holds the potential for producing high-energy-density biohydrogen through dark fermentation (Bhatia et al., 2021a). Yet its complex structure, containing lignin, poses challenges for enzyme access to hydrolyzable sugars, necessitating pretreatment methods like acids, alkalis, or advanced techniques such as ionic liquids or supercritical treatments. Overall, biohydrogen production from lignocellulosic biomass via dark fermentation offers a promising pathway toward sustainable energy generation, despite the challenges associated with biomass recalcitrance and pretreatment requirements (Kang et al., 2023).

Biomethane production from organic waste involves a series of steps in the anaerobic digestion process, including hydrolysis, acidogenesis, acetogenesis, and methanogenesis, each contributing to the conversion of complex organic compounds into methane-rich biogas (Saravanakumar et al., 2023). Initially, hydrolytic bacteria break down complex organic polymers such as carbohydrates, proteins, and lipids into simpler molecules like sugars, amino acids, and fatty acids through hydrolysis. These smaller molecules are then fermented by acidogenic bacteria during the acidogenesis phase, leading to the production of volatile fatty acids (VFAs), alcohols, and other organic acids. Subsequently, acetogenic bacteria convert VFAs into acetate, hydrogen, and carbon dioxide through acetogenesis. Finally, methanogenic archaea catalyze the conversion of acetate, hydrogen, and carbon dioxide into methane and carbon dioxide during methanogenesis (Bhatia et al., 2018). Throughout these sequential biochemical processes, biogas rich in methane is produced, which can be further purified to produce biomethane suitable for injection into natural gas pipelines or use as a transportation fuel. This biomethane is harvested from sources such as municipal solid waste, agricultural residues, or wastewater sludge. Biomethane production not only generates renewable energy but also mitigates greenhouse gas emissions and reduces reliance on finite fossil fuels, contributing to a more sustainable and environmentally friendly energy future (Haldar et al., 2023). Beyond its intrinsic energy benefits, biomethane production offers a sustainable solution to waste management, effectively reducing greenhouse gas emissions.

Hythane, a blend of hydrogen and methane, combines the advantages of both, offering improved combustion characteristics and reduced emissions compared to pure biomethane (Abdur Rawoof et al., 2021). Microbial production of hythane typically involves the anaerobic digestion of organic waste materials, similar to biomethane production, but with specific modifications to favor the coproduction of hydrogen and methane. By adjusting process parameters such as reactor configuration, substrate composition, and microbial consortium, microbial systems can be optimized to enhance the production of hythane (Si et al., 2020). The advantages of hythane over biomethane include its higher hydrogen content, which improves combustion efficiency and reduces emissions of pollutants such as carbon monoxide, and unburned hydrocarbons. Additionally, hythane can be used as a fuel for internal combustion engines with minimal modifications, making it a versatile and cost-effective alternative to conventional fuels. Moreover, the production of hythane from organic waste helps mitigate greenhouse gas emissions and contributes to waste management efforts, aligning with sustainability goals. Overall, microbial production of hythane offers a promising pathway toward cleaner and more sustainable energy generation, leveraging the synergistic benefits of hydrogen and methane in a single fuel blend.

An ideal aviation fuel must possess excellent cold stability, capable of withstanding temperatures ranging from –47°C to 40°C, particularly crucial for flights at high altitudes exceeding 30,000 feet. Additionally, it should boast sufficient energy density to meet the substantial energy demands characteristic of long-haul flights (Doliente et al., 2020). Bio-jet fuel, derived from renewable biomass sources through microbial processes, presents a sustainable alternative to conventional aviation fuels. Microbial fermentation transforms biomass feedstocks into medium- and long-chain fatty acids and some of them are able to convert these fatty acids into alkanes and/or alkenes which are then refined into bio-jet fuel (Jiménez-Díaz et al., 2017). Bio-jet fuel is composed primarily of hydrocarbons having properties simliar to traditional fuels, including energy content, density, and freezing point, ensuring compatibility with aircraft performance and safety requirements. One of its key advantages lies in its significantly reduced carbon dioxide emissions compared to conventional fuels, contributing to efforts to combat climate change. Compatible with existing aircraft engines and infrastructure, bio-jet fuel enables seamless integration into commercial flights without requiring costly modifications. Diverse feedstocks, such as non-food crops, algae, and waste oils, can be utilized, minimizing competition with food production and optimizing resource utilization. Moreover, ongoing advancements in biotechnology and process optimization enhance energy efficiency and scalability, further cementing its viability as a sustainable energy solution for the aviation industry. With its renewable origins, environmental benefits, and compatibility with existing infrastructure, bio-jet fuel produced using microbial processes represents a pivotal step toward a greener and more sustainable future for aviation.

Meanwhile, drop-in biofuels emerge as pivotal players in the transition toward sustainable energy systems. These biofuels, which offer compatible alternatives to traditional petroleum-based fuels, seamlessly integrate with existing infrastructure and vehicles, minimizing the need for costly infrastructure upgrades and facilitating a smoother transition to renewable energy sources (Martinez-Villarreal et al., 2022). Researchers are exploring a myriad of biomass-derived feedstocks and conversion technologies to develop drop-in biofuels such as renewable diesel and gasoline, further reducing humanity's dependence on finite fossil resources and mitigating the environmental impact associated with transportation.

Microorganisms can also be used to produce electricity. Microbial fuel cells (MFCs) utilize microbial cells to generate electricity, typically consisting of an anodic chamber where microbes grow and produce electrons through metabolic processes, and a cathodic chamber where electrons combine with protons and oxygen to form water (Vinayak et al., 2021). Various organic feedstocks, such as wastewater, organic waste, or agricultural residues, are metabolized by microbes and serve as electron donors (Gurav et al., 2019a, 2021; Lee et al., 2021). Some MFCs employ mediators, such as redox-active compounds, to facilitate electron transfer from the microbial cells to the electrode surface, enhancing efficiency (Gurav et al., 2020). Additionally, biochar electrodes play a crucial role by providing a conductive surface for microbial attachment and facilitating electron transfer. Moreover, the porous structure of biochar enables pollutant removal through adsorption, further enhancing the environmental benefits of MFC technology. This integrated approach demonstrates the potential of MFCs for sustainable electricity generation, coproduction of valuable products, and environmental remediation (Gurav et al., 2019b).

Expanding the narrative further, the exploration of bio-oil production unveils yet another dimension of the biofuel landscape. Through processes such as pyrolysis and hydrothermal liquefaction, biomass (microbial and plants) is converted into a liquid bio-oil rich in aliphatic and aromatic hydrocarbons, phenolics, long-chain fatty acids, and nitrogenous compounds (Paul et al., 2022; Sekar et al., 2021). This bio-oil, with its versatility and compatibility with existing infrastructure, offers a renewable substitute for petroleum-based fuels and chemicals, further reducing humanity's reliance on finite fossil resources and mitigating the environmental impact associated with conventional energy production (Shahi et al., 2020). In conclusion, we can predict that microbes have the potential to fulfill the future energy demand.

REFERENCES

Abdur Rawoof, S.A., Kumar, P.S., Vo, D.-V.N., Devaraj, T., Subramanian, S. 2021. Biohythane as a high potential fuel from anaerobic digestion of organic waste: A review. *Renewable and Sustainable Energy Reviews*, 152, 111700.

Ahuja, V., Palai, A.K., Kumar, A., Patel, A.K., Farooque, A.A., Yang, Y.-H., Bhatia, S.K. 2024. Biochar: Empowering the future of energy production and storage. *Journal of Analytical and Applied Pyrolysis*, 177, 106370.

Bhatia, S.K., Bhatia, R.K., Yang, Y.-H. 2017. An overview of microdiesel — A sustainable future source of renewable energy. *Renewable and Sustainable Energy Reviews*, 79, 1078–1090.

Bhatia, S.K., Jagtap, S.S., Bedekar, A.A., Bhatia, R.K., Rajendran, K., Pugazhendhi, A., Rao, C.V., Atabani, A.E., Kumar, G., Yang, Y.-H. 2021a. Renewable biohydrogen production from lignocellulosic biomass using fermentation and integration of systems with other energy generation technologies. *Science of the Total Environment*, 765, 144429.

Bhatia, S.K., Joo, H.-S., Yang, Y.-H. 2018. Biowaste-to-bioenergy using biological methods – A mini-review. *Energy Conversion and Management*, 177, 640–660.

Bhatia, S.K., Kant Bhatia, R., Jeon, J.-M., Pugazhendhi, A., Kumar Awasthi, M., Kumar, D., Kumar, G., Yoon, J.-J., Yang, Y.-H. 2021b. An overview on advancements in biobased transesterification methods for biodiesel production: Oil resources, extraction, biocatalysts, and process intensification technologies. *Fuel*, 285, 119117.

Bhatia, S.K., Mehariya, S., Bhatia, R.K., Kumar, M., Pugazhendhi, A., Awasthi, M.K., Atabani, A.E., Kumar, G., Kim, W., Seo, S.-O., Yang, Y.-H. 2021c. Wastewater based microalgal biorefinery for bioenergy production: Progress and challenges. *Science of The Total Environment*, 751, 141599.

Bhatia, S.K., Rajesh Banu, J., Singh, V., Kumar, G., Yang, Y.-H. 2023. Algal biomass to biohydrogen: Pretreatment, influencing factors, and conversion strategies. *Bioresource Technology*, 368, 128332.

Brar, K.K., Cortez, A.A., Pellegrini, V.O.A., Amulya, K., Polikarpov, I., Magdouli, S., Kumar, M., Yang, Y.-H., Bhatia, S.K., Brar, S.K. 2022. An overview on progress, advances, and future outlook for biohydrogen production technology. *International Journal of Hydrogen Energy*, 47(88), 37264–37281.

Cherwoo, L., Gupta, I., Flora, G., Verma, R., Kapil, M., Arya, S.K., Ravindran, B., Khoo, K.S., Bhatia, S.K., Chang, S.W., Ngamcharussrivichai, C., Ashokkumar, V. 2023. Biofuels an alternative to traditional fossil fuels: A comprehensive review. *Sustainable Energy Technologies and Assessments*, 60, 103503.

da Silva Fernandes, F., de Souza É.S., Carneiro, L.M., Alves Silva, J.P., de Souza, J.V.B., da Silva Batista, J. 2022. Current Ethanol Production Requirements for the Yeast Saccharomyces cerevisiae. *International Journal of Microbiology*, 2022, 7878830.

Doliente, S.S., Narayan, A., Tapia, J.F.D., Samsatli, N.J., Zhao, Y., Samsatli, S. 2020. Bio-aviation fuel: A comprehensive review and analysis of the supply chain components. *Frontiers in Energy Research*, 8, 110.

Gurav, R., Bhatia, S.K., Choi, T.-R., Jung, H.-R., Yang, S.-Y., Song, H.-S., Park, Y.-L., Han, Y.-H., Park, J.-Y., Kim, Y.-G., Choi, K.-Y., Yang, Y.-H. 2019a. Chitin biomass powered microbial fuel cell for electricity production using halophilic Bacillus circulans BBL03 isolated from sea salt harvesting area. *Bioelectrochemistry*, 130, 107329.

Gurav, R., Bhatia, S.K., Choi, T.-R., Kim, H.-J., Lee, H.-J., Cho, J.-Y., Ham, S., Suh, M.-J., Kim, S.-H., Kim, S.-K., Yoo, D.-W., Yang, Y.-H. 2021. Seafood processing chitin waste for electricity generation in a microbial fuel cell using halotolerant catalyst Oceanisphaera arctica YHY1. *Sustainability*, 13(15), 8508.

Gurav, R., Bhatia, S.K., Choi, T.-R., Kim, H.J., Song, H.-S., Park, S.-L., Lee, S.-M., Lee, H.-S., Kim, S.-H., Yoon, J.-J., Yang, Y.-H. 2020. Utilization of different lignocellulosic hydrolysates as carbon source for electricity generation using novel Shewanella marisflavi BBL25. *Journal of Cleaner Production*, 277, 124084.

Gurav, R., Bhatia, S.K., Moon, Y.-M., Choi, T.-R., Jung, H.-R., Yang, S.-Y., Song, H.-S., Jeon, J.-M., Yoon, J.-J., Kim, Y.-G., Yang, Y.-H. 2019b. One-pot exploitation of chitin biomass for simultaneous production of electricity, n-acetylglucosamine and polyhydroxyalkanoates in microbial fuel cell using novel marine bacterium Arenibacter palladensis YHY2. *Journal of Cleaner Production*, 209, 324–332.

Haldar, D., Bhattacharjee, N., Shabbirahmed, A.M., Anisha, G.S., Patel, A.K., Chang, J.-S., Dong, C.-D., Singhania, R.R. 2023. Purification of biogas for methane enrichment using biomass-based adsorbents: A review. *Biomass and Bioenergy*, 173, 106804.

Hwang, J.H., Kim, H.J., Kim, H.J., Shin, N., Oh, S.J., Park, J.-H., Cho, W.-D., Ahn, J., Bhatia, S.K., Yang, Y.-H. 2024. Galactose-based biohydrogen production from seaweed biomass by novel strain Clostridium sp. JH03 from anaerobic digester sludge. *Biotechnology and Bioprocess Engineering*, 29, 219–231.

Jiménez-Díaz, L., Caballero, A., Pérez-Hernández, N., Segura, A. 2017. Microbial alkane production for jet fuel industry: Motivation, state of the art and perspectives. *Microbial Biotechnology*, 10(1), 103–124.

Kang, B.-J., Jeon, J.-M., Bhatia, S.K., Kim, D.-H., Yang, Y.-H., Jung, S., Yoon, J.-J. 2023. Two-stage bio-hydrogen and polyhydroxyalkanoate production: Upcycling of spent coffee grounds. *Polymers*, 15(3), 681.

Khan, M.M., Sharma, R.P., Kadian, A.K., Hasnain, S.M.M. 2022. An assessment of alcohol inclusion in various combinations of biodiesel-diesel on the performance and exhaust emission of modern-day compression ignition engines – A review. *Materials Science for Energy Technologies*, 5, 81–98.

Kim, S.H., Kim, H.J., Kim, S.H., Jung, H.J., Kim, B., Cho, D.H., Jeon, J.M., Yoon, J.J., Kim, S.H., Park, J.H., Bhatia, S.K., Yang, Y.H. 2023. Unraveling biohydrogen production and sugar utilization systems in the Electricigen Shewanella marisflavi BBL25. *Journal of Microbiology and Biotechnology*, 33(5), 687–697.

Kumar, M.D., Kavitha, S., Tyagi, V.K., Rajkumar, M., Bhatia, S.K., Kumar, G., Banu, J.R. 2022. Macroalgae-derived biohydrogen production: Biorefinery and circular bioeconomy. *Biomass Conversion and Biorefinery*, 12(3), 769–791.

Lee, S.M., Lee, H.-J., Kim, S.H., Suh, M.J., Cho, J.Y., Ham, S., Song, H.-S., Bhatia, S.K., Gurav, R., Jeon, J.-M., Yoon, J.-J., Choi, K.-Y., Kim, J.-S., Lee, S.H., Yang, Y.-H. 2021. Engineering of Shewanella marisflavi BBL25 for biomass-based polyhydroxybutyrate production and evaluation of its performance in electricity production. *International Journal of Biological Macromolecules*, 183, 1669–1675.

Martinez-Villarreal, S., Breitenstein, A., Nimmegeers, P., Perez Saura, P., Hai, B., Asomaning, J., Eslami, A.A., Billen, P., Van Passel, S., Bressler, D.C., Debecker, D.P., Remacle, C., Richel, A. 2022. Drop-in biofuels production from microalgae to hydrocarbons: Microalgal cultivation and harvesting, conversion pathways, economics and prospects for aviation. *Biomass and Bioenergy*, 165, 106555.

Paul, T., Sinharoy, A., Baskaran, D., Pakshirajan, K., Pugazhenthi, G., Lens, P.N.L. 2022. Bio-oil production from oleaginous microorganisms using hydrothermal liquefaction: A biorefinery approach. *Critical Reviews in Environmental Science and Technology*, 52(3), 356–394.

Raj, T., Chandrasekhar, K., Naresh Kumar, A., Rajesh Banu, J., Yoon, J.-J., Kant Bhatia, S., Yang, Y.-H., Varjani, S., Kim, S.-H. 2022. Recent advances in commercial biorefineries for lignocellulosic ethanol production: Current status, challenges and future perspectives. *Bioresource Technology*, 344, 126292.

Saravanakumar, A., Sudha, M.R., Chen, W.-H., Pradeshwaran, V., Ashokkumar, V., Selvarajoo, A. 2023. Biomethane production as a green energy source from anaerobic digestion of municipal solid waste: A state-of-the-art review. *Biocatalysis and Agricultural Biotechnology*, 53, 102866.

Sekar, M., Mathimani, T., Alagumalai, A., Chi, N.T.L., Duc, P.A., Bhatia, S.K., Brindhadevi, K., Pugazhendhi, A. 2021. A review on the pyrolysis of algal biomass for biochar and bio-oil – Bottlenecks and scope. *Fuel*, 283, 119190.

Shahi, T., Beheshti, B., Zenouzi, A., Almasi, M. 2020. Bio-oil production from residual biomass of microalgae after lipid extraction: The case of Dunaliella Sp. *Biocatalysis and Agricultural Biotechnology*, 23, 101494.

Si, B., Yang, H., Huang, S., Watson, J., Zhang, Y., Liu, Z. 2020. An innovative multistage anaerobic hythane reactor (MAHR): Metabolic flux, thermodynamics and microbial functions. *Water Research*, 169, 115216.

Vinayak, V., Khan, M.J., Varjani, S., Saratale, G.D., Saratale, R.G., Bhatia, S.K. 2021. Microbial fuel cells for remediation of environmental pollutants and value addition: Special focus on coupling diatom microbial fuel cells with photocatalytic and photoelectric fuel cells. *Journal of Biotechnology*, 338, 5–19.

Xue, C., Zhao, J., Liu, F., Lu, C., Yang, S.-T., Bai, F.-W. 2013. Two-stage in situ gas stripping for enhanced butanol fermentation and energy-saving product recovery. *Bioresource Technology*, 135, 396–402.

Zhao, L., Wang, D. 2020. Combined effects of a biobutanol/ethanol–gasoline (E10) blend and exhaust gas recirculation on performance and pollutant emissions. *ACS Omega*, 5(7), 3250–3257.

2 Bioalcohol Production

Fermenting Nature's Bounty into Liquid Gold

Naimi Sirjohn, Ravi Kant Bhatia, and Pradeep Kumar

2.1 INTRODUCTION

The quest for sustainable energy sources has intensified due to the drive for improved living standards and the need to meet global energy demands (Bhatia et al., 2018). Conversely, the reliance on fossil fuels as primary energy sources has led to pervasive issues like environmental degradation and climate change on a global scale. Bioethanol, alternatively referred to as ethyl alcohol or C_2H_5OH (Azhar et al., 2017), can be employed in its pure form or combined with gasoline, forming what is known as "gasohol." It serves various purposes, acting as a gasoline enhancer or octane booster. Additionally, when incorporated into bioethanol-diesel blends, it aids in diminishing the emission of exhaust gases (Razak et al., 2021). The ongoing expansion of the global economy has led to a parallel increase in energy consumption, heightening concerns about the accumulation of atmospheric greenhouse gases (GHGs) and their consequential impact on climate change (Ulucak, 2021). To address these challenges, numerous countries are actively investing in renewable energy solutions, with a particular focus on biofuel production. Biofuel refers to any fuel sourced from biomass, which includes plant or algae material as well as animal waste. What sets bioalcohol apart is its renewable nature (Afolalu et al., 2021), as the feedstock material can be easily replenished. This stands in contrast to nonrenewable fossil fuels like petroleum, coal, and natural gas. The appeal of bioalcohol lies in its recognition as a form of renewable energy. In broad terms, bioalcohol stands as a liquid biofuel (Khan et al., 2021) that can be generated from various feedstocks and via diverse conversion techniques. Its appeal lies in its renewable nature and its capacity to diminish particulate emissions in compression-ignition engines. Noteworthy attributes of bioalcohol, such as high-octane number, wider flammability limits, increased flame speed, and vaporization heat, equip it to be a competitive counterpart to fossil fuels in terms of efficiency (Baeyens et al., 2015). However, bioalcohol does have drawbacks, including its pronounced corrosiveness, low flame luminosity, and challenges related to vapor pressure and water miscibility. Bioalcohol has several advantages over conventional fuels. It comes from a renewable resource (Jin et al., 2011), i.e., crops (like cereals, sugar beet, and maize). Another benefit of fossil fuels is the GHG emissions. The road transport network accounts for 22% of all GHG emissions, and through the use of

DOI: 10.1201/9781003585398-2

bioalcohol, some of these emissions will be reduced as the fuel crops absorb the CO_2 they emit through growing (Sydney et al., 2019).

The concept of bioalcohol production is not novel; it was initiated in response to the global fuel crisis during the 1970s. The capacity of ethanol production witnessed substantial growth, starting from less than one billion liters in 1975, reaching 39 billion liters in 2006, and anticipated to reach 100 billion liters by mid-2015 (Johnson et al., 2020). This growth is attributed to the widespread applications of bioalcohol in various sectors, especially in transportation. Bioalcohol production is broadly categorized into first, second, and third generations based on diverse feedstocks, production technologies, and their respective levels of development (Zabed et al., 2017). Ethanol derived from food crops such as corn, wheat, barley, and sweet sorghum is termed grain alcohol. In contrast, ethanol produced from lignocellulosic biomass and agricultural residues like rice straw, wheat straw, and grasses is referred to as biomass ethanol or bioethanol (Ibeto et al., 2011). Sugarcane juice is not a pristine liquid; it contains various impurities like minerals, salts, acids, dirt, and fiber, alongside water and sugars. To transform it into an efficient raw material for ethanol production through fermentation, these impurities must be eliminated. Therefore, sugarcane juice undergoes both physical and chemical treatments to remove undesirable components (Panigrahi et al., 2021).

This is particularly relevant given the escalating prices of petroleum and growing apprehensions about the environmental impact of fossil fuels, notably their contribution to global warming. Microbial production of bioalcohols from renewable sources holds great promise as an alternative to conventional fuels derived from fossil origins (Sahni et al., 2021). Ethanol, extensively produced from crops like cereals and sugarcane, has been utilized as a blend with gasoline or as a stand-alone biofuel. However, alcohols with longer carbon chains, such as butanol, present even more favorable properties and better align with existing fuel distribution (Ahmed et al., 2024) infrastructure. Biofuel is often promoted as a cost-effective and environmentally friendly alternative to traditional fossil fuels, aligning with the broader efforts to transition toward sustainable and eco-conscious energy sources (Celińska et al., 2009). Different kinds of bioethanol exist and can be categorized as follows.

2.2 DIFFERENT TYPES OF BIOALCOHOL

2.2.1 Ethanol

The simple sugars released in the hydrolysis step can be easily converted to bioethanol with the help of a few microorganisms. Bioethanol is the main product of fermentation, along with a few by-products such as CO_2 and water. The most commonly used microorganism for bioethanol production from sugar-containing feedstocks is *Saccharomyces cerevisiae*, due to its ability to degrade sucrose into hexoses (glucose and fructose) (Vohra et al., 2014). The conversion of glucose and galactose into ethanol involves distinct pathways: the Embden-Meyerhof pathway of glycolysis for glucose and the Leloir pathway for galactose. In the Embden-Meyerhof pathway, glucose undergoes two major stages. First, the sugar is converted to a common intermediate, glucose-6-phosphate, and then in the second stage, this intermediate is converted into pyruvate. The final product of this pathway varies based on the microorganism used. In the case of yeast (Shen et al., 2019), for example, pyruvate is reduced to alcohol (ethanol) and CO_2 through a two-step process catalyzed by enzymes, known as alcoholic fermentation. The bioethanol production process involves two main approaches: separated hydrolysis and fermentation (SHF) and simultaneous saccharification and fermentation (SSF). SHF employs a two-step process, starting with hydrolysis to break down the feedstock into sugars using enzymes (Szambelan et al., 2018), followed by fermentation. However, SHF faces challenges with end-product inhibition due to sugars produced during hydrolysis. On the other hand, SSF combines hydrolysis and fermentation in a single step, efficiently converting sugars into bioethanol and overcoming end-product inhibition (Sudiyani et al., 2019). Recent research in acid hydrolysis has focused on seaweed and microalgae species, with sulfuric acid as the hydrolyzing agent. While acid hydrolysis has limitations in bioethanol yield, enzymatic hydrolysis, especially in SSF, shows promise with significantly higher yields, playing a crucial role in advancing bioethanol production (Karimi et al., 2021).

In comparison to acid hydrolysis, enzymatic hydrolysis offers a more viable path for the generation of bioethanol in the upcoming years. Only 0.390 g/g of bioethanol was produced at the greatest acid hydrolysis yield in macroalgae, compared to 0.520 g/g in microalgae. However, employing SSF technology, the enzymatic hydrolysis of macroalgae produced a bioethanol yield of 0.909 g/g (Robak et al., 2018). To a greater extent, the application of SHF and SSF technologies in hydrolysis and fermentation contributed significantly to the generation of bioethanol. The potential of different algae species to generate bioethanol is a crucial factor to consider. Microalgae and macroalgae, owing to variations in their cell wall compositions, exhibit diverse bioethanol yields based on the sugars produced by each species (Ismail et al., 2020). In the realm of bioethanol production, macroalgae have garnered significant attention due to their advantageous characteristics. Notably, macroalgae's primary component is carbohydrates, encompassing various types of sugars conducive to fermentation for bioethanol production. Furthermore, the low lignin content in macroalgae presents a notable advantage (Kumar et al., 2021), making them well-suited for fermentation processes with high yields and enhanced conversion rates. This underscores the potential of macroalgae as a promising source for bioethanol production. This review assesses the potential of third-generation feedstock for bioethanol production, emphasizing its technological challenges for future commercial (Jambo et al., 2016) viability. The key attributes of sustainability and susceptibility to energy conversion make algal feedstock suitable for bioethanol production. The innovative use of micro- and macroalgae as feedstock represents a strategic move in the bioethanol industry to enhance commercial feasibility. Numerous studies underscore the suitability of microalgae, particularly due to their high lipid production (Zhang et al., 2024) capacity. Macroalgae, or seaweeds, have emerged as promising bioethanol feedstock, offering substantial carbohydrates and economic opportunities for coastal communities. Enzymatic hydrolysis is considered a promising method for bioethanol production due to its reasonable economic cost and minimal environmental impact. Additionally, the choice of microorganism strains (Rastogi et al., 2017) in fermentation significantly

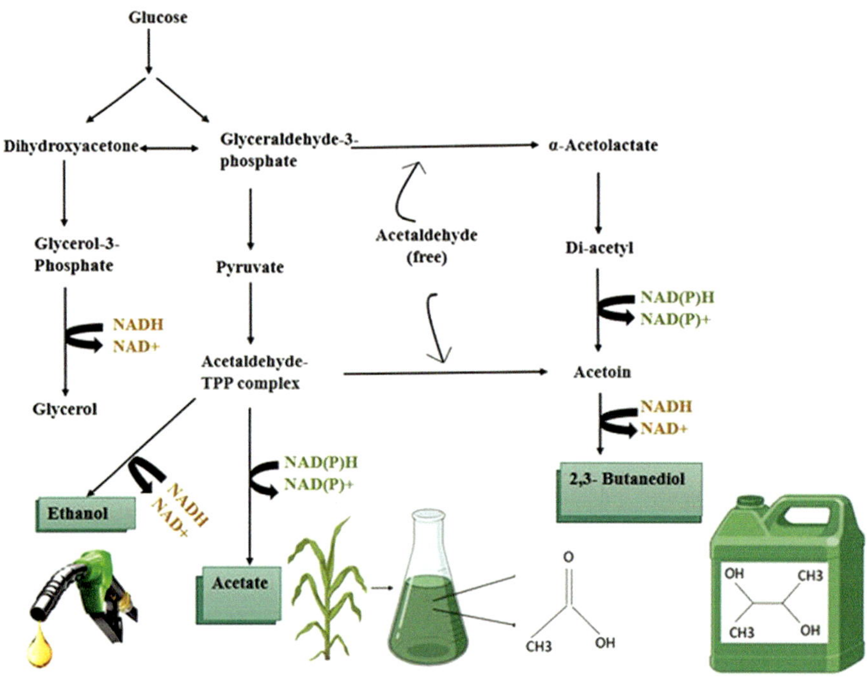

FIGURE 2.1 Production of different bioalcohols using glucose as substrate.

influences the conversion of sugars into bioethanol. Therefore, in-depth studies optimizing enzymatic hydrolysis and fermentation are essential for the development of an efficient and advanced bioethanol production process from third-generation feedstock. While the future prospects of algae in bioethanol production pose challenges that require time and effort for exploration, the anticipated contribution lies in reducing fossil fuel consumption, fostering a cleaner and more sustainable Earth in the future (Figure 2.1).

2.2.2 2,3-BD

2,3-Butanediol (2,3-BD) holds significance as a fundamental chemical compound in numerous industrial contexts. Its versatility extends to serving as a foundational element for the production of essential chemicals like methyl ethyl ketone, gamma-butyrolactone, and 1,3-butadiene. Furthermore, the optically active form of 2,3-BD holds specific significance in the asymmetric synthesis of valuable chiral specialty chemicals and functions as an antifreeze agent (Celińska and Grajek, 2009). With a heating value of 27,200 J/g, 2,3-BD has potential applications as a liquid fuel or fuel additive. Nonetheless, the growing energy crisis has led to increased expenses in the chemical production of 2,3-BD from nonrenewable resources (Xie et al., 2022), prompting a shift toward the biological production of 2,3-BD from renewable resources. Its widespread availability and ability to renew have sparked interest in it as an alternative source for bioprocesses (Gawal et al., 2023). Breaking down lignocellulose results in a sugar mixture primarily composed of glucose and xylose (Zhou et al., 2021). Enterobacteriaceae members like *Klebsiella pneumoniae*, *Klebsiella oxytoca*, *Enterobacter cloacae*, and *Serratia marcescens* are among the most favorable microorganisms for the biological synthesis of 2,3-BD (Celińska and Grajek, 2009). These strains exhibit swift growth in basic media and efficiently metabolize principal lignocellulose-derived sugars such as glucose and xylose into 2,3-BD. In bacterial metabolism, 2,3-BD is generated via what is known as the mixed-acid fermentation pathway. Alongside 2,3-BD, pyruvate derived from monosaccharides also undergoes conversion into a blend of acetate (Sharma et al., 2018), lactate, succinate, acetoin, and ethanol. Minimizing the release of these by-products during 2,3-BD fermentation is crucial to enhancing the economic viability of the production process.

2.2.3 n-Butanol

Bioalcohols with longer carbon chains (Balat et al., 2009), such as butanol, possess more favorable properties and would be better suited to the existing fuel distribution infrastructure. Consequently, breeding technologies, genetic engineering, and the exploration of undiscovered species offer promising avenues (Maurya et al., 2021) to develop microorganisms with elevated alcohol productivity and yields. The microorganisms such as *Saccharomyces cerevisiae*, and bacterial species such as *Zymomonas mobilis*, would be capable of converting all lignocellulosic sugars, utilizing carbon dioxide or monoxide, exhibiting high resistance (Zabed et al., 2017) to inhibitors and fermentation products, and being easily cultivable in large-scale bioreactors. Several mesophiles, such as *Escherichia coli*, *Bacillus subtilis*, *Pseudomonas putida*, and *Lactobacillus brevis*, have undergone genetic engineering to enable the production of n-butanol through the heterologous expression of genes sourced from *Clostridium acetobutylicum* (Mishra et al., 2023). The *C. acetobutylicum* and other similar clostridia, which have been utilized in the Weizmann process since the 1920s, have the most extensively researched n-butanol route.

2.3 DIFFERENT GENERATIONS OF BIOALCOHOLS

Biofuels are categorized into different generational groups, primarily determined by the raw materials employed in their production. First-generation biofuels (Naik et al., 2010) utilize sugars, edible oils, and starch. Second-generation biofuels are derived from non-edible biomass. Third-generation biofuels, known as algal biofuels, are produced using algae. The latest advancement, fourth-generation biofuels (Abdullah et al., 2019), involves innovative technologies such as capturing CO_2. As of now, the majority of biofuels in production belong to the first generation. These are directly sourced from food crops. For instance, countries like the United States and Brazil utilize corn and sugarcane, while European nations employ wheat and barley (Kim et al., 2004) for ethanol production. This categorization highlights the evolution of biofuel technologies and the ongoing shift toward more sustainable and advanced raw materials. The current demand for eco-friendly transportation fuels derived from renewable sources is already significant and is expected to see a substantial rise in the future. Presently, the biofuel sector predominantly manufactures ethanol using corn starch or sugarcane, while biodiesel is crafted from vegetable oils and animal fats.

2.3.1 First-Generation Bioalcohols

The ascendancy of first-generation bioethanol is accompanied by a growing awareness of its long-term sustainability challenges, bringing to the forefront critical issues such as the repercussions on land use, water resources, potential soil contamination arising from distillation residues, and the escalating competition for resources in food and feed production. In response to these sustainability concerns, ongoing research and development efforts in fuel ethanol are diligently working to address these negative externalities. The paramount role of process design is evident in the pursuit of modifying major pathways in first-generation ethanol synthesis, striving for enhanced efficiency and reduced environmental impact. This involves leveraging the central role of improved enzymes (Susmozas et al., 2020) and microorganisms, exemplified by innovations like SSF in first-generation facilities, particularly when utilizing starchy materials. A crucial facet of this evolutionary process is the compensatory strategy of offsetting ethanol production costs through the integrated valorization of energy and by-products. This entails maximizing the utility of by-products for applications in feed and green chemistry (Severo et al., 2019) within the broader biorefinery concept. The outcomes of real-scale experiments with first-generation ethanol underscore the potential for this compensatory approach. Rather than viewing first-generation bioethanol as a misstep, it should be recognized as an instrumental initial phase that provided indispensable experience.

This foundational knowledge has proven invaluable for the successful implementation of greener biofuels in subsequent generations, starting with second-generation lignocellulosic bioethanol, which is now making its presence felt in the market. Within this dynamic context, the emergence of integrated biorefineries stands out as a promising avenue. These facilities offer diversification of usable feedstocks (Yue et al., 2014), resulting in reduced facility sizes, optimized supply chains, and efficient valorization of resources such as bagasse from sugarcane, corn stover, and the untapped potential of microalgae for capturing carbon dioxide during fermentation steps. First-generation bioalcohol production typically involves the conversion of sugars from crops or starch-containing materials into bioalcohols such as ethanol (Suraiya et al., 2015). The common feedstocks include corn, sugarcane, or other crops with high sugar or starch content. The raw materials are harvested and processed to extract sugars or starch. The ethanol production process begins with saccharification, where starch-based (Bušić et al., 2018) feedstocks such as corn transform into sugars. Subsequently, fermentation takes place, during which sugars from either sugar-containing crops or saccharified starch are converted into ethanol by yeast or bacteria. Following fermentation, distillation separates ethanol from water and other components, concentrating and purifying it based on boiling points (Vane et al., 2008). Optional steps include dehydration to further concentrate the ethanol and denaturing if the ethanol is intended for industrial or fuel use. The final ethanol product is then stored and distributed for various applications, ranging from fuel blending to industrial processes. First-generation bioethanol production, despite its contribution to renewable energy, is associated with several disadvantages (Ayodele et al., 2020). One significant concern is the competition for resources with food production, as bioethanol is often derived from crops like corn and sugarcane. This competition raises apprehensions about potential impacts on food prices and availability. Furthermore, the cultivation of biofuel crops can lead to deforestation, changes in land use, and habitat loss, contributing to environmental (Ravindranath et al., 2011) degradation. The process of converting these crops into bioethanol is energy-intensive, requiring substantial inputs of water, fertilizers, and energy, which may compromise the overall environmental sustainability.

2.3.1.1 Corn as a Substate for Bioethanol Production

Corn, a staple crop with diverse applications, holds significant global importance. The majority of corn production, approximately 64%, is dedicated to animal feed, highlighting its crucial role in livestock agriculture (Mumm et al., 2014). Human consumption follows closely, accounting for around 19% of global corn production. However, the utilization of corn varies across regions, with Africa and Central America primarily using corn for human food, while other regions emphasize its role in animal feed (Murdia et al., 2019). Despite its critical role, a notable portion of corn, approximately 5% globally, is lost as waste. Waste, as defined by FAOSTAT, encompasses losses occurring from the farm to the household level during handling, storage, and transport (Fabi et al., 2021). In Central America, the loss rate is particularly high, averaging over 9% of corn production (Ranum et al., 2014). Examining the potential of bioethanol production from corn reveals intriguing possibilities. If the wasted corn were fully utilized for bioethanol production, approximately 9.3 billion gallons of bioethanol could be generated. This could replace 6.7 billion gallons of gasoline if bioethanol, particularly E85, is adopted as an alternative vehicle fuel (Monroe et al., 2019). Furthermore, employing the corn dry milling process, which produces dry distillers' dried grains and solubles (DDGS) as a coproduct, presents an additional avenue. About 11 million metric tons of DDGS could be available for animal feed, potentially replacing 13 million metric tons of corn used in animal feed. If the replaced corn is redirected to bioethanol production, an extra 5.1 billion gallons of bioethanol (equivalent to 3.7 billion gallons of gasoline in a midsize passenger car fueled by E85) could be produced (Mohamaddi et al., 2021). The utilization of wasted corn could contribute to a reduction of approximately 0.93% in global gasoline consumption annually.

Genetic engineering technology undoubtedly holds significant potential for the future of agriculture and biofuel production, as outlined previously (Kour et al., 2019). Nonetheless, the acceptance of biotech-derived crops has encountered skepticism and regulatory obstacles in numerous countries. Maize, classified as a

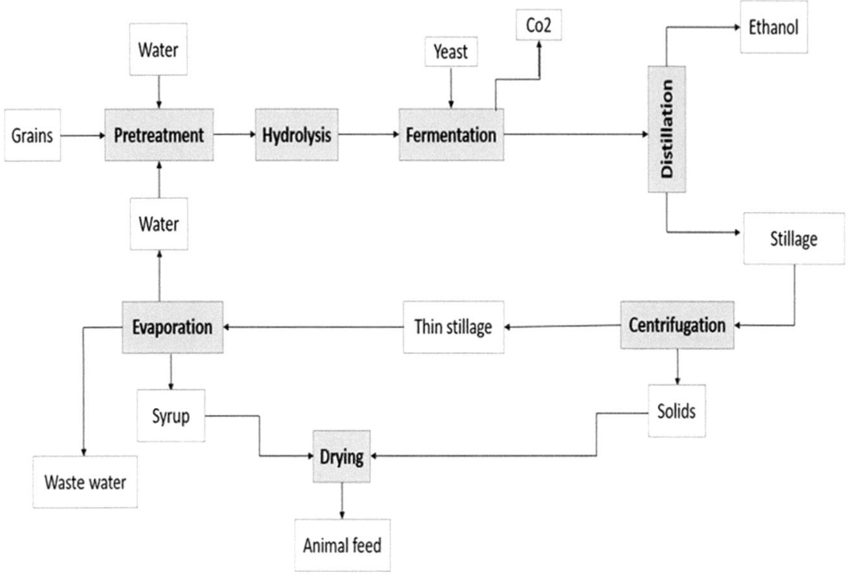

FIGURE 2.2 A simplified depiction of second-generation bioethanol production involves utilizing grain feedstock with yeast as the inoculum.

C4 plant, possesses a compartmentalized photosynthetic system employing phosphoenolpyruvate carboxylase (PEPC) as its primary carboxylase. Studies have indicated that transgenic maize engineered to overexpress PEPC exhibits enhanced CO_2 fixation rates and compensation points (Sakamoto et al., 2006), increased fresh and dry weight, augmented leaf surface area and stomatal density, along with improved resistance to water stress. Rearranging the architecture of plants becomes another way to improve photosynthesis in order to accommodate the large planting densities that are frequently used in agriculture (Warnasooriya et al., 2014). Recent research on rice has shown that an erect leaf phenotype is caused by lowering plant hormone brassinosteroid levels or the number of brassinosteroid receptors. These erect-leaf rice plants—which were created by genetic engineering (Sinclair et al., 1999) or mutagenesis—produce more grain and biomass when grown in high-density planting situations without the need for more fertilizer. This illustration will delve into the role of yeasts in bioethanol fermentation and explore immobilization techniques, aiming to enhance ethanol production for the betterment of humankind (Figure 2.2).

2.3.2 Second-Generation Bioalcohols

The primary goals of initiating the pretreatment process for biofuel production include: (i) reducing the size of the raw materials through a physical method, (ii) exposing components such as hemicellulose, cellulose, and starch before hydrolysis to enhance the production of reducing sugars, (iii) facilitating better access for

TABLE 2.1
Various Feedstock along with Their Constitutional Values for Second-Generation Feedstock

Dry Matter (%)	Lignin (%)	Carbohydrates (%)	Ethanol Yield (L/kg of Dry Biomass)	Reference
Bagasse	14.5	67.15	0.28	(Huang et al., 2020)
Sugarcane	–	67.00	0.50	(Oliveira et al., 2014)
Oat	4.0	65.60	0.41	(Bharathiraja et al., 2014)
Rice	–	87.50	0.48	(Binod et al., 2010)
Sorghum	1.40	71.60	0.44	(Stamenković et al., 2020)
Wheat	–	35.85	0.40	(Panahi et al., 2022)
Corn	0.60	73.70	0.46	(Swain et al., 2019)

enzymes during subsequent enzymatic hydrolysis to break down carbohydrates into fermentable sugars, and (iv) decreasing the crystallinity degree of the cellulose matrix. It is frequently emphasized that the pretreatment process is highly recommended as it directly yields fermentable sugars (Kucharska et al., 2018), prevents premature degradation of sugars, inhibits the formation of inhibitors before hydrolysis and fermentation, reduces processing costs, and decreases the demand for conventional energy overall. In lignocellulosic materials, pretreatment serves to deform rigid components like lignin, cellulose, and hemicellulose, degrading crystallinity for optimal hydrolysis conditions (Saadon et al., 2022). While mechanical reduction alone is effective, combining it with chemical pretreatment creates a smoother environment for subsequent hydrolysis, leading to enhanced yields of reducing sugars. The composition of lignocellulosic material is contingent upon factors such as species, variety, growth conditions, and maturity (Zhang et al., 2015). The ethanol yield and conversion productivity are influenced by the biomass type, necessitating a high content of cellulose and hemicellulose while maintaining low lignin content. Aditiya explored diverse pretreatment methods for bioethanol production from typical lignocellulosic materials (Aditiya et al., 2016), aiming to bring celluloses into an amorphous form for efficient enzymatic consumption, resulting in more sugar monomers.

Various pretreatment methods are categorized as biological, chemical, physical, and physicochemical (Das et al., 2021), each contributing to the overall efficiency of the production process. Biological pretreatment utilizes microorganisms to carry out the initial treatment, and these microorganisms are specifically equipped to break down the lignocellulosic components of the feedstock into an amorphous form. The primary goals of initiating the pretreatment process for biofuel production include: (i) reducing the size of the raw materials through a physical method, (ii) exposing components such as hemicellulose, cellulose, and starch before hydrolysis to enhance the production of reducing sugars, (iii) facilitating better access for enzymes during subsequent enzymatic hydrolysis to break down carbohydrates into fermentable sugars, and (iv) decreasing the crystallinity degree of the cellulose matrix. It is frequently emphasized that the pretreatment process is highly recommended as it directly yields fermentable sugars (Kucharska et al., 2018), prevents premature degradation of sugars, inhibits the formation of inhibitors before hydrolysis and fermentation, reduces processing costs, and decreases the demand for conventional energy overall. In lignocellulosic materials, pretreatment serves to deform rigid components like lignin, cellulose, and hemicellulose, degrading crystallinity for optimal hydrolysis conditions (Saadon et al., 2022). While mechanical reduction alone is effective, combining it with chemical pretreatment creates a smoother environment for subsequent hydrolysis, leading to enhanced yields of reducing sugars. The composition of lignocellulosic material is contingent upon factors such as species, variety, growth conditions, and maturity (Zhang et al., 2015). The ethanol yield and conversion productivity are influenced by the biomass type, necessitating a high content of cellulose and hemicellulose while maintaining low lignin content. Aditiya explored diverse pretreatment methods for bioethanol production from typical lignocellulosic materials (Aditiya et al., 2016), aiming to bring celluloses into an amorphous form for efficient enzymatic consumption, resulting in more sugar monomers.

Yeast, particularly *Saccharomyces cerevisiae*, has played a pivotal role in alcohol production for millennia, especially in brewing and winemaking (Walker et al., 2018). Its use in distillation keeps costs low while ensuring high ethanol yields, productivity, and resilience to high ethanol concentrations. In the modern context, yeasts are harnessed for generating fuel bioethanol from renewable sources. Various yeast strains, such as *Pichia stipitis*, *S. cerevisiae*, and *Kluyveromyces fragilis* (Ndubuisi et al., 2023), have demonstrated proficiency in ethanol production from different sugars. *S. cerevisiae* stands out as the most widely employed yeast in industrial ethanol production due to its ability to tolerate a broad pH range, making the process less susceptible to infections (Oliva-Neto et al., 2013). While baker's yeast has traditionally been used in ethanol production for its low cost and easy availability, challenges arose as wild-type yeasts outcompeted it, leading to contamination in industrial processes. Stressful conditions like increased ethanol concentration (Gibson et al., 2007), temperature, osmotic stress, and bacterial contamination can compromise yeast survival during fermentation.

To address these challenges, flocculent yeasts have been utilized in ethanol production, aiding downstream processing and enabling operation at high cell density. Common challenges during sugar fermentation include elevated temperatures (35–45°C) and high ethanol concentrations (over 20%) (Lin et al., 2012). Ethanol-tolerant and thermotolerant yeast strains, isolated from natural resources, offer solutions to these challenges. For instance, *K. marxianus* is a thermotolerant yeast capable of co-fermenting both hexose and pentose sugars (Leonel et al., 2021). Overcoming the issue of pentose fermentation involves employing hybrid, genetically engineered, or co-culture approaches. Hybrid yeast strains, developed by fusing protoplasts of *S. cerevisiae* and xylose-fermenting yeasts, enable simultaneous fermentation of pentose and hexose sugars (Zhang et al., 2012). Genetically engineered *S. cerevisiae* strains, incorporating genes from other yeast species, demonstrate the ability to ferment xylose. Co-culture of different yeast strains in the same reactor enhances ethanol production by efficiently utilizing both hexose and pentose sugars (Sandri et al., 2023). These advancements contribute to the industrial-scale efficiency of ethanol production, offering sustainable solutions for the bioethanol industry. Microorganisms, particularly yeasts, play a crucial role in bioethanol production by efficiently fermenting a diverse range of sugars into ethanol. Their utilization in industrial plants is attributed to several valuable properties that contribute to the success of ethanol production. These properties include achieving high ethanol yields (Branco et al., 2018) (exceeding 90.0% of theoretical yield), demonstrating tolerance to high ethanol concentrations (greater than 40.0 g/L), displaying high ethanol productivity (exceeding 1.0 g/L/h), thriving in simple and cost-effective growth media, and exhibiting resilience in undiluted fermentation broth. Additionally, yeasts exhibit resistance to inhibitors and possess the ability to suppress contaminants, making them well-suited for growth in various conditions. Yeasts, as the primary actors in the fermentation process, significantly influence the amount of ethanol produced.

Brown rot, soft rot, and white rot fungi are common microorganisms employed in this method, acting as liberators of the intricate lignocellulosic structure. Brown rot fungi specialize in cellulose degradation, whereas both soft and white rot fungi can break down both cellulose and lignin (Sista et al., 2018). Because biological pretreatment is sustainable, it is eco-friendly. It does not require a lot of energy as it runs in mild circumstances. Because fungi naturally break down lignocellulosic walls, no additional chemicals are needed, making the process safe and non-toxic to the environment. The same goal is achieved in chemical pretreatment when particular supportive chemicals (Den et al., 2018) are added to lignocellulosic or starchy materials in order to get them ready for the hydrolysis process that follows in the production sequence. Although there isn't a single pretreatment technique that works well for all situations, diluted acid pretreatment is frequently thought to be the most practical (Canam et al., 2013) for use in industrial settings. The benefits of chemical pretreatment include chemical material accessibility, affordability, resistance to technical advancements (unlike designed enzymes), ease of storage, and durability.

Through direct chemical reactions, the chemical compounds help break down lignocellulosic walls (Dutta et al., 2022) and complicated carbohydrate chains. This process requires less energy in the form of heat, but it takes longer and produces less sugar overall (especially in the case of alkali pretreatment). Another way of pretreating biomass is mechanical pretreatment, which involves physically reducing the biomass's size using chopping, cutting, or material-breaking techniques. Similar to other pretreatment techniques, mechanical pretreatment aims to reduce the crystallinity of biomass, hence improving the remaining bioethanol production processes. Traditionally, wet disc milling, ball milling, vibratory ball milling (Mood et al., 2013), compression milling, hammer milling, and roll milling can be used to reduce physical size. The most well-known issue with mechanical pretreatment is how much power it uses. It has been stated that one-third of the energy used in the overall bioethanol manufacturing process (Aditiya et al., 2016) is used in the physical reduction strategy via mechanical pretreatment later leading to hydrolysis. Hydrolysis includes microorganisms (that are employed in the later process of fermentation) that are able to digest only simpler sugar forms derived from complex carbohydrates (Balat et al., 2008) of biomass; the hydrolysis process—which is typically catalyzed by an enzyme or acid—separates a long chain of carbohydrates (from cellulose or starch) with the addition of water molecules. This stage is critical in the production of bioethanol since the quality of the hydrolysate will affect the subsequent fermentation process, which is connected to the quality of the ethanol as the end product.

A proposed solution to effectively utilize unfermented substrate without compromising DDGS quality involves employing food-related strains of Zygomycetes and Ascomycetes filamentous fungi. Strains like *Rhizopus* sp., *Fusarium venenatum*, *Aspergillus oryzae*, *Neurospora intermedia*, and *Monascus purpureus* (Lennartsson et al., 2014) have been confirmed to grow on mostly wheat-based thin stillage in aerobic conditions. This growth results in the production of 11–19 g/L fungal biomass and 0.9–4.7 g/L ethanol, based on unpublished data. The process involves the easy separation of fungal biomass, characterized by its filamentous nature (Nair et al., 2016), from the liquid, and subsequent drying. The ethanol remains in the fermented broth, which is directed to the evaporators. The volatile ethanol naturally joins the outgoing steam, following the current approach used in first-generation plants. This eliminates the need for additional process steps to separate the ethanol. In the context of pentose utilization (Ferreira et al., 2018) and second-generation processes, the research focus has primarily been on Zygomycetes. Initiatives, such as those by Ram et al. (2020), explored the use of sulfite liquor from the paper pulp industry as a substrate for *Rhizopus* (Ram et al., 2020). Various publications highlight the potential of food-related Zygomycetes. While the ethanol yield from xylose is often limited (around 0.2 g/g), the production of fungal biomass has shown more promising results (approximately 0.35 g/g). The inherent limitation in anaerobic fermentation of xylose by Zygomycetes, attributed to an imbalance among the redox carriers (Lynd et al., 2003) without access to oxygen, imposes a natural constraint on ethanol production. This limitation, especially in industrial-scale operations, hinders micro-adjustments in the oxygen level to achieve high ethanol yields (Lennartsson et al., 2014). Despite challenges, the production of fungal biomass in aerobic conditions can likely be optimized from pentoses by adjusting process parameters and feed composition. However, the utilization of pentose sugars by fungi is generally a slower process than that of hexose sugars and has not been reported at high hexose concentrations for these filamentous fungi. Aeration plays a crucial role in various aspects of the process (Seiboth et al., 2007), including pentose utilization for biomass production, decomposition of carbohydrate polymers in thin stillage, and the utilization of metabolites and unfermented sugars. Zygomycetes, known for their ability to produce enzymes like amylases, cellulases, proteases, and lipases, can effectively utilize most substrates (Kereluik et al., 2022), further emphasizing the importance of aeration. Cultivating filamentous fungi presents its challenges, particularly in mixing due to the broth's viscosity caused by the filamentous nature of the cells. The fungi may also attach to equipment inside the reactor. Two approaches to counteract these issues involve adjusting process conditions to control growth morphology, such as forming pellets or modifying the cultivation vessel to suit the growth of filamentous fungi (Gomes et al., 2023). Reactor types like air-lift and bubble-column reactors, lacking internal moving parts and relying on aeration for mixing, have shown promising performance in fungal cultivations on thin stillage in aerobic conditions, based on unpublished data (Liu et al., 2011). This approach also offers the benefit of relatively low energy demand for mixing.

Rice straw, a plentiful crop residue globally, yields approximately 731 million tons annually, distributed across Africa, Asia, Europe, and America. This volume has the potential to generate around 205 billion liters of bioethanol (Singh et al., 2023) each year. In Asia, it constitutes a significant field-based residue, contributing 667.59 million MT, with a theoretical ethanol production potential of 282 billion liters from the total 668 million MT (Binod et al., 2010). Unfortunately, an increasing proportion of rice straw undergoes field burning, a wasteful practice considering the high fuel prices and the pressing need to reduce GHG emissions and air pollution. Rice cultivation produces two main residues, straw and husk, both holding energy potential (Lim et al., 2012). While the use of rice husk is well-established in many Asian countries, rice straw remains largely untapped as a renewable energy source. The preference for husk stems from its easy availability at rice mills, whereas collecting rice straw is labor-intensive and limited to harvest time. Although baling could enhance collection logistics, the expensive equipment makes it economically impractical for most rice farmers (Bhattacharyya et al., 2021). Consequently, technologies for utilizing rice straw for energy purposes must be exceptionally efficient to offset the high costs associated with straw collection. Rice straw comprises a complex mix of carbohydrate polymers, with cellulose and hemicellulose tightly bound by layers of lignin, acting as a barrier against enzymatic hydrolysis (Sun et al., 2010). To enable efficient enzymatic action, a pretreatment step is essential to break the lignin seal, exposing cellulose and hemicellulose. The primary objectives of pretreatment include reducing cellulose crystallinity, increasing biomass surface area, eliminating hemicellulose, and breaking the lignin seal (Mosier et al., 2005). Through pretreatment, cellulose becomes more accessible to enzymes, facilitating the rapid and high-yield conversion of carbohydrate polymers into fermentable sugars. Various methods, such as physical, chemical, and thermal techniques, alone or in combination, are employed in pretreatment. However, it's worth noting that pretreatment is often considered one of the costliest stages in the conversion of cellulosic biomass to fermentable sugars (Kumar et al., 2017). Despite facing challenges such as its intricate nature and high lignin and ash content, ongoing efforts are focused on developing an economically viable technology for rice straw. Efficient pretreatment methods are being researched to eliminate undesired components, leading to successful outcomes. Current statistics indicate that rice straw could meet the bioethanol demand for the transport sector. To overcome hurdles like xylose and glucose co-fermentation, improvements in both process and strain engineering are essential to enhance system efficiency. The optimal combination of pretreatment, hydrolysis, and fermentation processes is crucial for maximum efficacy (Huang et al., 2019). With advancements like genetically modified yeast, synthetic hydrolyzing enzymes, and other sophisticated technologies, when efficiently integrated, the bioethanol production process using rice straw is poised to become a viable technology very shortly.

The implementation of second-generation bioethanol on an industrial scale represents a strategic approach that caters to the demands of both the energy and agricultural sectors. This advanced bioethanol production addresses the energy industry's needs, particularly in the renewable energy sector, by providing a sustainable biofuel product (Zabed et al., 2017). Simultaneously, it contributes to the agricultural sector by utilizing biomass as a value-added product. Bioethanol has gained widespread popularity in developed and developing countries alike, primarily driven by environmental concerns, air quality improvement (Shahare et al., 2017), and the desire to reduce reliance on fossil fuels. While first-generation bioethanol production is well-established in many countries to meet the demands of the transportation sector, concerns are arising regarding its sustainability due to its reliance on edible feedstocks (Ahorsu et al., 2018). This dependence raises ethical questions, as using food crops for fuel seems incongruous when considering the global issue of malnutrition. The advent of second-generation bioethanol production addresses these concerns by utilizing non-edible feedstocks derived from agricultural and forestry biomass (Jusakulvijit et al., 2021), presenting a more sustainable and ethical alternative to conventional bioethanol production. The success of second-generation biofuels hinges on research investment and the application of knowledge gleaned from the development of first-generation bioethanol. Integrating new technologies into first-generation bioethanol should be aligned with feedstock availability (Kirshner et al., 2022). Sugarcane offers several advantages as a feedstock compared to other crops like corn, sugar beet, and wheat. One key advantage is its perennial nature, allowing for six or seven harvests before replanting and reducing field costs. Adapted well to soil and climate in certain Brazilian regions, sugarcane has replaced low-productivity pastures, transforming the agribusiness strategy of many cities (Jonson et al., 2022). However, sugarcane harvesting is limited to 6–8 months per year, and its perishable nature prevents storage. In contrast, corn, used in the USA, can be transported and stored, enabling distilleries to remain active almost year-round. This difference means that distilleries are idle for 4–6 months annually, not producing bioethanol during that period (Mayer et al., 2016).

Second-generation bioethanol, often termed "advanced biofuels," distinguishes itself through innovative processes primarily utilizing lignocellulosic feedstocks and agricultural forest residues. The notable advantage of these feedstocks lies in their easy availability (Valdivia et al., 2016) without competing with food resources, thereby minimizing their environmental impact. Despite these merits, the industrial scale-up of second-generation bioethanol has encountered significant challenges, primarily stemming from technological issues. High production costs and moderate yields of bioethanol are notable hurdles attributed to the lignin composition of these feedstocks (Vohra et al., 2014). A key goal is to streamline the fermentation process to effectively convert all sugars, including pentoses and hexoses, released during pretreatment and hydrolysis into ethanol. Technical hurdles in second-generation biofuel production encompass the varied composition of biomass, the formation of inhibitors during pre-saccharification treatment, challenges with end-product inhibition, osmotic and oxidative stress (da Silva et al., 2020), and ethanol accumulation. Additionally, the adoption of advanced technologies and the need for specialized facilities to facilitate the conversion process present further challenges to second-generation bioethanol production (Zuliani et al., 2021). Overcoming these technological and economic barriers is crucial for the widespread adoption of advanced biofuels and their integration into the renewable energy landscape. Comparing the third generation of bioethanol to the first and second, there are greater advantages. The utilization of marine organisms like algae is the main focus of third-generation bioethanol.

2.3.3 Third-Generation Bioalcohols: Algal Biofuels

2.3.3.1 Algae

Algae, a diverse group of photosynthetic organisms, faces classification controversies, notably regarding Cyanobacteria. Algae can be microalgae (unicellular) or macroalgae (multicellular), with microalgae often floating on water surfaces (phytoplankton) due to their lipid content, while macroalgae (seaweeds) are typically attached to rocks (Heimann et al., 2015). Microalgae have garnered global attention from biofuel researchers, given their potential as a renewable energy source. More recently, macroalgae have emerged as a third-generation feedstock, sparking increased research interest. Microalgae types include dinoflagellates, green algae, golden algae, and diatoms (Rajkumar et al., 2013). They exhibit varying protein, carbohydrate, and lipid content, with some species storing lipids exceeding 70% of dry biomass. Factors like light, temperature, and nutrient levels influence these contents. Microalgae's cell wall components, such as cellulose and proteins, can be converted into bioethanol through hydrolysis (Kusmiyati et al., 2023). Research on liquid fuel from microalgae began in the 1980s, exploring various species under specific growth conditions. Microalgae, as an energy source, offers significant fuel yields from small crop areas, with high photosynthetic efficiency aiding in global warming mitigation. Beyond bioethanol, microalgal extracts find applications in diverse sectors like food, pharmaceuticals, fertilizers, lubricants, and cosmetics (Rajvanshi et al., 2019).

2.3.3.2 Microalgae

Microalgae are celebrated for their adaptability to diverse environmental conditions, thriving in extreme habitats. Major groups like *Diatoms*, *Cyanoprocaryota*, *Euglenophycota*, *Cryptophycophyta*, and *Chlorophycophyta* showcase this resilience (El-Sheekh et al., 2021). Specific species, including *Chlorella* sp. and *Scenedesmus obliquus* sp., exhibit the ability to flourish in alkaline water environments. Remarkably, microalgae have found success in being cultivated in industrial liquid wastes, tapping into them as a nutrient source (Liu et al., 2023). Unlike conventional agricultural plants, microalgae excel in diverse water environments, providing a non-competitive alternative. One significant advantage over plant-based crops is the swift doubling of biomass in less than a day under favorable conditions, enabling year-round harvesting. Lam and Lee propose that microalgal biomass production can range from 15 to 25 tons annually (Lam et al., 2012). Microalgae cultivation harnesses solar energy to combine water with CO_2, presenting nearly carbon-neutral biomass production. Various methods, including open or covered ponds and closed photobioreactors, are employed for cultivation (Gupta et al., 2015). Harvesting techniques involve centrifugation, foam fractionation, flocculation, membrane filtration, and ultrasonic separation. The concentration of microalgal biomass in the upper zone enhances harvest efficiency and economic feasibility. The rate of microalgal growth in the culture system is influenced by factors such as light intensity, photoperiod, temperature, and nutrient availability.

Crucial environmental factors for microalgae cultivation include light intensity, pH, salinity, and temperature, while nutritional factors encompass nitrogen, carbon, phosphorus, sulfur, and iron availability and source types. Genera such as *Scenedesmus*, *Chlorella*, *Chlorococcum*, and *Tetraselmis* from the *Clorophyta* division (Lakatos et al., 2019), along with cyanobacteria like *Synechococcus*, have undergone extensive examination as potential sources for bioethanol production. Generally, cultivation under high light intensity (ranging from 150 to 450 $\mu m^{-2} s^{-1}$), utilizing a CO_2-air mix of 2–5% and maintaining mesophilic temperatures (20–30°C), achieves (Brown et al., 1999) approximately 50% carbohydrate content under nutrient starvation, particularly nitrogen.

2.3.3.3 Macroalgae

Extensive research has been dedicated to biofuels derived from microalgae, primarily concentrating on biodiesel and biogas production, with additional consideration given to bioethanol and biohydrogen (Kumar et al., 2016). The production methods and operational parameters differ for each biofuel. Numerous studies have proven the feasibility of industrial processes for biodiesel production, and some propose anaerobic digestion following lipid extraction from algal biomass (Menegazzo et al., 2019). However, the quest for a refined and efficient process for bioethanol production is an ongoing endeavor. Macroalgae, commonly known as seaweeds, have been prevalent for centuries and utilized as a marine vegetable in Asian countries such as China, Japan, and Korea. They play a vital ecological role by serving as habitat and substrata for various marine life forms, including invertebrates, fish, mammals, and birds (Levin et al., 2001). Additionally, macroalgae contribute to coastal protection against erosion and have a substantial impact on the marine carbon cycle. Seaweeds exhibit regional specificity, with certain species confined to specific areas. Currently, 221 seaweed species are commercially utilized, with 145 species for food purposes and 110 species for phycocolloid production (Rao et al., 2018). Despite the discovery that seaweeds are rich in proteins, vitamins, amino acids, growth hormones, and minerals, and are deemed suitable for consumption, they are not widely adopted as a primary source of energy in daily life. Seaweeds can be classified into three primary groups: brown (Phaeophyceae), red (Rhodophyceae), and green (Chlorophyceae) (Baweja et al., 2016). These groups possess varying compositions, encompassing carbohydrates, proteins, and lipids, with notable differences between them. The cultivation of seaweeds is dominated by five genera, namely, *Saccharina*, *Undaria*, *Porphyra*, *Eucheuma/Kappaphycus*, and *Gracilaria* (Pereira et al., 2024), which collectively contribute to about 98% of the world's cultivated seaweed production. Notable species include kelps such as *Saccharina japonica* and *Undaria pinnatifida*, tropical red algal species like *Kappaphycus* and *Eucheuma*, nori species including *Porphyra* and *Pyropia*, and the red algal agarophyte *Gracilaria*.

The structural cell wall of seaweeds typically comprises a matrix of linear sulfated galactan polymers. Researchers have identified seaweeds as promising candidates for bioethanol conversion (Chen et al., 2015) due to their known low or absent lignin content. Brown seaweeds contain carbohydrates like laminarin, mannitol, fucoidan, cellulose, and alginates. In red seaweeds, the cell wall is composed of polysaccharides such as agar, cellulose, xylene, mannan, and carrageenan, while green seaweeds' cell walls consist (Jambo et al., 2016) of cellulose, mannose, and xylene. Notably, red seaweed exhibits the highest carbohydrate composition among the three types. Seaweeds' distinct biochemical characteristics make them potentially valuable resources for bioethanol production, offering an alternative and sustainable feedstock for the biofuel industry. Seaweeds contain a significant amount of carbohydrates (Yanagisawa et al., 2013), so during the production stage, it is essential to convert this composition into bioethanol. On the other hand, the process of making bioethanol also requires its capacity to store enough carbon sources. The following step of the process, hydrolysis followed by fermentation, often uses the powdered form and slurry of algae.

2.3.4 Fourth-Generation Bioalcohols

Fourth-generation biofuels utilize genetically modified microalgae as their primary feedstocks. These microorganisms are subjected to genetic modifications aimed at enhancing their capacity for CO_2 assimilation during photosynthesis, thereby serving as effective carbon sinks, while simultaneously augmenting their biofuel production capabilities (Priyadharsini et al., 2022). Several strains of microalgae, including *Chlamydomonas reinhardtii*, *Phaeodactylum tricornutum*, and *Thalassiosira pseudonana*, have undergone genetic alterations to improve their growth rates and adaptability to nutrient-deficient (Singh et al., 2016) environments. Experimental studies have documented notable enhancements in lipid productivity and carbon fixation within these genetically modified microalgae. Environmental advantages associated with these modifications include enhanced CO_2 (Barati et al., 2021) sequestration, utilization as a medium for wastewater treatment, and mitigation of GHG emissions. The overarching objective of fourth-generation biofuels is to achieve the lowest possible environmental footprint compared to preceding generations. However, a comprehensive evaluation of the environmental ramifications of genetic modification procedures remains a focal point of current research endeavors in this field. Genetically modified microalgae can be grown in either closed or open systems (Beacham et al., 2017). Closed systems provide more security by shielding the cultivation process from external influences and reducing the risk of contamination. However, their higher operational costs make them less economically viable. In contrast, open systems are prone to leakage, potentially leading to the escape or unintentional release of microalgae into the environment (Abdullah et al., 2019). A research investigation focused on assessing the environmental consequences of cultivating genetically modified microalgae in an open system. Findings from the study indicated that over the experimental duration, the cultivation of genetically modified microalgae did not cause any adverse effects (Henley et al., 2013) on the environment or native species. The introduction of genetically modified algae is viewed as a promising avenue for acquiring customized algae traits. However, it is crucial to acknowledge the uncertainty surrounding the altered species and the potential unknown threats they may pose to the ecosystem.

TABLE 2.2
Bioethanol Production Progression from First to Fourth Generations

Generation	Feedstocks	Processing Technology	Reference
First	Food crops, animal fat, edible oil seeds	Esterification and transesterification, along with fermentation of sugars, are chemical processes crucial in the production of biofuels	(Nigam et al., 2011)
Second	Cereal straw, forest residues, lignocellulosic feedstock material, and non-edible seeds	Pretreatment of feedstock and fermentation using physical, biological, chemical, and thermochemical processes	(Li et al., 2016)
Third	Algae	Cultivation of algae, harvesting, oil extraction, transesterification, or fermentation	(Ahmad et al., 2013)
Fourth	Algae and other microbes	Altering the metabolism of algae to enhance carbon capture, the process of cultivation, harvesting, transesterification, and fermentation is undertaken in the context of biofuel production	(Choi et al., 2019)

2.4 PRETREATMENT OF BIOMASS

The central challenge encountered in biofuel production lies in the pretreatment of biomass. Lignocellulosic biomass, characterized by its tripartite composition of hemicellulose, lignin, and cellulose, necessitates pretreatment methods aimed at the dissolution and separation of these constituents (Balasubramani et al., 2024). This process enhances the accessibility of the remaining solid biomass to subsequent chemical or biological treatments. The lignocellulosic structure consists of a complex matrix (Singhvi et al., 2019) wherein cellulose and lignin are interconnected by hemicellulose chains. Pretreatment procedures are designed to disrupt this matrix, leading to a reduction in the crystallinity of cellulose and an increase in the proportion of amorphous cellulose (Ling et al., 2017), thereby optimizing enzymatic hydrolysis. Pretreatment renders lignocellulosic biomass more susceptible to rapid hydrolysis, resulting in heightened yields of monomeric sugars. Various steps for pretreatment can be discussed as follows.

2.4.1 Acid Hydrolysis

The enhancement of anaerobic digestibility through ambient temperature acid pretreatment of lignocellulose is well-established. Dilute acid pretreatment primarily targets hemicellulose, leaving lignin degradation minimally affected (Chandra et al., 2007). This process involves the solubilization of hemicellulose, rendering cellulose more accessible to enzymes, particularly cellulase. Mineral acids such as HCl and H_2SO_4 are commonly employed in acid pretreatment. Typically, dilute acid pretreatment is a straightforward single-stage process, where biomass undergoes treatment with dilute sulfuric acid at appropriate (Ab Rasid et al., 2021) concentrations and temperatures for a specified duration. In an effort to reduce enzyme requirements, a two-stage process was devised at the National Renewable Energy Laboratory (NREL) in Golden, Colorado (Davis et al., 2013). Literature on dilute acid hydrolysis of rice straw is scarce, primarily due to the process's incapacity to remove lignin and its resulting low sugar yield. However, acid hydrolysis proves effective in releasing simple sugars from polysaccharides in macroalgae. Sulfuric acid particularly (Jönsson et al., 2020) serves as a common catalyst in this process. The acid disrupts bonds in polysaccharide chains, inducing amorphousness and high susceptibility to hydrolysis. Functioning as a catalyst, acid facilitates the cleavage of polysaccharides through glycosidic bond hydrolysis. Dilute H_2SO_4 treatment (Rahmati et al., 2020) at elevated temperatures ensures the swift and complete hydrolysis of polysaccharides into monosaccharides. Nevertheless, acid pretreatment leads to the generation of diverse inhibitors such as acetic acid, furfural, and 5-hydroxymethylfurfural. These compounds act as growth inhibitors for microorganisms. Consequently, hydrolysates intended for fermentation require detoxification prior to use.

2.4.2 Pretreatment with an Oxidizing Agent

Pretreatment involving oxidizing agents is a method where an oxidizing compound, such as hydrogen peroxide or peracetic acid, is introduced to biomass suspended in water. This process aims to remove hemicellulose and lignin, enhancing cellulose accessibility (Chaturvedi et al., 2013). Various reactions, including electrophilic substitution, side chain displacement, alkyl aryl ether linkage cleavage, or oxidative cleavage of aromatic nuclei, can occur during oxidative pretreatment. Hydrogen peroxide pretreatment utilizes oxidative delignification (Mittal et al., 2017) to detach and solubilize lignin, loosening the lignocellulosic matrix and improving enzyme digestibility. Wei investigated the impact of hydrogen peroxide pretreatment on structural features and enzymatic hydrolysis of rice straw. Parameters like lignin content, weight loss, accessibility for Cadoxen, water-holding capacity, and crystallinity were monitored during pretreatment to assess the modification of the lignocellulosic structure (Wei et al., 1985). Pretreatment at 60°C for 5 h in a solution with 1% (w/w) H_2O_2 and NaOH resulted in 60% delignification, 40% weight loss, a fivefold increase in Cadoxen accessibility, a one-time increase in water-holding capacity, and only a slight decrease in crystallinity compared to untreated straw. Enhancements to the pretreatment effect could be achieved by increasing the initial alkalinity and pretreatment temperature of the hydrogen peroxide solution (Yan et al., 2020). A saturated improvement in structural features occurred when the weight ratio of hydrogen peroxide to straw exceeded 0.25 g H_2O_2/g straw in an alkaline H_2O_2 solution with 1% (w/w) NaOH at 32°C. The rates and extents of hydrolysis, cellulase adsorption, and cellobiose accumulation during hydrolysis were enhanced with improved structural features of the pretreated straw (Zhang et al., 2021). A fourfold increase in the extent of enzymatic hydrolysis of straw for 24 hours was attributed to alkaline hydrogen peroxide pretreatment. There are reports on using peracetic acid for rice straw pretreatment. Quantitative changes in the composition of treated straw, crystallinity of treated straw and extracted cellulose, and susceptibility of treated straw with peracetic acid resulted in a slight loss of hemicellulose and cellulose in the straw (Pascoli et al., 2022). Peracetic acid treatments caused little or no breakdown of the crystalline structure of cellulose in the straw. The first degree of enzymatic solubilization relative to the amount of residual straw was 42% after treatment with 20% peracetic acid (Taniguchi et al., 1982).

2.4.3 Ionic Liquid Method

Among various pretreatment methods, ionic liquid pretreatment stands out as a highly promising strategy for lignin removal and enhancing the solubility of carbohydrates. Ionic liquids (ILs) are unique organic salts with a low melting point (below 100°C) and have garnered significant attention due to their exceptional properties, including very low vapor pressure, non-flammability, high chemical and thermal stability, relative non-toxicity, high polarity, and high conductivity (Ziaei-Rad et al., 2021). Imidazolium-based ILs, in particular, are potent and contentious components utilized in pretreatment processes. Additionally, acid ILs such as 1-ethyl-3-methylimidazolium hydrogen sulfate and 1-butyl-3-methylimidazolium hydrogen sulfate have been employed for biomass fractionation (da Costa Lopes et al., 2013), yielding high glucose yields of up to 90% following enzymatic saccharification. Furthermore, efficient separation of biomass-derived xylose and [emim][HSO_4] from biomass mixtures has been achieved by da Costa Lopes et al., with recovery yields of 90.8 and 98.1 wt% for IL and xylose, respectively.

Despite their considerable potential, the widespread application of ILs faces significant hurdles primarily due to their high costs. The estimated production cost of [emim][OAc] ranges from $20 to $101 per kilogram, while also exhibiting low thermal stability and moisture tolerance (Brandt-Talbot et al., 2017). The high moisture content of fresh biomass, reaching up to 50%, along with the hygroscopic nature of ILs, presents considerable challenges in maintaining low water content during biomass treatment.

2.4.4 Enzymatic Hydrolysis

Recent advancements in enzyme technology have significantly propelled lignocellulosic ethanol research forward. Enzymatic hydrolysis typically occurs under mild conditions, involving low pressure and extended retention time, especially concerning hemicellulose hydrolysis. Studies on rice straw hydrolysis using sulfuric acid at varying concentrations and temperatures have been conducted, demonstrating the potential for high sugar yields (Zhu et al., 2015). The kinetics of glucose production from rice straw by *Aspergillus niger* revealed dependencies on particle size, temperature, pH, substrate concentration, and cell loading. Various pretreatment methods, including alkali-assisted photocatalysis and ammonia treatment, have shown enhanced hydrolysis efficiency. Combining enzymes like cellulase, xylanases, and pectinases has been explored to improve hydrolysis efficiency, although it comes with increased process costs (Bhati et al., 2021). Cellulases, including endo-glucanases, exo-glucanases, and β-glucosidase, play crucial roles. Endo-glucanases attack the amorphous region of cellulose, while exo-glucanases cleave cellobiose units from cellulose fibers. β-Glucosidase further splits cellobiase residues into two glucose units. This enzymatic approach, achieving over 80% conversion rates, is attractive for bioethanol production. Addressing factors such as substrate structure and enzyme interaction is vital (Rastogi et al., 2017). Algal-based feedstocks, with their superior porosity, enhance enzyme contact during hydrolysis. The utilization of surfactants, which are surface-active substances like polyethylene glycol and

Tween, can significantly enhance the efficiency of enzymatic hydrolysis. This improvement arises from the adsorption of surfactants onto the surface of lignin instead of the enzymes (Li et al., 2017). As a consequence of this interaction, enzymes are safeguarded from inactivation, leading to a more effective enzymatic hydrolysis process. Cellulase and β-glucosidase, either individually or in combination, stand out as the preferred enzymes for enzymatic hydrolysis, primarily owing to their ability to effectively break down polysaccharides. Additionally, the reactions are carried out under mild conditions, optimizing the enzyme's performance (Rahman et al., 2018). The fermentation process involves the use of microorganisms such as *Saccharomyces cerevisiae*, *Escherichia coli*, and *Zymomonas mobilis*. The potential of bioethanol production through enzymatic hydrolysis has shown a more promising trajectory compared to acid hydrolysis in recent studies. While acid hydrolysis in macroalgae reported a maximum bioethanol yield of only 0.390 g/g and 0.520 g/g in microalgae, enzymatic hydrolysis demonstrated significantly higher yields (Mohapatra et al., 2019). Macroalgae achieved an impressive bioethanol yield of 0.909 g/g through SSF technology, and microalgae exhibited a yield of 0.890 g/g (Ray et al., 2018). The integration of SSF and separate hydrolysis and fermentation (SHF) technologies played a crucial role in enhancing bioethanol production. However, each process has its own set of advantages and disadvantages. SHF allows independent operation at optimum temperatures but is susceptible to end-product inhibition and contamination issues. On the other hand, SSF offers cost reduction, requiring a smaller amount of enzymes with fewer contamination and inhibitory effects. The main challenge with SSF lies in the higher optimum temperature for enzymatic hydrolysis (El-Bakry et al., 2015), posing difficulties in process control. Despite this, literature suggests that SSF is generally preferred over SHF due to its ability to lower costs and achieve higher production rates.

2.5 FACTORS AFFECTING BIOALCOHOL PRODUCTION

Several factors significantly impact bioethanol production, including temperature, sugar concentration, pH, fermentation time, agitation rate, and inoculum size. Elevated temperatures can denature enzymes, diminishing their activity and affecting the overall yield. The optimal sugar concentration for bioethanol production is typically achieved at around 150 g/L (Moonsamy et al., 2021). The pH of the fermentation broth plays a crucial role in influencing bacterial contamination, yeast growth, fermentation rate, and by-product formation. For biomass fermentation using *Saccharomyces cerevisiae*, the optimal pH range is 4.0–5.0 (Lin et al., 2012). Deviating from this range may necessitate longer incubation periods or result in a substantial reduction in ethanol concentration. Agitation rate is another key factor for optimizing bioethanol yield. Higher agitation rates generally lead to increased ethanol production. A commonly utilized agitation rate for yeast cell fermentation is in the range of 150–200 rpm (Rodmui et al., 2008). However, excessive agitation may impede the metabolic activities of the cells. In summary, carefully controlling these factors is essential for maximizing the efficiency of bioethanol production processes. The pH of the fermentation broth plays a crucial role in ethanol production (Brexo et al., 2017), impacting bacterial contamination, yeast growth, fermentation rate, and the formation of by-products.

The concentration of H^+ in the fermentation broth influences the permeability of essential nutrients into cells. Yeast survival and growth are particularly affected by the pH, with an optimal range of 2.75–4.25. Specifically for ethanol production using *S. cerevisiae* (Arroyo et al., 2009), the ideal pH range is 4.0–5.0. If the pH falls below 4.0, it extends the incubation period without a significant reduction in ethanol concentration. On the other hand, a pH above 5.0 leads to a substantial decrease in ethanol concentration. The duration of fermentation significantly (Liu et al., 2015) influences microbial growth. A shorter fermentation time can lead to inefficient fermentation due to inadequate microorganism growth. Conversely, an extended fermentation time, especially in batch mode (Brexo et al., 2017), can have a toxic effect on microbial growth, primarily due to the elevated concentration of ethanol in the fermented broth. Achieving complete fermentation at lower temperatures is possible with longer fermentation times, albeit resulting in the lowest ethanol yield. The agitation rate plays a crucial role in controlling the permeability of nutrients into cells from the fermentation broth and the removal of ethanol (Zhu et al., 2007) from cells to the fermentation broth. Higher agitation rates correspond to increased ethanol production, enhanced sugar consumption, and a reduction in the inhibitory effects of ethanol on cells. The optimal agitation rate for yeast cell fermentation typically ranges between 150 and 200 rpm (Boswell et al., 2002). Excessive agitation rates, however, are not conducive to smooth ethanol production, as they impose limitations on the metabolic activities of the cells.

2.6 DOWNSTREAM PROCESS AND PURIFICATION

The purification step in bioethanol production involves various techniques, including rectification, distillation, and dehydration, each significantly influencing the final products (Bušić et al., 2018). Distillation stands out as the most widely employed technique in the purification stage, despite its high energy consumption. The fundamental principle of distillation revolves around separating mixtures based on component volatilities, necessitating careful observation of the resultant concentration of the content (Sørensen et al., 2014). For bioethanol to be considered for commercialization, it must meet all the required standards established by international organizations such as ASTM and ANP. These standards play a crucial role in ensuring the quality and compliance of bioethanol with industry benchmarks. A distillation unit typically comprises (1) feed (the ethanol to be purified), (2) an energy source (usually steam), (3) overhead, (4) bottom product, and (5) a condenser (Lei et al., 2003). However, when aiming for high purity in bioethanol production, the system often undergoes modifications aligned with advancements in engineering technology, focusing on producing high-grade bioethanol with reduced energy consumption. The distillation process facilitates mass transfer between different components in a counter-current fashion. During this process, two distinct zones are formed based on the volatility of the components (Weeranoppanant et al., 2017), with more volatile components in the vapor-rich region and less volatile components in the liquid-rich region. Ultimately, the end product is drawn off from the system, ready to be blended with gasoline fuel or used directly as fuel. This stage plays a critical role in achieving the desired purity and quality standards for bioethanol (Figures 2.3 and 2.4).

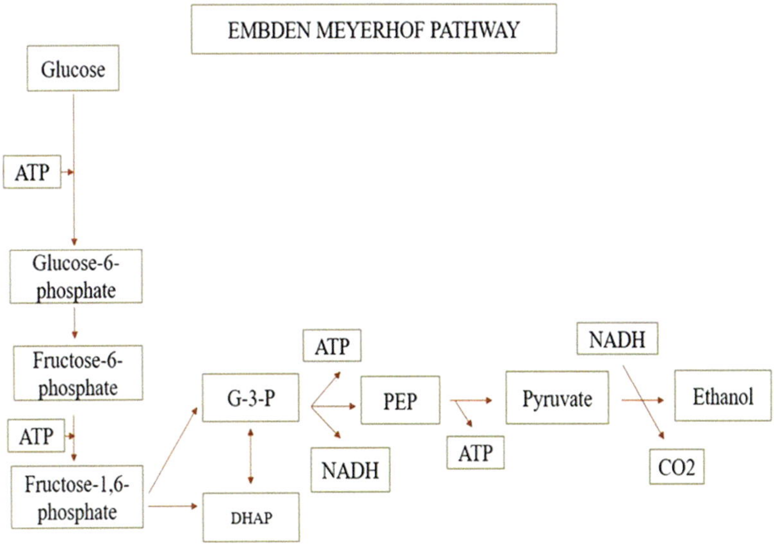

FIGURE 2.3 This figure illustrates the formation of ethanol from glucose under anaerobic conditions using Embden-Meyerhof pathway.

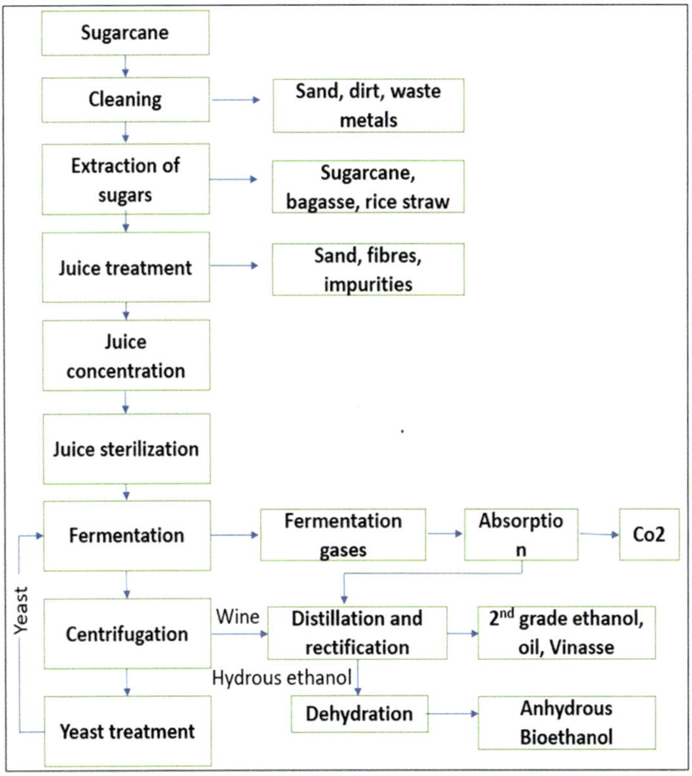

FIGURE 2.4 A schematic block flow diagram delineating the conventional bioethanol production process from sugarcane.

2.7 CHALLENGES

2.7.1 Global Supply and Demand

In the future, biofuels are poised to play a crucial role in addressing the global energy demand. This chapter delves into the four generations of bioalcohols, exploring their biomass sources, conversion technologies, environmental implications, and economic viability (Hoang et al., 2021). Each bioalocohol generation presents distinct advantages and challenges. Ensuring an uninterrupted supply of raw materials is paramount to meet the escalating energy needs. Achieving self-sufficiency in raw oil supply requires producing major bialcohol feedstocks domestically without compromising food availability (Noorollahi et al., 2021). Geographical location, economic conditions, and food-fuel demand influence the availability and production of biofuel feedstocks. This underscores the unsuitability of first-generation biofuels, as they trigger debates surrounding food-fuel competition. In contrast, the third and fourth generations prove more suitable, avoiding such controversies. Crucial to the economic (Luque et al., 2008) and environmental sustainability of biofuels is technological advancement in their production. Establishing cost-effective, high-yield conversion systems is imperative for profitable commercial biofuel production. Metabolic engineering tools offer avenues for improving conversion yield by modifying (Jagadevan et al., 2018) existing biological pathways, enhancing both the quantity and quality of biofuels. This tool also plays a pivotal role in modifying feedstocks or enhancing microbes for superior conversion. Further research is essential to achieve high-yield, cost-effective conversion processes. Regarding GHG mitigation, the second and third generations (Nanda et al., 2018) of biofuels have demonstrated superiority over the first generation. Anticipated improvements in GHG mitigation can be expected from the fourth generation of biofuels. From an economic standpoint, the first-generation biofuel currently stands out as the most cost-effective fuel (Darda et al., 2019). However, its production capacity is constrained to specific countries due to its high land intensity. In contrast, second- and third-generation biofuels face higher costs, primarily attributed to substantial investment (Alalwan et al., 2019) requirements and relatively low conversion efficiencies of feedstock into biofuel. Despite these challenges, ongoing efforts in optimization and advancements in conversion technology hold the promise of rendering second- and third-generation bioethanol and biodiesel production more economically viable. In summary, this chapter suggests that the future of biofuels may entail a combination of some or all four generations, rather than relying solely on one generation. This approach acknowledges the unique strengths and challenges of each generation, aiming for a more diversified and sustainable biofuel landscape.

2.7.2 Microbial Contamination

Industrial fermentations often face contamination challenges from bacteria and wild yeast, including both *Saccharomyces* and non-*Saccharomyces* species. These contaminants compete with selected yeast strains in fermenters, and difficulties in sterilizing large volumes of juice and water during yeast cell recycling contribute to their entry into the process (Bonatelli et al., 2019). Gram-positive bacteria like *Lactobacillus* and *Bacillus* are common contaminants, controlled through acid treatment, antibiotics, hop products, and chemical biocides. However, some bacteria resistant to these methods can be challenging to control during recycling, causing significant bioethanol losses (Chen et al., 2021). High bacterial contamination can lead to yeast flocculation and fermentation inhibition. In processes involving yeast cell recycling by centrifugation, flocculant yeast strains pose problems. The yeast cream, diluted with water and subjected to acid treatment in the Brazilian alcoholic fermentation process, may become less concentrated due to flocculation, reducing centrifugation efficiency (Reis et al., 2013). This results in more wine going through acid treatment, requiring additional sulfuric acid expenditure. The intensified acid treatment can negatively impact yeast activity, and flocculant cells may settle at the bottom of fermentors, leaving residual sugars unfermented and prolonging fermentation time. Addressing these challenges (Eggleston et al., 2013) is essential for efficient bioethanol production. In recent times, distilleries have streamlined procedures to identify the most effective antibiotic for controlling bacterial contamination within a mere 6 hours. However, the primary challenge lies in preventing bacterial contamination in industrial fermentations (Seo et al., 2020) that involve yeast cell recycling. Some distilleries used to eliminate a portion of yeast cells, drying them for sale as animal feed. These distilleries refrained from using traditional antibiotics to avoid leaving residues (Pikkemaat et al., 2011) in the dry yeast. There is a need for the development of new, safer, and environmentally friendly compounds for bacterial control, particularly for distilleries engaged in the sale of dry yeast.

2.7.3 Vinasse Production

Addressing the scientific challenge of reducing the volume of vinasse and exploring new uses and products from it is an ongoing concern. Vinasse, the residue obtained after distillation and ethanol removal from wine (Fuess et al., 2022), is rich in minerals like potassium, calcium, magnesium, nitrogen, and phosphorus. While vinasse has found utility as a fertilizer in sugarcane fields, contributing to water savings and soil improvement, its application requires careful procedures to prevent water table (Ortegón et al., 2016) contamination. Additionally, exploring innovative applications for vinasse remains an active area of research. Currently, distilleries generate approximately 12 liters of vinasse for every liter of ethanol produced. This ratio is influenced by the alcohol content in the wine at the end of fermentation (Varela et al., 2015), typically reaching 8–9%. The challenge lies in the fact that higher alcohol concentrations in the wine diminish the viability of yeast cells. Sustaining cellular viability is crucial for fermentation processes involving yeast cell recycling (Davey et al., 2012), where yeast cells need to remain alive until the conclusion of the fermentation process. This is essential as these cells are reused multiple times over a period of 6–8 months.

2.7.4 Waste Water Production from Distilleries

Bioethanol production facilities generate a wastewater by-product known by various names such as stillage, thin stillage, distillery wastewater, distillery slop, dunder, spent wash, and vinasse. Typically, 8–20 liters of stillage are produced per liter of ethanol. The characteristics of stillage, including high biological oxygen demand (BOD) (Wilkie et al., 2000), high chemical oxygen demand (COD), and low pH, vary based on feedstock and processing methods. Although stillage from sugar and starch feedstocks has been well studied, information on variations in stillage from different cellulosic feedstocks (Mojović et al., 2009) and production processes, especially at a commercial scale, is less comprehensive. Recent technological advances and initiatives to increase cellulosic ethanol production indicate a potential surge in stillage production from such materials. Stillage by-products from sugar and starch feedstocks exhibit similar values for BOD and COD (Zhang et al., 2016) with pH values ranging from 3.8 to 5.5. Reports on stillage characteristics from cellulosic feedstocks are limited, but BOD and COD values fall within the lower range of those reported for sugar and starch feedstocks. Nutrient values in stillage from sugar and starch feedstocks vary widely. The treatment of such wastewater to meet acceptable levels poses economic and logistical challenges.

2.8 ECONOMIC SURVEY FOR BIOALCOHOL PRODUCTION

While the land cost for algae biomass production is relatively low, the expenses associated with infrastructure and mixing are higher. Despite these advantages, the elevated costs of energy recovery in algae production offset a significant portion of the benefits. In the case of rapeseed biodiesel, the impact of food prices and GHG (Judd et al., 2017) emissions stand out as prominent costs. Research by Talebi delves into the cost efficiency of biomethane production from various sources, including the organic fraction of municipal solid waste (OFMSW), slaughterhouse waste (SHW), and grass (Talebi et al., 2022) and slurry. OFMSW emerges as the most cost-effective source, followed by SHW and grass and slurry, with biomethane from the latter falling within the price range of petroleum-derived transport fuels (Sikarwar et al., 2017). As of January 2007, estimates by the International Energy Agency (IEA) indicate that the cost of sugarcane ethanol in Brazil is \$0.30 per liter gasoline-equivalent (lge), while ethanol from maize, sugar beet, and wheat ranges from \$0.6 to \$0.8 per lge, excluding subsidies (Cazzola et al., 2013). Lignocellulosic ethanol, produced at the pilot scale, costs around \$1.0 per lge. Various Brazilian countries have now started to utilize various alternative feedstock residues from bioethanol for the production of bioelectricity (Khatiwada et al., 2016). The efforts underscore a commitment to advancing the sustainability of biofuel production in Brazilian distilleries, with a dual aim of conserving critical natural resources—such as water and land—and minimizing the combustion of sugarcane, which in turn reduces the release of GHGs into the atmosphere (Amorim et al., 2019).

To broaden the scope of sustainable practices, exploration into alternative feedstocks, including sorghum bagasse, corn stover, grass, algae, and more, is underway. The evaluation of these diverse materials aims not only to diversify feedstock sources but also to contribute to the global effort to mitigate GHG emissions on a larger scale (Amorim et al., 2010). It's worth noting that, in this pursuit, the bioethanol production processes based on sugarcane in Brazil stand out as exemplary. The GHG emissions associated with these processes are considered superior when compared to the emissions from other major substrates (Toor et al., 2020) currently utilized in bioethanol production. This distinction underscores the significant strides made in the Brazilian biofuel industry toward sustainability and environmental responsibility.

At present, about 90% of Brazilian distilleries are affiliated with sugar factories, engaging in energy-intensive processes such as sugarcane milling (Leal et al., 2013), juice clarification, evaporation, and distillation. Various stages of sugar and alcohol production require substantial steam, which distilleries generate by using sugarcane bagasse. The challenge lies in enhancing the efficiency of steam usage, allowing more bagasse to be repurposed. Currently, distilleries primarily use sugarcane bagasse for steam production. However, the development of high-efficiency boilers enables (Martinez-Hernandez et al., 2018) the burning of less bagasse, creating opportunities for alternative uses such as bioelectricity and second-generation bioethanol. Comparing the global energy matrix to Brazil's, a notable contrast emerges. While the world relies on only 12% renewable energy, Brazil has surpassed 45% in the use of renewable energies, with a significant contribution from distilleries (Bernal et al., 2017). In 2009, electricity generated from bagasse combustion accounted for 3% of the total electricity produced in the country, presenting new prospects for distilleries in the realm of sustainable energy.

2.9 FUTURE TRENDS

The existing primary energy source, fossil fuels, is finite, underscoring the necessity to supplement it with renewable alternatives like agricultural biomass. Maize, a major biofuel crop in the United States, accounts for 31% of global cereal production and utilizes over one-fifth of dedicated cereal cultivation land worldwide (Torney et al., 2007). As the second-largest biotech crop globally, around 10% of maize cultivation involves biotech varieties, primarily after soybean. The current genetic engineering of maize has mainly focused on a few transformable genotypes, lacking consistent desired agronomic traits (Yadava et al., 2017). Enhancing the ability to directly introduce transgenes into inbred or elite genetic backgrounds is crucial for efficient bioethanol production, reducing the time needed for transgene integration into elite maize lines. Ongoing technological advancements aim to improve transgene expression quality through tissue or developmental stage-specific expression, stringent regulation, induced gene expression, site-specific integration, and gene stacking (sequential addition of transgenes in a genome) (Liu et al., 2013). Genetic engineering aimed at enhancing biomass yield is a multifaceted endeavor. While several biotechnological crop lines designed for increased yield are undergoing testing, many of the genes involved in this trait remain unidentified. The primary concern in biobutanol (butanol and 2,3-butanediol) production lies in the cost of feedstock. Consequently, the high expenses associated with traditional substrates have steered research attention toward renewable alternatives. Renewable feedstocks primarily consist of biomass rich in hemicellulose, which can serve as viable fermentation substrates.

The augmentation and stabilization of biomass yield hinge on comprehending and improving mechanisms related to stress tolerance and carbohydrate metabolism (Raza et al., 2020). Stress tolerance in plants has primarily been achieved through the manipulation of effector genes, involving ion transporters and biosynthetic enzymes, and regulatory genes such as transcription factors and signal transduction components. An exemplary case is the use of transgenic maize expressing d-endotoxins from *Bacillus thuringiensis* (Bt) to confer resistance to biotic stress. Strategies for next-generation insect-resistant crops explore diverse approaches, including broad-spectrum (Bartsch et al., 2016) insecticidal proteins and novel proteins from various sources. While insect damage accounts for a notable percentage of crop loss, abiotic stress has a more significant role. Reactive oxygen species (ROS) accumulation under (Sachdev et al., 2021) stress conditions is a common denominator, leading to deleterious effects. Mitogen-activated protein kinases (MAPKs), associated with both biotic and abiotic stress responses, offer a promising avenue for enhancing stress tolerance in plants. Additionally, the manipulation of antioxidant enzymes like superoxide dismutase (SOD) has proven effective. Photosynthesis improvements (Kwon et al., 2001) in C4 plants like maize involve enhancing enzymes like PEPC. Transgenic maize overexpressing PEPC has demonstrated improved CO_2 fixation, increased fresh and dry weight, and enhanced resistance to water stress (Qian et al., 2015). Modifying plant architecture, such as reducing brassinosteroid levels, can further enhance photosynthesis and biomass production.

Addressing grain yield involves targeting enzymes like ADP-glucose pyrophosphorylase (AGP), a key regulator of starch biosynthesis. Deregulation of AGP through genetic engineering has shown promising results in increasing seed and biomass yield (Torney et al., 2007). In conclusion, genetic engineering holds great potential for advancing agriculture and biofuel production by addressing complex traits related to stress tolerance, photosynthesis, and grain yield. Despite its potential, public acceptance and regulatory challenges remain significant hurdles that need careful consideration. Developing multiple transgene strategies is crucial for tackling complex traits and ensuring the broad applicability and acceptance of genetic engineering technologies in bioethanol production (Dixit et al., 2023). The development of an integrated agri-biotechnology system for food, feed, and fuel production is anticipated to be a regulatory challenge but a potential future for maize in bioethanol production.

Over the past decade, federal policies have been implemented to boost ethanol production, responding to both favorable and challenging economic conditions from 2000 to 2010. These conditions have had a significant impact on the ethanol production landscape, and despite the initial impetus for development (Robertson et al., 2017), the growth of cellulosic ethanol production has not matched early predictions. The United States and Brazil have been global leaders in ethanol production since the 1970s. While Brazil has a long-standing history of ethanol production for transportation (Sajid et al., 2021), the United States initially showed minimal inclination to significantly increase ethanol use in the transportation sector. Around the mid-2000s, in 2005, U.S. ethanol's share in gasoline blends was a mere 2.8% of total fuel use. However, by 2008, this figure had risen to nearly 7%. Early governmental support for corn ethanol involved a partial exemption from the federal gasoline excise tax for the fuel containing a minimum of 10% biomass-derived ethanol (Shinn et al., 2011), a fuel blender's tax credit, and a small ethanol producer tax credit. In 2004, the Volumetric Ethanol Excise Tax Credit (VEETC) transitioned to a volume-based system. The Energy Policy Act of 2005 introduced the Renewable Fuel Standard (RFS), establishing initial production goals for cellulosic ethanol (Solomon et al., 2015). Other critical components of the Energy Policy Act included amendments to the Biomass Research and Development Act and the initiation of the systems biology and bioenergy program. The bioenergy program empowered the U.S. Department of Energy (USDOE) to collaborate with industrial and academic institutions to propel the advancement of biofuels.

2.10 CONCLUSION AND PERSPECTIVES

In summary, the journey toward harnessing lignocellulosic biomass, particularly the abundant and compositionally appealing rice straw, unveils substantial potential for bioalcohol production. The biological transformation of rice straw into fermentable sugars, facilitated by hydrolysing enzymes, emerges as an environmentally compelling alternative, notwithstanding the intricate challenges posed by its complex composition, high lignin, and ash content. The ongoing strides in devising efficient pretreatment methods signify considerable success, offering a glimpse into the prospect of meeting the bioethanol demands of the transportation sector. The imperative remains on process and strain engineering to surmount hurdles related to xylose and glucose co-fermentation, elevating the overall system's efficiency. A judicious blend of pretreatment, hydrolysis, and fermentation processes stands as a linchpin for maximizing efficacy. The integration of genetically modified yeast, synthetic hydrolyzing enzymes, and other avant-garde technologies holds the tantalizing promise of rendering bioethanol production from rice straw a plausible reality in the foreseeable future. The vast potential of genetic engineering technology in shaping the future of agriculture and biofuel production, as elucidated earlier, is indisputable. Nevertheless, the global acceptance of crops derived from biotechnology has encountered skepticism and regulatory challenges across various nations. A significant public apprehension revolves around the management of pollen dissemination, especially concerning wind-pollinated crops like maize. Plastid genome transformation emerges as a strategic approach, offering the advantage of constraining the transmission of the transgene through pollen, thereby preserving the plant's fertility and enabling enhanced production of transgenic products. While the transformation of plastid genomes has been successfully demonstrated in a few plant species, its broader implementation requires careful consideration and scrutiny.

The exploration into traditional fermentation methods underscores the significance of augmenting carbohydrate content and biomass productivity, necessitating thorough technical and economic evaluations to navigate the energy-related challenges inherent in utilizing microbiological biomass for ethanol production. Simultaneously, the allure of engineered cyanobacteria in industrial bioethanol processes beckons, yet in-depth investigations into their structural, metabolic, and genetic facets are imperative for comprehending both their industrial applicability and the potential environmental risks they might pose. A meticulous numerical scrutiny of production costs becomes paramount when juxtaposing genetically engineered microorganisms with traditional processes that rely on enzymes and yeasts for efficient bioethanol production. Despite bottlenecks in biofuel production from microalgae and cyanobacteria, the ongoing exploration of hydrolysis, fermentation technologies, and the nuanced understanding of genetically modified cyanobacteria remains pivotal. The evaluation of costs associated with microalgae/cyanobacteria cultivation necessitates careful consideration, especially when juxtaposed with fossil fuels, which, despite environmental and climatic advantages, currently pose economic challenges. The strategic integration of second-generation processes with existing first-generation ethanol processes emerges as an appealing avenue, albeit with a caveat for prudently weighing the economic viability of existing ethanol production, particularly when tethered to valuable by-products like animal feed. Looking toward the horizon, the unfolding narrative of genetic engineering technology presents a tableau of immense potential for agriculture and biofuel production. Despite encountering headwinds in the form of public skepticism and regulatory hurdles, ingenious strategies such as plastid genome transformation and engineered male sterility offer pathways to address concerns surrounding pollen dissemination in wind-pollinated crops like maize. These pathways pave the road toward a sustainable and responsible future in biofuel production, navigating the intricate landscape of challenges with innovative and conscientious solutions. While challenges persist, the exploration of traditional and novel fermentation methods, along with the continuous evolution of technological approaches, highlights the ongoing quest to maximize carbohydrate content, biomass productivity, and overall system efficiency. With the integration of genetically modified organisms, advanced enzymatic processes, and sustainable production methods, bioethanol from lignocellulosic biomass holds substantial promise for meeting the world's energy needs while minimizing environmental impact.

ACKNOWLEDGMENTS

The authors would like to thank Prof. Prem Kumar Khosla, Chancellor, Shoolini University of Biotechnology and Management Sciences, Solan, Himachal Pradesh, India, and the School of Biotechnology here for providing the laboratory and technical facilities in the field of Nano-Biotechnology.

BIBLIOGRAPHY

Ab Rasid, N.S., Shamjuddin, A., Rahman, A.Z.A. and Amin, N.A.S., 2021. Recent advances in green pre-treatment methods of lignocellulosic biomass for enhanced biofuel production. *Journal of Cleaner Production*, *321*, p.129038.

Abdullah, B., Muhammad, S. A. F. A. S., Shokravi, Z., Ismail, S., Kassim, K. A., Mahmood, A. N., & Aziz, M. M. A. (2019). Fourth generation biofuel: A review on risks and mitigation strategies. *Renewable and sustainable energy reviews*, *107*, 37–50.

Aditiya, H.B., Mahlia, T.M.I., Chong, W.T., Nur, H. and Sebayang, A.H., 2016. Second generation bioethanol production: A critical review. *Renewable and Sustainable Energy Reviews*, *66*, pp.631–653.

Afolalu, S.A., Yusuf, O.O., Abioye, A.A., Emetere, M.E., Ongbali, S.O. and Samuel, O.D., 2021, March. Biofuel; A sustainable renewable source of energy-a review. In *IOP Conference Series: Earth and Environmental Science* (Vol. 665, No. 1, p. 012040). IOP Publishing.

Ahmad, F., Khan, A.U. and Yasar, A., 2013. Transesterification of oil extracted from different species of algae for biodiesel production. *African Journal of Environmental Science and Technology*, *7*(6), pp.358–364.

Ahmed, S., Lockwood, K., Miller, J.H., Huq, N., Luecke, J., Labbe, N. and Foust, T., 2024. Understanding fundamental effects of biofuel structure on ignition and physical fuel properties. *Fuel*, *358*, p.129999.

Ahorsu, R., Medina, F. and Constantí, M., 2018. Significance and challenges of biomass as a suitable feedstock for bioenergy and biochemical production: A review. *Energies*, *11*(12), p.3366.

Alalwan, H.A., Alminshid, A.H. and Aljaafari, H.A., 2019. Promising evolution of biofuel generations. Subject review. *Renewable Energy Focus*, 28, pp.127–139.

Arroyo-López, F.N., Orlić, S., Querol, A. and Barrio, E., 2009. Effects of temperature, pH and sugar concentration on the growth parameters of Saccharomyces cerevisiae, S. kudriavzevii and their interspecific hybrid. *International Journal of Food Microbiology*, *131*(2–3), pp.120–127.

Ayodele, B.V., Alsaffar, M.A. and Mustapa, S.I., 2020. An overview of integration opportunities for sustainable bioethanol production from first-and second-generation sugar-based feedstocks. *Journal of Cleaner Production*, *245*, p.118857.

Azhar, S. H. M., Abdulla, R., Jambo, S. A., Marbawi, H., Gansau, J. A., Faik, A. A. M., & Rodrigues, K. F. (2017). Yeasts in sustainable bioethanol production: A review. *Biochemistry and Biophysics Reports*, *10*, 52–61.

Baeyens, J., Kang, Q., Appels, L., Dewil, R., Lv, Y. and Tan, T., 2015. Challenges and opportunities in improving the production of bioethanol. *Progress in Energy and Combustion Science*, *47*, pp.60–88.

Balasubramani, V., Nagarajan, K. J., Karthic, M., & Pandiyarajan, R. (2024). Extraction of lignocellulosic fiber and cellulose microfibrils from agro waste-palmyra fruit peduncle: Water retting, chlorine-free chemical treatments, physio-chemical, morphological, and thermal characterization. *International Journal of Biological Macromolecules*, *259*, 129273.

Balat, M. and Balat, H., 2009. Recent trends in global production and utilization of bio-ethanol fuel. *Applied Energy*, *86*(11), pp.2273–2282.

Balat, M., Balat, H. and Öz, C., 2008. Progress in bioethanol processing. *Progress in Energy and Combustion Science*, *34*(5), pp.551–573.

Barati, B., Zeng, K., Baeyens, J., Wang, S., Addy, M., Gan, S. Y., & Abomohra, A. E. F. (2021). Recent progress in genetically modified microalgae for enhanced carbon dioxide sequestration. *Biomass and Bioenergy*, *145*, 105927.

Bartsch, D., Gathmann, A., Saeglitz, C. and Sinha, A., 2016. Field testing of transgenic plants. *Plant Biotechnology and Genetics: Principles, Techniques, and Applications*, pp.333–346.

Baweja, P., Kumar, S., Sahoo, D., & Levine, I. (2016). Biology of seaweeds. In *Seaweed in health and disease prevention* (pp. 41–106). Academic Press.

Beacham, T. A., Sweet, J. B., & Allen, M. J. (2017). Large scale cultivation of genetically modified microalgae: A new era for environmental risk assessment. *Algal Research*, *25*, 90–100.

Bernal, A. P., dos Santos, I. F. S., Silva, A. P. M., Barros, R. M., & Ribeiro, E. M. (2017). Vinasse biogas for energy generation in Brazil: An assessment of economic feasibility, energy potential and avoided CO2 emissions. *Journal of Cleaner Production*, *151*, 260–271.

Bharathiraja, B., Jayamuthunagai, J., Praveenkumar, R., VinothArulraj, J., Vinoshmuthukumar, P. and Saravanaraj, A., 2014. Bioethanol production from lignocellulosic materials–an overview. *The SciTech Journal*, *1*(7), pp.28–36.

Bhati N. Shreya and Sharma, A.K., 2021. Cost-effective cellulase production, improvement strategies, and future challenges. *Journal of Food Process Engineering*, *44*(2), p.e13623.

Bhattacharyya, P., Bisen, J., Bhaduri, D., Priyadarsini, S., Munda, S., Chakraborti, M., Adak, T., Panneerselvam, P., Mukherjee, A.K., Swain, S.L. and Dash, P.K., 2021. Turn the wheel from waste to wealth: economic and environmental gain of sustainable rice straw management practices over field burning in reference to India. *Science of the Total Environment*, *775*, p.145896.

Bonatelli, M.L., Ienczak, J.L. and Labate, C.A., 2019. Sugarcane must fed-batch fermentation by Saccharomyces cerevisiae: impact of sterilized and non-sterilized sugarcane must. *Antonie van Leeuwenhoek*, *112*, pp.1177–1187.

Boswell, C.D., Nienow, A.W. and Hewitt, C.J., 2002. Studies on the effect of mechanical agitation on the performance of brewing fermentations: Fermentation rate, yeast physiology, and development of flavor compounds. *Journal of the American Society of Brewing Chemists*, *60*(3), pp.101–106.

Branco, R.H., Serafim, L.S. and Xavier, A.M., 2018. Second generation bioethanol production: on the use of pulp and paper industry wastes as feedstock. *Fermentation*, *5*(1), p.4.

Brandt-Talbot, A., Gschwend, F. J., Fennell, P. S., Lammens, T. M., Tan, B., Weale, J., & Hallett, J. P. (2017). An economically viable ionic liquid for the fractionation of lignocellulosic biomass. *Green Chemistry*, *19*(13), 3078–3102.

Brown, M.R., Mular, M., Miller, I., Farmer, C. and Trenerry, C., 1999. The vitamin content of microalgae used in aquaculture. *Journal of Applied Phycology*, *11*, pp.247–255.

Bušić, A., Marđetko, N., Kundas, S., Morzak, G., Belskaya, H., Ivančić Šantek, M., Komes, D., Novak, S. and Šantek, B., 2018. Bioethanol production from renewable raw materials and its separation and purification: a review. *Food Technology and Biotechnology*, *56*(3), pp.289–311.

Canam, T., Town, J., Iroba, K., Tabil, L. and Dumonceaux, T., 2013. Pretreatment of lignocellulosic biomass using microorganisms: approaches, advantages, and limitations. *Sustainable degradation of lignocellulosic biomass-techniques, applications and commercialization*, pp.181–206.

Cazzola, P., Morrison, G., Kaneko, H., Cuenot, F., Ghandi, A. and Fulton, L., 2013. Production costs of alternative transportation fuels. Influence of Crude Oil Price and Technology Maturity.

Celińska, E. and Grajek, W., 2009. Biotechnological production of 2, 3-butanediol—current state and prospects. *Biotechnology Advances*, *27*(6), pp.715–725.

Chandra, R.P., Bura, R., Mabee, W.E., Berlin, D.A., Pan, X. and Saddler, J.N., 2007. Substrate pretreatment: the key to effective enzymatic hydrolysis of lignocellulosics? *Biofuels*, pp.67–93.

Chaturvedi, V. and Verma, P., 2013. An overview of key pretreatment processes employed for bioconversion of lignocellulosic biomass into biofuels and value added products. *3 Biotech*, *3*, pp.415–431.

Chen, J., Bai, J., Li, H., Chang, C. and Fang, S., 2015. Prospects for bioethanol production from macroalgae. *Trends in Renewable Energy*, *1*(3), pp.185–197.

Chen, J., Zhang, B., Luo, L., Zhang, F., Yi, Y., Shan, Y., Liu, B., Zhou, Y., Wang, X. and Lü, X., 2021. A review on recycling techniques for bioethanol production from lignocellulosic biomass. *Renewable and Sustainable Energy Reviews*, *149*, p.111370.

Choi, Y.Y., Patel, A.K., Hong, M.E., Chang, W.S. and Sim, S.J., 2019. Microalgae Bioenergy with Carbon Capture and Storage (BECCS): An emerging sustainable bioprocess for reduced CO2 emission and biofuel production. *Bioresource Technology Reports*, *7*, p.100270.

da Costa Lopes, A. M., João, K. G., Morais, A. R. C., Bogel-Łukasik, E., & Bogel-Łukasik, R. (2013). Ionic liquids as a tool for lignocellulosic biomass fractionation. *Sustainable Chemical Processes*, *1*, 1–31.

da Silva, A.S.A., Espinheira, R.P., Teixeira, R.S.S., de Souza, M.F., Ferreira-Leitão, V. and Bon, E.P., 2020. Constraints and advances in high-solids enzymatic hydrolysis of lignocellulosic biomass: a critical review. *Biotechnology for Biofuels*, *13*(1), pp.1–28.

Darda, S., Papalas, T. and Zabaniotou, A., 2019. Biofuels journey in Europe: Currently the way to low carbon economy sustainability is still a challenge. *Journal of Cleaner Production*, *208*, pp.575–588.

Davey, H.M., Cross, E.J., Davey, C.L., Gkargkas, K., Delneri, D., Hoyle, D.C., Oliver, S.G., Kell, D.B. and Griffith, G.W., 2012. Genome-wide analysis of longevity in nutrient-deprived Saccharomyces cerevisiae reveals importance of recycling in maintaining cell viability. *Environmental Microbiology*, *14*(5), pp.1249–1260.

Davis, R., Tao, L., Tan, E.C.D., Biddy, M.J., Beckham, G.T., Scarlata, C., Jacobson, J., Cafferty, K., Ross, J., Lukas, J. and Knorr, D., 2013. *Process design and economics for the conversion of lignocellulosic biomass to hydrocarbons: dilute-acid and enzymatic deconstruction of biomass to sugars and biological conversion of sugars to hydrocarbons* (No. NREL/TP-5100–60223). National Renewable Energy Lab.(NREL), Golden, CO (United States).

Den, W., Sharma, V.K., Lee, M., Nadadur, G. and Varma, R.S., 2018. Lignocellulosic biomass transformations via greener oxidative pretreatment processes: access to energy and value-added chemicals. *Frontiers in Chemistry*, *6*, p.141.

Dixit, Y., Yadav, P., Sharma, A.K., Pandey, P. and Kuila, A., 2023. Multiplex genome editing to construct cellulase engineered Saccharomyces cerevisiae for ethanol production from cellulosic biomass. *Renewable and Sustainable Energy Reviews*, *187*, p.113772.

Dutta, N., Usman, M., Luo, G. and Zhang, S., 2022. An insight into valorization of lignocellulosic biomass by optimization with the combination of hydrothermal (HT) and biological techniques: A review. *Sustainable Chemistry*, *3*(1), pp.35–55.

Eggleston, G., Cole, M. and Andrzejewski, B., 2013. New commercially viable processing technologies for the production of sugar feedstocks from sweet sorghum (Sorghum bicolor L. Moench) for manufacture of biofuels and bioproducts. *Sugar Tech*, *15*, pp.232–249.

El-Bakry, M., Abraham, J., Cerda, A., Barrena, R., Ponsá, S., Gea, T. and Sánchez, A., 2015. From wastes to high value added products: novel aspects of SSF in the production of enzymes. *Critical Reviews in Environmental Science and Technology*, *45*(18), pp.1999–2042.

El-Sheekh, M. and Abomohra, A.E.F. eds., 2021. *Handbook of algal biofuels: aspects of cultivation, conversion, and biorefinery*. Elsevier.

Fabi, C., Cachia, F., Conforti, P., English, A. and Moncayo, J.R., 2021. Improving data on food losses and waste: From theory to practice. *Food Policy*, *98*, p.101934.

Ferreira, J.A., Brancoli, P., Agnihotri, S., Bolton, K. and Taherzadeh, M.J., 2018. A review of integration strategies of lignocelluloses and other wastes in 1st generation bioethanol processes. *Process Biochemistry*, *75*, pp.173–186.

Fuess, L.T., Lens, P.N., Garcia, M.L. and Zaiat, M., 2022. Exploring potentials for bioresource and bioenergy recovery from Vinasse, the "New" Protagonist in Brazilian Sugarcane Biorefineries. *Biomass*, *2*(4), pp.374–411.

Gawal, P.M. and Subudhi, S., 2023. Advances and challenges in bio-based 2, 3-BD downstream purification: A comprehensive review. *Bioresource Technology Reports*, 24, 101638.

Gibson, B.R., Lawrence, S.J., Leclaire, J.P., Powell, C.D. and Smart, K.A., 2007. Yeast responses to stresses associated with industrial brewery handling. *FEMS Microbiology Reviews*, *31*(5), pp.535–569.

Gomes, D.G., Coelho, E., Silva, R., Domingues, L. and Teixeira, J.A., 2023. Bioreactors and engineering of filamentous fungi cultivation. In *Current Developments in Biotechnology and Bioengineering* (pp. 219–250). Elsevier.

Gupta, P.L., Lee, S.M. and Choi, H.J., 2015. A mini review: photobioreactors for large scale algal cultivation. *World Journal of Microbiology and Biotechnology*, *31*, pp.1409–1417.

Heimann, K. and Huerlimann, R., 2015. Microalgal classification: major classes and genera of commercial microalgal species. In *Handbook of marine microalgae* (pp. 25–41). Academic Press.

Henley, W. J., Litaker, R. W., Novoveská, L., Duke, C. S., Quemada, H. D., & Sayre, R. T. (2013). Initial risk assessment of genetically modified (GM) microalgae for commodity-scale biofuel cultivation. *Algal Research*, *2*(1), 66–77.

Hoang, A.T., Ong, H.C., Fattah, I.R., Chong, C.T., Cheng, C.K., Sakthivel, R. and Ok, Y.S., 2021. Progress on the lignocellulosic biomass pyrolysis for biofuel production toward environmental sustainability. *Fuel Processing Technology*, *223*, p.106997.

Huang, J., Khan, M.T., Perecin, D., Coelho, S.T. and Zhang, M., 2020. Sugarcane for bioethanol production: Potential of bagasse in Chinese perspective. *Renewable and Sustainable Energy Reviews*, *133*, p.110296.

Huang, S., Liu, T., Peng, B. and Geng, A., 2019. Enhanced ethanol production from industrial lignocellulose hydrolysates by a hydrolysate-cofermenting Saccharomyces cerevisiae strain. *Bioprocess and biosystems engineering*, *42*, pp.883–896.

Ibeto, C.N., Ofoefule, A.U. and Agbo, K.E., 2011. A global overview of biomass potentials for bioethanol production: a renewable alternative fuel. *Trends in Applied Sciences Research*, *6*(5), p.410.

Ismail, M.M., Ismail, G.A. and El-Sheekh, M.M., 2020. Potential assessment of some micro- and macroalgal species for bioethanol and biodiesel production. *Energy sources, part a: recovery, utilization, and environmental effects*, pp.1–17.

Jagadevan, S., Banerjee, A., Banerjee, C., Guria, C., Tiwari, R., Baweja, M. and Shukla, P., 2018. Recent developments in synthetic biology and metabolic engineering in microalgae towards biofuel production. *Biotechnology for Biofuels*, *11*, pp.1–21.

Jambo, S.A., Abdulla, R., Azhar, J.A., 2016. A review on third generation bioethanol feedstock. *Renewable and Sustainable Energy Reviews*, *65*, pp.756–769.

Jin, C., Yao, M., Liu, H., Chia-Fon, F.L. and Ji, J., 2011. Progress in the production and application of n-butanol as a biofuel. *Renewable and Sustainable Energy Reviews*, *15*(8), pp.4080–4106.

Johnson, C., Milbrandt, A., Zhang, Y., Hardison, R. and Sharpe, A., 2020. *Jamaican Domestic Ethanol Fuel Feasibility and Benefits Analysis* (No. NREL/TP-5400–76011). National Renewable Energy Lab. (NREL), Golden, CO (United States).

Jonson, S., 2020. Multifunctional production systems in Brazil: Opportunities, barriers, and implementation.

Jönsson, M., Allahgholi, L., Sardari, R.R., Hreggviðsson, G.O. and Nordberg Karlsson, E., 2020. Extraction and modification of macroalgal polysaccharides for current and next-generation applications. *Molecules*, *25*(4), p.930

Judd, S.J., Al Momani, F.A.O., Znad, H. and Al Ketife, A.M.D., 2017. The cost benefit of algal technology for combined CO2 mitigation and nutrient abatement. *Renewable and Sustainable Energy Reviews*, *71*, pp.379–387.

Jusakulvijit, P., Bezama, A. and Thrän, D., 2021. The availability and assessment of potential agricultural residues for the regional development of second-generation bioethanol in Thailand. *Waste and Biomass Valorization*, 12, 1–28.

Karimi, F., Mazaheri, D., Saei Moghaddam, M., Mataei Moghaddam, A., Sanati, A.L. and Orooji, Y., 2021. Solid-state fermentation as an alternative technology for cost-effective production of bioethanol as useful renewable energy: a review. *Biomass Conversion and Biorefinery*, pp.1–17.

Kereluik, M., 2022. *A Review of the Enzymes Secreted from Fungi and Evaluation of their Range of Commercial Applications* (Doctoral dissertation).

Khan, M.A.H., Bonifacio, S., Clowes, J., Foulds, A., Holland, R., Matthews, J.C., Percival, C.J. and Shallcross, D.E., 2021. Investigation of biofuel as a potential renewable energy source. *Atmosphere*, *12*(10), p.1289.

Khatiwada, D., Leduc, S., Silveira, S. and McCallum, I., 2016. Optimizing ethanol and bioelectricity production in sugarcane biorefineries in Brazil. *Renewable Energy*, *85*, pp.371–386.

Kim, S. and Dale, B.E., 2004. Global potential bioethanol production from wasted crops and crop residues. *Biomass and Bioenergy*, *26*(4), pp.361–375.

Kirshner, J., Brown, E., Dunlop, L., Cairo, J.P.F., Redeker, K., Veneu, F., Brooks, S., Kirshner, S. and Walton, P.H., 2022. "A future beyond sugar": Examining second-generation biofuel pathways in Alagoas, northeast Brazil. *Environmental Development*, *44*, p.100739.

Kour, D., Rana, K.L., Yadav, N., Yadav, A.N., Rastegari, A.A., Singh, C., Negi, P., Singh, K. and Saxena, A.K., 2019. Technologies for biofuel production: current development, challenges, and future prospects. *Prospects of renewable bioprocessing in future energy systems*, pp.1–50.

Kucharska, K., Rybarczyk, P., Hołowacz, I., Łukajtis, R., Glinka, M. and Kamiński, M., 2018. Pretreatment of lignocellulosic materials as substrates for fermentation processes. *Molecules*, *23*(11), p.2937.

Kumar, K., Ghosh, S., Angelidaki, I., Holdt, S.L., Karakashev, D.B., Morales, M.A. and Das, D., 2016. Recent developments on biofuels production from microalgae and macroalgae. *Renewable and Sustainable Energy Reviews*, *65*, pp.235–249.

Kumar, D., and Pugazhendi, A., 2021. Biofuel production from Macroalgae: present scenario and future scope. *Bioengineered*, *12*(2), p.9216.

Kumar, A.K. and Sharma, S., 2017. Recent updates on different methods of pretreatment of lignocellulosic feedstocks: a review. *Bioresources and bioprocessing*, *4*(1), pp.1–19.

Kusmiyati, K., Hadiyanto, H. and Fudholi, A., 2023. Treatment updates of microalgae biomass for bioethanol production: A comparative study. *Journal of Cleaner Production*, *383*, p.135236.

Kwon, S.Y., Lee, H.S. and Kwak, S.S., 2001. Development of environmental stress-tolerant plants by gene manipulation of antioxidant enzymes. *The Plant Pathology Journal*, *17*(2), pp.88–93.

Lakatos, G.E., Ranglová, K., Manoel, J.C., Grivalský, T., Kopecký, J. and Masojídek, J., 2019. Bioethanol production from microalgae polysaccharides. *Folia Microbiologica*, *64*, pp.627–644.

Lam, M.K. and Lee, K.T., 2012. Microalgae biofuels: a critical review of issues, problems and the way forward. *Biotechnology Advances*, *30*(3), pp.673–690.

Lei, Z., Li, C. and Chen, B., 2003. Extractive distillation: a review. *Separation & Purification Reviews*, *32*(2), pp.121–213.

Lennartsson, P.R., Erlandsson, P. and Taherzadeh, M.J., 2014. Integration of the first and second generation bioethanol processes and the importance of by-products. *Bioresource Technology*, *165*, pp.3–8.

Leonel, L.V., Arruda, P.V., Chandel, A.K., Felipe, M.G.A. and Sene, L., 2021. Kluyveromyces marxianus: A potential biocatalyst of renewable chemicals and lignocellulosic ethanol production. *Critical Reviews in Biotechnology*, *41*(8), pp.1131–1152.

Levin, L.A., Boesch, D.F., Covich, A., Dahm, C., Erséus, C., Ewel, K.C., Kneib, R.T., Moldenke, A., Palmer, M.A., Snelgrove, P. and Strayer, D., 2001. The function of marine critical transition zones and the importance of sediment biodiversity. *Ecosystems*, *4*, pp.430–451.

Li, C., Aston, J.E., Lacey, J.A., Thompson, V.S. and Thompson, D.N., 2016. Impact of feedstock quality and variation on biochemical and thermochemical conversion. *Renewable and Sustainable Energy Reviews*, *65*, pp.525–536.

Li, X. and Zheng, Y., 2017. Lignin-enzyme interaction: Mechanism, mitigation approach, modeling, and research prospects. *Biotechnology Advances*, *35*(4), pp.466–489.

Lim, J.S., Manan, Z.A., Alwi, S.R.W. and Hashim, H., 2012. A review on utilisation of biomass from rice industry as a source of renewable energy. *Renewable and Sustainable Energy Reviews*, *16*(5), pp.3084–3094.

Lin, Y., Zhang, W., Li, C., Sakakibara, K., Tanaka, S. and Kong, H., 2012a. Factors affecting ethanol fermentation using Saccharomyces cerevisiae BY4742. *Biomass and Bioenergy*, *47*, pp.395–401.

Lin, Y., Zhang, W., Li, C., Sakakibara, K., Tanaka, S. and Kong, H., 2012b. Factors affecting ethanol fermentation using Saccharomyces cerevisiae BY4742. *Biomass and Bioenergy*, *47*, pp.395–401.

Ling, Z., Chen, S., Zhang, X., & Xu, F. (2017). Exploring crystalline-structural variations of cellulose during alkaline pretreatment for enhanced enzymatic hydrolysis. *Bioresource Technology*, *224*, 611–617.

Liu, Z., Bai, H., Lee, J. and Sun, D.D., 2011. A low-energy forward osmosis process to produce drinking water. *Energy & Environmental Science*, *4*(7), pp.2582–2585.

Liu, Z., Hao, N., Hou, Y., Wang, Q., Liu, Q., Yan, S., Chen, F. and Zhao, L., 2023. Technologies for harvesting the microalgae for industrial applications: Current trends and perspectives. *Bioresource Technology*, p.129631.

Liu, X., Jia, B., Sun, X., Ai, J., Wang, L., Wang, C., Zhao, F., Zhan, J. and Huang, W., 2015. Effect of initial pH on growth characteristics and fermentation properties of Saccharomyces cerevisiae. *Journal of Food Science*, *80*(4), pp.M800–M808.

Liu, W., Yuan, J.S. and Stewart Jr, C.N., 2013. Advanced genetic tools for plant biotechnology. *Nature Reviews Genetics*, *14*(11), pp.781–793.

Luque, R., Herrero-Davila, L., Campelo, J.M., Clark, J.H., Hidalgo, J.M., Luna, D., Marinas, J.M. and Romero, A.A., 2008. Biofuels: a technological perspective. *Energy & Environmental Science*, *1*(5), pp.542–564.

Lynd, L.R., Weimer, P.J., Van Zyl, W.H. and Pretorius, I.S., 2002. Microbial cellulose utilization: fundamentals and biotechnology. *Microbiology and Molecular Biology Reviews*, *66*(3), pp.506–577.

Martinez-Hernandez, E., Amezcua-Allieri, M.A., Sadhukhan, J. and Anell, J.A., 2018. Sugarcane bagasse valorization strategies for bioethanol and energy production. *Sugarcane-Technology and Research.*

Maurya, D.K., Kumar, A., Chaurasiya, U., Hussain, T. and Singh, S.K., 2021. Modern era of microbial biotechnology: opportunities and prospects. In *Microbiomes and plant health* (pp. 317–343). Academic Press.

Mayer, F.D., Brondani, M., Hoffmann, R., Feris, L.A., Marcilio, N.R. and Baldo, V., 2016. Small-scale production of hydrous ethanol fuel: Economic and environmental assessment. *Biomass and Bioenergy*, *93*, pp.168–179.

McGinn, P.J., Dickinson, K.E., Bhatti, S., Frigon, J.C., Guiot, S.R. and O'Leary, S.J., 2011. Integration of microalgae cultivation with industrial waste remediation for biofuel and bioenergy production: opportunities and limitations. *Photosynthesis Research*, *109*, pp.231–247.

Menegazzo, M.L. and Fonseca, G.G., 2019. Biomass recovery and lipid extraction processes for microalgae biofuels production: A review. *Renewable and Sustainable Energy Reviews*, *107*, pp.87–107.

Mishra, R., Raj, A. and Saurabh, S., 2023. Present Status and Future Prospect of Butanol Fermentation. *Production of Biobutanol from Biomass*, pp.105–131.

Mittal, A., Katahira, R., Donohoe, B. S., Black, B. A., Pattathil, S., Stringer, J. M., & Beckham, G. T. (2017). Alkaline peroxide delignification of corn stover. *ACS Sustainable Chemistry & Engineering*, *5*(7), 6310–6321.

Mohammadi Shad, Z., Venkitasamy, C. and Wen, Z., 2021. Corn distillers dried grains with solubles: production, properties, and potential uses. *Cereal Chemistry*, *98*(5), pp.999–1019.

Mohapatra, S., Ray, R.C. and Ramachandran, S., 2019. Bioethanol from biorenewable feedstocks: technology, economics, and challenges. In *Bioethanol production from food crops* (pp. 3–27). Academic Press.

Mojović, L., Pejin, D., Grujić, O., Markov, S., Pejin, J., Rakin, M., Vukašinović, M., Nikolić, S. and Savić, D., 2009. Progress in the production of bioethanol on starch-based feedstocks. *Chemical Industry and Chemical Engineering Quarterly/CICEQ*, *15*(4), pp.211–226.

Monroe, R., Kass, M., & McConnell, S. (2019). Potential Impacts of Increased Ethanol Blend-Level in Gasoline on Distribution and Retail Infrastructure.

Mood, S.H., Golfeshan, A.H., Tabatabaei, M., Jouzani, G.S., Najafi, G.H., Gholami, M. and Ardjmand, M., 2013. Lignocellulosic biomass to bioethanol, a comprehensive review with a focus on pretreatment. *Renewable and Sustainable Energy Reviews*, *27*, pp.77–93.

Moonsamy, T.A., 2021. *Techno-economic analysis and benchmarking of integrated first and second generation biorefinery scenarios annexed to a typical sugar mill for bioethanol production* (Doctoral dissertation, Stellenbosch: Stellenbosch University).

Mosier, N., Wyman, C., Dale, B., Elander, R., Lee, Y.Y., Holtzapple, M. and Ladisch, M., 2005. Features of promising technologies for pretreatment of lignocellulosic biomass. *Bioresource Technology*, *96*(6), pp.673–686.

Mumm, R.H., Goldsmith, P.D., Rausch, K.D. and Stein, H.H., 2014. Land usage attributed to corn ethanol production in the United States: sensitivity to technological advances in corn grain yield, ethanol conversion, and co-product utilization. *Biotechnology for Biofuels*, *7*(1), pp.1–17.

Murdia, L.K., Wadhwani, R., Wadhawan, N., Bajpai, P. and Shekhawat, S., 2016. Maize utilization in India: an overview. *American Journal of Food and Nutrition*, *4*(6), pp.169–176.

Naik, S.N., Goud, V.V., Rout, P.K. and Dalai, A.K., 2010. Production of first and second generation biofuels: a comprehensive review. *Renewable and Sustainable Energy Reviews*, *14*(2), pp.578–597.

Nair, R.B. and Taherzadeh, M.J., 2016. Valorization of sugar-to-ethanol process waste vinasse: a novel biorefinery approach using edible ascomycetes filamentous fungi. *Bioresource Technology*, *221*, pp.469–476.

Nanda, S., Rana, R., Sarangi, P.K., Dalai, A.K. and Kozinski, J.A., 2018. A broad introduction to first-, second-, and third-generation biofuels. *Recent advancements in biofuels and bioenergy utilization*, pp.1–25.

Ndubuisi, I.A., Amadi, C.O., Nwagu, T.N., Murata, Y. and Ogbonna, J.C., 2023. Non-conventional yeast strains: Unexploited resources for effective commercialization of second generation bioethanol. *Biotechnology Advances*, 63, 108100.

Nigam, P.S. and Singh, A., 2011. Production of liquid biofuels from renewable resources. *Progress in Energy and Combustion Science*, *37*(1), pp.52–68.

Noorollahi, Y., Janalizadeh, H., Yousefi, H. and Jahangir, M.H., 2021. Biofuel for energy self-sufficiency in agricultural sector of Iran. *Sustainable Energy Technologies and Assessments*, *44*, p.101069.

Oliva-Neto, P., Dorta, C., Carvalho, A.F.A., Lima, V.D. and Silva, D.D., 2013. The Brazilian technology of fuel ethanol fermentation—yeast inhibition factors and new perspectives to improve the technology. *Materials and Processes for Energy: Communicating Current Research and Technological Developments*, *1*, pp.371–379.

Oliveira, L.R., Nascimento, V.M., Goncalves, A.R. and Rocha, G.J., 2014. Combined process system for the production of bioethanol from sugarcane straw. *Industrial Crops and Products*, *58*, pp.1–7.

Ortegón, G.P., Arboleda, F.M., Candela, L., Tamoh, K. and Valdes-Abellan, J., 2016. Vinasse application to sugar cane fields. Effect on the unsaturated zone and groundwater at Valle del Cauca (Colombia). *Science of the Total Environment*, *539*, pp.410–419.

Panahi, H.K.S., Dehhaghi, M., Guillemin, G.J., Gupta, V.K., Lam, S.S., Aghbashlo, M. and Tabatabaei, M., 2022. Bioethanol production from food wastes rich in carbohydrates. *Current Opinion in Food Science*, *43*, pp.71–81.

Panigrahi, C., Shaikh, A. E. Y., Bag, B. B., Mishra, H. N., & De, S. (2021). A technological review on processing of sugarcane juice: Spoilage, preservation, storage, and packaging aspects. *Journal of Food Process Engineering*, *44*(6), e13706.

Pascoli, D.U., Dichiara, A., Roumeli, E., Gustafson, R. and Bura, R., 2022. Lignocellulosic nanomaterials production from wheat straw via peracetic acid pretreatment and their application in plastic composites. *Carbohydrate Polymers*, *295*, p.119857.

Pereira, R., Yarish, C., & Critchley, A. T. (2024). Seaweed aquaculture for human foods in land based and IMTA systems. In *Applications of Seaweeds in Food and Nutrition* (pp. 77–99). Elsevier.

Pikkemaat, M.G. and van Egmond, H.J., 2011. Microbial screening methods for antibiotic residues. In *Responsible Use of Antibiotics in Animals* (pp. 120–120).

Priyadharsini, P., Nirmala, N., Dawn, S. S., Baskaran, A., SundarRajan, P., Gopinath, K. P., & Arun, J. (2022). Genetic improvement of microalgae for enhanced carbon dioxide sequestration and enriched biomass productivity: review on CO2 bio-fixation pathways modifications. *Algal Research*, *66*, 102810.

Qian, B., Li, X., Liu, X., Chen, P., Ren, C. and Dai, C., 2015. Enhanced drought tolerance in transgenic rice over-expressing of maize C4 phosphoenolpyruvate carboxylase gene via NO and Ca2+. *Journal of Plant Physiology*, *175*, pp.9–20.

Rahman, N.F.A., Harun, S., Sajab, M.S., Zubairi, S.I., Markom, M., Jahim, J.M., Tusirin, M., Nor, M., Abdullah, M.A. and Hashim, N., 2018. Boosting enzymatic hydrolysis of pressurized ammonium hydroxide pretreated empty fruit bunch using response surface methodology. *Journal of Engineering Science and Technology*, *13*(8), pp.2421–2445.

Rahmati, S., Doherty, W., Dubal, D., Atanda, L., Moghaddam, L., Sonar, P., Hessel, V. and Ostrikov, K.K., 2020. Pretreatment and fermentation of lignocellulosic biomass: reaction mechanisms and process engineering. *Reaction Chemistry & Engineering*, *5*(11), pp.2017–2047.

Rajkumar, R. and Yaakob, Z., 2013. The biology of microalgae. *Biotechnological Applications of Microalgae: Biodiesel and Value-Added Products*, pp.7–16.

Rajvanshi, M., Sagaram, U.S., Subhash, G.V., Kumar, G.R.K., Kumar, C., Govindachary, S. and Dasgupta, S., 2019. Biomolecules from microalgae for commercial applications. In *Sustainable downstream processing of microalgae for industrial application* (pp. 3–38). CRC Press.

Ram, C., Rani, P., Gebru, K.A. and Mariam Abrha, M.G., 2020. Pulp and paper industry wastewater treatment: use of microbes and their enzymes. *Physical Sciences Reviews*, *5*(10), p.20190050.

Ranum, P., Peña-Rosas, J.P. and Garcia-Casal, M.N., 2014. Global maize production, utilization, and consumption. *Annals of the New York Academy of Sciences*, *1312*(1), pp.105–112.

Rao, P.S., Periyasamy, C., Kumar, K.S., Rao, A.S. and Anantharaman, P., 2018. Seaweeds: distribution, production and uses. *Bioprospecting of algae. Society for Plant Research*, pp.59–78.

Rastogi, M. and Shrivastava, S., 2017. Recent advances in second generation bioethanol production: An insight to pretreatment, saccharification and fermentation processes. *Renewable and Sustainable Energy Reviews*, *80*, pp.330–340.

Ravindranath, N.H., Lakshmi, C.S., Manuvie, R. and Balachandra, P., 2011. Biofuel production and implications for land use, food production and environment in India. *Energy Policy*, *39*(10), pp.5737–5745.

Ray, R.C. and Ramachandran, S. eds., 2018. *Bioethanol production from food crops: sustainable sources, interventions, and challenges*. Academic Press.

Raza, A., Salehi, H., Rahman, M.A., Zahid, Z., Madadkar Haghjou, M., Najafi-Kakavand, S., Charagh, S., Osman, H.S., Albaqami, M., Zhuang, Y. and Siddique, K.H., 2022. Plant hormones and neurotransmitter interactions mediate antioxidant defenses under induced oxidative stress in plants. *Frontiers in Plant Science*, *13*.

Razak, N. H., Hashim, H., Yunus, N. A., & Klemeš, J. J. (2021). Reducing diesel exhaust emissions by optimisation of alcohol oxygenates blend with diesel/biodiesel. *Journal of Cleaner Production*, *316*, 128090.

Reis, V.R., Bassi, A.P.G., Silva, J.C.G.D. and Ceccato-Antonini, S.R., 2013. Characteristics of Saccharomyces cerevisiae yeasts exhibiting rough colonies and pseudohyphal morphology with respect to alcoholic fermentation. *Brazilian Journal of Microbiology*, *44*, pp.1121–1131.

Robak, K. and Balcerek, M., 2018. Review of second generation bioethanol production from residual biomass. *Food technology and biotechnology*, *56*(2), p.174.

Robertson, G.P., Hamilton, S.K., Barham, B.L., Dale, B.E., Izaurralde, R.C., Jackson, R.D., Landis, D.A., Swinton, S.M., Thelen, K.D. and Tiedje, J.M., 2017. Cellulosic biofuel contributions to a sustainable energy future: Choices and outcomes. *Science*, *356*(6345), p.eaal2324.

Rodmui, A., Kongkiattikajorn, J. and Dandusitapun, Y., 2008. Optimization of agitation conditions for maximum ethanol production by coculture. *Agriculture and Natural Resources*, *42*(5), pp.285–293.

Saadon, S.Z.A.H., Osman, N.B. and Yusup, S., 2022. Pretreatment of fiber-based biomass material for lignin extraction. In *Value-Chain of Biofuels* (pp. 105–135). Elsevier.

Sachdev, S., Ansari, S.A., Ansari, M.I., Fujita, M. and Hasanuzzaman, M., 2021. Abiotic stress and reactive oxygen species: Generation, signaling, and defense mechanisms. *Antioxidants*, *10*(2), p.277.

Sahni, S., Singh, M.K. and Narang, A., 2021. Sustainable Solution for Future Energy Challenges Through Microbes. *Energy: Crises, Challenges and Solutions*, pp.231–249.

Sajid, Z., da Silva, M.A.B. and Danial, S.N., 2021. Historical analysis of the role of governance systems in the sustainable development of biofuels in Brazil and the United States of America (USA). *Sustainability*, *13*(12), p.6881.

Sakamoto, T., Morinaka, Y., Ohnishi, T., Sunohara, H., Fujioka, S., Ueguchi-Tanaka, M., Mizutani, M., Sakata, K., Takatsuto, S., Yoshida, S. and Tanaka, H., 2006. Erect leaves caused by brassinosteroid deficiency increase biomass production and grain yield in rice. *Nature Biotechnology*, *24*(1), pp.105–109.

Sandri, J.P., Milessi, T.S., Zangirolami, T.C. and Mussatto, S.I., 2023. Screening of yeast co-culture using crude hydrolysate for co-fermentation of pentose and hexose. *Biofuels, Bioproducts and Biorefining*, *17*(6), pp.1639–1653.

Seiboth, B., Pakdaman, B.S., Hartl, L. and Kubicek, C.P., 2007. Lactose metabolism in filamentous fungi: how to deal with an unknown substrate. *Fungal Biology Reviews*, *21*(1), pp.42–48.

Seo, S.O., Park, S.K., Jung, S.C., Ryu, C.M. and Kim, J.S., 2020. Anti-contamination strategies for yeast fermentations. *Microorganisms*, *8*(2), p.274.

Severo, I.A., Siqueira, S.F., Depra, M.C., Maroneze, M.M., Zepka, L.Q. and Jacob-Lopes, E., 2019. Biodiesel facilities: What can we address to make biorefineries commercially competitive?s *Renewable and Sustainable Energy Reviews*, *112*, pp.686–705.

Shahare, V.V., Kumar, B. and Singh, P., 2017. Biofuels for sustainable development: a global perspective. In: Singh, R., Kumar, S. (eds). *Green technologies and environmental sustainability*, pp.67–89.

Sharma, A., Nain, V., Tiwari, R., Singh, S. and Nain, L., 2018. Optimization of fermentation condition for co-production of ethanol and 2, 3-butanediol (2, 3-BD) from hemicellolosic hydrolysates by Klebsiella oxytoca XF7. *Chemical Engineering Communications*, *205*(3), pp.402–410.

Shen, L., 2019. *Sugar metabolism: from enzyme cascades towards physiology and application* (Doctoral dissertation, Universitaet Duisburg-Essen (Germany)).

Shinn, T., 2011. An idea whose time had come: an exploratory analysis of ethanol's rise to agenda prominence in the United States.

Sikarwar, V.S., Zhao, M., Fennell, P.S., Shah, N. and Anthony, E.J., 2017. Progress in biofuel production from gasification. *Progress in Energy and Combustion Science*, *61*, pp.189–248

Sinclair, T.R. and Sheehy, J.E., 1999. Erect leaves and photosynthesis in rice. *Science*, *283*(5407), pp.1455–1455.

Singh, P., Kumari, S., Guldhe, A., Misra, R., Rawat, I., & Bux, F. (2016). Trends and novel strategies for enhancing lipid accumulation and quality in microalgae. *Renewable and Sustainable Energy Reviews*, *55*, 1–16.

Singh, K., Singh, S., Kumar, V., Khandai, S., Kumar, A., Bhowmick, M.K., Kumar, V., Srivastava, A. and Hellin, J., 2023. Rice Straw Management: Energy Conservation and Climate Change Mitigation. In *Handbook of Energy Management in Agriculture* (pp. 1–25). Singapore: Springer Nature Singapore.

Singhvi, M. S., & Gokhale, D. V. (2019). Lignocellulosic biomass: hurdles and challenges in its valorization. *Applied Microbiology and Biotechnology*, *103*(23), 9305–9320.

Sista Kameshwar, A.K. and Qin, W., 2018. Comparative study of genome-wide plant biomass-degrading CAZymes in white rot, brown rot and soft rot fungi. *Mycology*, *9*(2), pp.93–105.

Solomon, B.D., Banerjee, A., Acevedo, A., Halvorsen, K.E. and Eastmond, A., 2015. Policies for the sustainable development of biofuels in the Pan American region: A review and synthesis of five countries. *Environmental Management*, *56*, pp.1276–1294.

Sørensen, E., Lam, K.F. and Sudhoff, D., 2014. Special distillation applications. In *Distillation* (pp. 367–401). Academic Press.

Stamenković, O.S., Siliveru, K., Veljković, V.B., Banković-Ilić, I.B., Tasić, M.B., Ciampitti, I.A., Đalović, I.G., Mitrović, P.M., Sikora, V.Š. and Prasad, P.V., 2020. Production of biofuels from sorghum. *Renewable and Sustainable Energy Reviews*, *124*, p.109769.

Sudiyani, Y., Dahnum, D., Burhani, D. and Putri, A.M.H., 2019. Evaluation and comparison between simultaneous saccharification and fermentation and separated hydrolysis and fermentation process. In *Second and Third Generation of Feedstocks* (pp. 273–290). Elsevier.

Sun, R., 2010. *Cereal straw as a resource for sustainable biomaterials and biofuels: chemistry, extractives, lignins, hemicelluloses and cellulose*. Elsevier.

Surriya, O., Saleem, S. S., Waqar, K., Gul Kazi, A., & Öztürk, M. (2015). Bio-fuels: a blessing in disguise. *Phytoremediation for Green Energy*, 11–54.

Susmozas, A., Martín-Sampedro, R., Ibarra, D., Eugenio, M.E., Iglesias, R., Manzanares, P. and Moreno, A.D., 2020. Process strategies for the transition of 1G to advanced bioethanol production. *Processes*, *8*(10), p.1310.

Swain, M.R., Singh, A., Sharma, A.K. and Tuli, D.K., 2019. Bioethanol production from rice-and wheat straw: an overview. *Bioethanol Production from Food Crops*, pp.213–231.

Sydney, E.B., Letti, L.A.J., Karp, S.G., Sydney, A.C.N., de Souza Vandenberghe, L.P., de Carvalho, J.C., Woiciechowski, A.L., Medeiros, A.B.P., Soccol, V.T. and Soccol, C.R., 2019. Current analysis and future perspective of reduction in worldwide greenhouse gases emissions by using first- and second-generation bioethanol in the transportation sector. *Bioresource Technology Reports*, *7*, p.100234.

Szambelan, K., Nowak, J., Szwengiel, A., Jeleń, H. and Łukaszewski, G., 2018. Separate hydrolysis and fermentation and simultaneous saccharification and fermentation methods in bioethanol production and formation of volatile by-products from selected corn cultivars. *Industrial Crops and Products*, *118*, pp.355–361.

Talebi, S., Edalatpour, A. and Tavakoli, O., 2022. Algal biorefinery: a potential solution to the food–energy–water–environment nexus. *Sustainable Energy & Fuels*, *6*(11), pp.2623–2664.

Taniguchi, M., Tanaka, M., Matsuno, R. and Kamikubo, T., 1982. Evaluation of chemical pretreatment for enzymatic solubilization of rice straw. *European Journal of Applied Microbiology and Biotechnology*, *14*, pp.35–39.

Toor, M., Kumar, S.S., Malyan, S.K., Bishnoi, N.R., Mathimani, T., Rajendran, K. and Pugazhendhi, A., 2020. An overview on bioethanol production from lignocellulosic feedstocks. *Chemosphere*, *242*, p.125080.

Torney, F., Moeller, L., Scarpa, A. and Wang, K., 2007. Genetic engineering approaches to improve bioethanol production from maize. *Current Opinion in Biotechnology*, *18*(3), pp.193–199.

Ulucak, R., 2021. A revisit to the relationship between financial development and energy consumption: Is globalization paramount? *Energy*, *227*, p.120337.

Valdivia, M., Galan, J.L., Laffarga, J. and Ramos, J.L., 2016. Biofuels 2020: biorefineries based on lignocellulosic materials. *Microbial Biotechnology*, *9*(5), pp.585–594.
Vane, L. M. (2008). Separation technologies for the recovery and dehydration of alcohols from fermentation broths. *Biofuels, Bioproducts and Biorefining*, 2(6), 553–588.
Vohra, M., Manwar, J., Manmode, R., Padgilwar, S. and Patil, S., 2014. Bioethanol production: Feedstock and current technologies. *Journal of Environmental Chemical Engineering*, *2*(1), pp.573–584.
Walker, G.M. and Walker, R.S., 2018. Enhancing yeast alcoholic fermentations. *Advances in Applied Microbiology*, *105*, pp.87–129.
Warnasooriya, S.N. and Brutnell, T.P., 2014. Enhancing the productivity of grasses under high-density planting by engineering light responses: from model systems to feedstocks. *Journal of Experimental Botany*, *65*(11), pp.2825–2834.
Weeranoppanant, N., Adamo, A., Saparbaiuly, G., Rose, E., Fleury, C., Schenkel, B. and Jensen, K.F., 2017. Design of multistage counter-current liquid–liquid extraction for small-scale applications. *Industrial & Engineering Chemistry Research*, *56*(14), pp.4095–4103.
Wei, C. J., & Cheng, C. Y. (1985). Effect of hydrogen peroxide pretreatment on the structural features and the enzymatic hydrolysis of rice straw. *Biotechnology and Bioengineering*, *27*(10), 1418–1426.
Wilkie, A.C., Riedesel, K.J. and Owens, J.M., 2000. Stillage characterization and anaerobic treatment of ethanol stillage from conventional and cellulosic feedstocks. *Biomass and Bioenergy*, *19*(2), pp.63–102.
Xie, S., Li, Z., Zhu, G., Song, W. and Yi, C., 2022. Cleaner production and downstream processing of bio-based 2, 3-butanediol: A review. *Journal of Cleaner Production*, *343*, p.131033.
Yadava, P., Abhishek, A., Singh, R., Singh, I., Kaul, T., Pattanayak, A. and Agrawal, P.K., 2017. Advances in maize transformation technologies and development of transgenic maize. *Frontiers in Plant Science*, *7*, p.1949.
Yan, X., Cheng, J.R., Wang, Y.T. and Zhu, M.J., 2020. Enhanced lignin removal and enzymolysis efficiency of grass waste by hydrogen peroxide synergized dilute alkali pretreatment. *Bioresource Technology*, *301*, p.122756.
Yanagisawa, M., Kawai, S. and Murata, K., 2013. Strategies for the production of high concentrations of bioethanol from seaweeds: production of high concentrations of bioethanol from seaweeds. *Bioengineered*, *4*(4), pp.224–235.
Yue, D., You, F. and Snyder, S.W., 2014. Biomass-to-bioenergy and biofuel supply chain optimization: Overview, key issues and challenges. *Computers & Chemical Engineering*, *66*, pp.36–56.
Zabed, H., Sahu, J.N., Suely, A., Boyce, A.N. and Faruq, G., 2017. Bioethanol production from renewable sources: Current perspectives and technological progress. *Renewable and Sustainable Energy Reviews*, *71*, pp.475–501.
Zhang, W. and Geng, A., 2012. Improved ethanol production by a xylose-fermenting recombinant yeast strain constructed through a modified genome shuffling method. *Biotechnology for Biofuels*, *5*, pp.1–11.
Zhang, H., Han, L. and Dong, H., 2021. An insight to pretreatment, enzyme adsorption and enzymatic hydrolysis of lignocellulosic biomass: Experimental and modeling studies. *Renewable and sustainable energy reviews*, *140*, p.110758.
Zhang, K., Johnson, L., Prasad, P.V., Pei, Z. and Wang, D., 2015. Big bluestem as a bioenergy crop: A review. *Renewable and Sustainable Energy Reviews*, *52*, pp.740–756.
Zhang, M. Y., Xu, X. R., Zhao,. (2024). Mechanism of enhanced microalgal biomass and lipid accumulation through symbiosis between a highly succinic acid-producing strain of Escherichia coli SUC and Aurantiochytrium sp. SW1. *Bioresource Technology*, *394*, 130232.

Zhou, Z., Liu, D. and Zhao, X., 2021. Conversion of lignocellulose to biofuels and chemicals via sugar platform: an updated review on chemistry and mechanisms of acid hydrolysis of lignocellulose. *Renewable and Sustainable Energy Reviews*, *146*, p.111169.
Zhu, Y., 2007. Immobilized cell fermentation for production of chemicals and fuels. *Bioprocessing for value-added products from renewable resources*, pp.373–396.
Zhu, S., Huang, W., Huang, W., Wang, K., Chen, Q. and Wu, Y., 2015. Pretreatment of rice straw for ethanol production by a two-step process using dilute sulfuric acid and sulfomethylation reagent. *Applied Energy*, *154*, pp.190–196.
Ziaei-Rad, Z., Fooladi, J., Pazouki, M., & Gummadi, S. N. (2021). Lignocellulosic biomass pre-treatment using low-cost ionic liquid for bioethanol production: An economically viable method for wheat straw fractionation. *Biomass and Bioenergy*, *151*, 106140.
Zuliani, L., Serpico, A., De Simone, M., Frison, N. and Fusco, S., 2021. Biorefinery gets hot: Thermophilic enzymes and microorganisms for second-generation bioethanol production. *Processes*, *9*(9), p.1583.

3 Biodiesel Production
Microbial Alchemy for Sustainable Transportation

Falak Shaheen, Rohit Topkar, Ranjit Gurav, Yung-Hun Yang, Jyoti Jadhav, and Rahul Jadhav

3.1 INTRODUCTION

Energy has historically been regarded as a crucial element in supporting a nation's economic growth, with fossil fuels serving as the predominant means of fulfilling this energy demand since their initial discovery (Bateni et al., 2017). However, environmental difficulties caused by the burning of fossil fuels, rising prices, and the depletion of natural energy sources such as petroleum reserves prompted scientists to seek alternate, affordable, sustainable, renewable, and reliable energy sources. Researchers have focused on biodiesel as a potential substitute for petroleum fuel among the possibilities. As per the report of the Energy Information Administration (EIA) (2016), there will be an increase in energy consumption between the years 2000 and 2030 by 71% worldwide and it is expected that the increase in emission of carbon dioxide will also rise by approximately 35%. Numerous waste biomass has been created in greater quantities in tandem with the world's population rise. As a result, a practical trash disposal solution is desperately needed. Repurposing waste for energy generation is a sustainable energy development; however, it is not economical to invest energy in waste disposal. To this purpose, the production and use of fuels produced from organic waste that are plant-based or biodegradable have drawn the attention of researchers (Zabermawi et al., 2022). Biodiesel (the most appealing biofuel) has attracted significant attention over the past few years as an environmentally friendly, sustainable, clean-burning, and non-toxic transportation fuel (. Indeed, as a result of the establishment of concessions and tax reductions, this green fuel has steadily become more affordable and extensively utilized globally. To increase the financial advantages of the biodiesel industry, the main by-product of biodiesel manufacturers, glycerol, which accounts for approximately 10% of the total volume produced, can be valorized into combustion improvers for diesel/biodiesel, such as solketal, solketalacetin, and acetins (Mortaza et al., 2018). Governments have implemented policies to boost biodiesel production due to its benefits, and by 2050, biodiesel should supply 27% of the world's transportation fuel needs. Various biofuels are available at different phases of advancement, including lignocellulosic-based fuels, hydrotreated vegetable oils and fats, biodiesel, bioethanol, and biomethane.

DOI: 10.1201/9781003585398-3

TABLE 3.1
A List of Oleaginous Microorganisms and Their Lipid Content

Microalgae	
Scenedesmus sp.	34.1
Chlorella protothecoides	49
Tetraselmis elliptica	14
C. Vulgaris NIES-227	89
Auxenochlorella protothecoides	66–63
Botryococcus braunii	28
Chlamydomonas reinhardtii, CC1010	59
Yeasts	
Cryptococcus sp. (KCTC 27583)	34
Rhodosporidium kratochvilovae HIMPA1	53–70
Trichosporon fermentans CICC 1368	36–62
Rhodosporidium toruloides Y4	48–67.5
Rhodosporidium toruloides	56–67
Rhodosporidium toruloides AS 2.1389	66
Rhodosporidium diobovatum	50
Lipomyces starkeyi	48–61
Rhodotorula glutinis	20
Cryptococcus curvatus	53–70
Lipomyces starkeyi CBS 1807	30
Fusarium oxysporum	22–53
Fusarium equiseti UMN-1	56
Sarocladium kiliense ADH17	33
Mortierella alpina LP M 301	31
Mortierella isabellina	50–83
Mucorales fungi	42.7–65.8
Candida sp. LEB-M3	50
Cunninghamella echinulata	57.5
Kodamaea Ohmeri BY 4-523	53
Colletotrichum	73
Bacteria	
Rhodococcus opacus DSM 1609	4
R. opacus PD630	14–70
R. opacus DSM 43205	66
R. opacus PD630	>50
Gordonia sp. DG	13–52
Nocardia globerula 432	>49.7
Streptomyces coelicolor TR0958	83
Streptomyces coelicolor TR0123	64

Source: Adopted from Yellapu et al., 2018; Patel et al., 2018; Shaheen et al., 2023.

3.2.2 Yeast and Filamentous Fungi (Molds)

Oleaginous yeast has a fast growth rate and abundant oil content comprising mainly C16 and C18 fatty acids, and hence it is preferred for biodiesel production. Studies have been performed for their genetic improvement as on average the lipid content in their cell wall is up to 40% of the cellular biomass (Uthi et al., 2022). Over 30 species of yeast are known as oleaginous yeast (Chintagunta et al., 2021); most studied microorganisms include *Yarrowia*, *Candida*, *Rhodosporidium*, *Rhizopus*, *Rhodotorula*, *Lipomyces*, *Trichosporon*, and *Cryptococcus* (Patel et al., 2020; Uthi et al., 2022). Under nutrient-limited conditions, oleaginous yeast can store lipids up to 70–80% of their cellular biomass (Chintagunta et al., 2021; Uthi et al., 2022). Lipid content and fatty acid profiles vary according to the species and the cultivation conditions. *Rhodotorula* sp. (*R. Glutinis*, *R. toruloides*, and *R. graminis*) are shown to have similar fatty acid profiles but different lipid content (Uthi et al., 2022). They are observed to grow on diverse carbon sources such as monosodium glutamate, glycerol, arabinose, xylose, glucose, and SCB hydrolysate as well as non-food sources such as lignocellulosic biomass (LCB) (Patel et al., 2020; Uthi et al., 2022). Oleaginous filamentous fungi are known to synthesis unusually high amounts of exclusive fatty acid comprising of γ-linolenic acid (GLA) unknown in other OMs (Patel et al., 2020). Molds or filamentous fungi, 80% of the biomass is stored lipids or fatty acids (Chintagunta et al., 2021). Low-cost substrates such as agricultural residues, molasses water, glycerol, sludge from sewage, and monosodium glutamate wastewater are used for the cultivation of fungi. A consortium of filamentous fungi, *Mortierella elongata* along with marine microalgae *Nannochloropsis oceanica*, when co-cultivated yielded a high volume of TAGs and PUFAs as well as total lipids as a result of enhanced bio-flocculation. *Cunninghamella echinulate* and *Mortierella alpina* LPM 301 when cultivated on glucose as a substrate along with orange peel and potassium nitrate, respectively, produced a high yield of total lipids of 46.6% and 31.1% including 14% of γ-linolenic acid (GLA) and 60.4% of arachidonic acid (ARA) (Patel et al., 2020). Further, quite opposite to microalgae, they can be cultivated in traditional bioreactors, causing a severe decline in production cost (Chintagunta et al., 2021) (Figure 3.1).

3.2.3 Microalgae

Microalgae exhibit a high biomass abundance, and a high photosynthetic rate, and are also known to amass a high volume of neutral lipid which may vary between 20% and 80% and may show an increase of up to 90% upon providing certain growth conditions (Chintagunta et al., 2021; Uthi et al., 2022). The yield per unit area of cultivation of oil crops (e.g., corn, soybean) is lower than the yield by microalgae (Cho and Park, 2018). Organic waste or diverse types of wastewater such as industrial, municipal, and agricultural wastewater are suitable substrates for microalgae. They are classified as heterotrophic, autotrophic, mixotrophic, and photoheterotrophic based on their mode of nutrition, further leading to different metabolic pathways. Organic compounds act as carbon sources for heterotrophic microalgae whereas CO_2 from the atmosphere acts as the carbon source for autotrophic microalgae.

FIGURE 3.1 Schematic diagram of biodiesel production from the oleaginous yeast. (Adapted with permission (5777170676126) from Patel et al., 2017).

The mixotrophic mode of nutrition requires both organic carbon sources and CO_2 for photosynthetic pathways and cellular respiration pathways (Patel et al., 2020). Chloroplast and endoplasmic reticulum are the main sites of TAG's synthesis and accumulation. Individual stress factors such as physical, chemical, environmental, or a combination can alter the metabolic pathways of formation and accumulation of neutral lipids in microalgae. Heterotrophic cultivation of microalgae is favored over autotrophic as it is economical and relatively easier to maintain. As the heterotrophic microalgae culture does not need a light source, its cultivation does not require a photobioreactor or illumination, hence reducing the cost of equipment and overall production. *Chlorella*, *Neochloris*, *Nannochloropsis*, *Scenedesmus*, *Chlamydomonas*, *Chlorococcum*, *Isochrysis*, *Cylindrotheca*, *Tetraselmis*, and *Auxenoch* are some of the microalgae with high lipid content. The microbial consortia of microalgae along with bacteria or yeast have shown higher yield in comparison to individual strains (Valdés et al., 2020).

3.3 SUBSTRATE USED AND CULTIVATION METHODS

At the industrial level, the major share of biodiesel is produced from edible vegetable oils, such as palm oil, soybean oil, and rapeseed oil. Recent trends show increased use of waste oils and animal fats. OMs convert a significant amount of carbon from the substrates, mainly carbohydrates to lipids, utilized in the regulatory and energetic functions. Simple sugars are the most preferred carbon source; however, the use of pure sugars as a substrate for the growth of oleaginous microbes for higher biomass is not economically viable.

3.3.1 Pure Sugar

Carbohydrates are the main components in biodiesel production as they supply energy for the cultivation of microbes responsible for lipid biosynthesis. Sugars like glucose, fructose, and sucrose are easy sources of carbon for microbe cultivation. It is reported that different fermentable sugars produce different biodiesel profiles. Most of the microbes prioritize utilizing glucose as a substrate for the production of lipids. However, the utilization of pure sugars is not economically favorable for the fermentation process of biodiesel manufacturing.

3.3.2 Waste Cooking Oil/Food Waste/VFA

Feedstocks as raw materials are a great source of triglycerides (triacylglycerides) for the production of biodiesel. The fatty acid is provided by feedstocks such as waste cooking oil, waste food, animal fats, and edible and non-edible oils, which on transesterification reaction produce biodiesel (Bhatia et al., 2020; Ramos et al., 2019). Characteristics of biodiesel vary depending on the feedstock used since the composition of fatty acids differs for each feedstock (Valdés et al., 2020). Using waste oils is a cheap and alternative method to produce biodiesel. These oils have lost their original properties due to repeated use and contain impurities (Chintagunta et al., 2021). Waste oils are generally produced from three different sources: the food industry, the non-food industry, and households or restaurants. The utilization of waste frying oils (WFO) reduces the cost of biodiesel production by 60–90% (Rezania et al., 2021). However, waste oil disposal is an environmental concern that can be resolved by using these oils to produce biodiesel. The drawback is that impurities in used cooking oils can resist the formation of biodiesel and require additional processing (Singh et al., 2019; Chintagunta et al., 2021). A newly discovered inexpensive method for the production of biodiesel is executed by a group of microbes called OMs. OMs utilize volatile fatty acids (VFA) for their optimal growth and convert them into microbial lipids. VFAs are short carbon-chain acids that are formed as an intermediate compound by anaerobic digestion of organic waste in bioreactors. As these microorganisms contain lipids as a major component of their cellular biomass, they are utilized as a substrate for biodiesel production (Bhatia et al., 2019b). The use of OMs for the manufacturing of biodiesel can help resolve the environmental problem of waste management from a variety of sectors (Patel et al., 2021; Zhang et al., 2021). Residuals from food industries and food businesses provide large amounts of waste as a cheap substrate in the manufacturing of biodiesels. Around 6 mt of coffee waste is generated each year and disposal of such waste is environmentally hazardous as it contains tannins, caffeine, and polyphenols. Diversion of coffee wastes in the production of a variety of biofuels is demonstrated in earlier studies as coffee residues are rich in carbon content and low in nitrogen content, making it an ideal substrate for the production of OMs (Kavitha et al., 2019). Similarly, the brewing industry waste has large organic content with low pH providing suitable conditions for the culturing of lipid-producing microbes, thus employing the waste in generating sustainable bioenergy and helping to achieve the goal of a "zero-waste society" (Nair et al., 2022).

3.3.3 Lignocellulosic Biomass

Lignocellulosic biomass (LCB) is the most abundant, easily available, and cheap substrate that can be utilized for biodiesel production. Around 180 billion metric tons of biomass are generated each year by the living world, from which around 75% of biomass contains carbohydrates as a major part of their composition (Sasmal and Mohanty, 2018). In plants, lignocellulose protects from biotic and abiotic factors and provides rigidity and strength. Lignocellulose is structurally composed of carbohydrates mainly consisting of cellulose, hemicelluloses, and lignin polymers with small quantities of starch and pectin. The composition of lignocellulose varies depending on location, age, species, and parts of the plant (Valdés et al., 2020). LCB can be obtained from agricultural waste, grasses and weeds, and forest residues (Kim et al., 2022). LCB is majorly acquired from agriculture residue which includes, sugarcane bagasse, wheat straws, rice straws, and corn residues (Lee et al., 2021b). In India, close to 500 metric tons of agricultural residue are generated each year. Manufacturing biodiesel from LCB is a tedious and complex process, as it requires a few operation stages; pretreatment and saccharification of LCB, culturing the OMs for the production of lipids, and transesterification for the production of biodiesel (Bhuvaneshwari et al., 2019; Chintagunta et al., 2021). To obtain sugars from LCB, pre-processing of LCB is mandatory, which includes enzymatic hydrolysis, application of mechanical stress, and physicochemical pretreatment (Gurav et al., 2022). Mechanical methods are utilized in order to decrease the particle size and increase the total surface subjected to extraction. The structural characters of the biomass utilized determine the mechanical process to be utilized. Based on the main forms of forces used, mechanical forces are of two types – solid shear forces and liquid shear forces. High-pressure homogenization, oil expeller, ultrasonication, autoclaving, and microwave irradiation are some of the common mechanical treatment methods. Non-mechanical pretreatment methods include physicochemical and biological or enzymatic methods. Ammonia fiber explosion (AFEX) exposes the biomass to hot liquid ammonia at high pressure for a short span releasing the pressure, causing structural changes in the biomass which increases its water retention capacity and digestibility. The biomass is also subjected to acid and alkali pretreatment, based on the structural characters of the biomass. Acid in diluted or concentrated form breaks down the glycosidic bonds, and bases aim to break the glycosidic bond and ester linkages. The biomass can also be pretreated using steam or hot water under high pressure. Other commonly used chemical treatments are CO_2 explosion, ozonolysis, and organosolvent pretreatment. Biological pretreatment methods utilize organisms that release enzymes such as laccase, lignin peroxidases, manganese-dependant peroxidases, and polyphone oxidases which degrade the biomass. However, biological methods have shown a lower rate of carbohydrate conversion in comparison to chemical and mechanical pretreatment methods. Pre-processing of LCB helps in reducing the size of particles and increases the surface area and porosity of biomass. Also, pretreatment helps in the degradation of lignin, which causes easy access of enzymes to cellulose and hemicellulose (. Further, pretreated LCB is hydrolyzed by enzymes and acid treatment to acquire constituent monomeric sugars, such as glucose, xylose, galactose, arabinose, and mannose (Valdés et al., 2020). The pretreatment methods are selected based on their efficiency to break down cell walls and release the trapped lignin. The methods selected for the pretreatment should release fewer toxic products during the downstreaming process. In addition to releasing fermentable sugars, the pretreatment process also generates by-products such as furfural, 5-hydroxymethylfurfural, vanillin, acetate, and formic acids. Consequently, detoxifying plant biomass hydrolysates is essential before utilizing them as feedstock for microbial fermentation to enhance growth and productivity (Bhatia et al., 2023; Lee et al., 2021a). OMs utilize glucose and other carbohydrates as a carbon source and ultimately convert sugars to pyruvate. Produced pyruvate is further used for *de novo* synthesis of lipids. Lastly, accumulated fatty acids in microbes are extracted and exploited to produce biofuels. The fuels produced using LCB as a substrate are environmentally friendly, as LCB has low sulfur and nitrogen content (Sasmal and Mohanty, 2018; Chintagunta et al., 2021) (Figure 3.2).

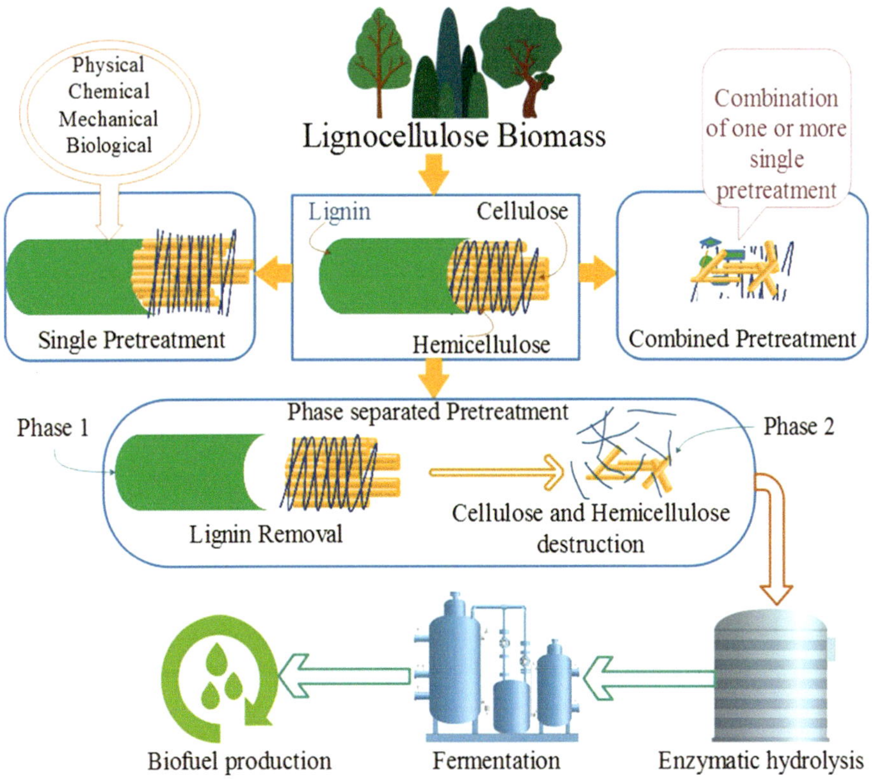

FIGURE 3.2 Different pretreatments used for lignocellulosic biomass to obtain hydrolysate for biodiesel production under fermentation (Figure available under Creative Commons CC-BY license was adopted from Preethi et al., 2021).

3.3.4 Industrial Wastewater

Wastewater is an unavoidable waste generated by various industries. The generated waste is not environmentally friendly and requires more attention in management. The majority of agro-industrial wastewater contains an adequate quantity of sugars that can be diverted for the production of biofuels. The wastewater from the dairy industry chiefly contains lactose as a carbohydrate, along with proteins, lipids, and minerals like NaCl and KCl. The presence of high organic content in dairy wastewater makes it a great source of contamination and is not easy to dispose of. Utilization of such wastewater as the energy source for culturing lipid-producing microbes can greatly eliminate the problem of waste management and help in lowering the economic cost of biodiesel production by implementing a waste-to-treasure approach (Iragavarapu et al., 2023; Nair et al., 2022). Breweries consume large quantities of water for the production of clear beer and support the operations of brewing procedures, cooling, steam-raising packaging, and cleaning, generating large volumes of effluent. An average of 3–10 L of wastewater is released for the production of 1 L of beer. The partially or untreated water released causes nutrient pollution and eutrophication of water bodies. Recent studies on the treatment of brewery wastewater by microalgae lead to higher biomass which is further used to extract bioactive compounds and produce biofuels through biochemical and thermochemical methods (Ferreira et al., 2019). Municipal wastewater is the water that remains after the sewage treatment process. After the process, the sludge obtained has a high biomass of lipids, making it a favorable feedstock for biodiesel production. Depending upon the source, sludge is divided into primary, secondary, tertiary, and blended sludge. The lipids mainly found in sludge (>54%) are ditriglycerides, triglycerides, monoglycerides, sterols, phospholipids, and FFAs. Primary treated water used as a substrate for the cultivation of OMs leads to storage of 9–32% of lipid content, whereas cultivation on predigested sludge can lead to 65% of lipid and 56% of FAME content by dried weight. The use of wastewater from different sources as a substrate for the cultivation of OMs can end the debate of food versus fuel, as biodiesel is considered a potential alternative to fossil fuels.

3.3.5 Gaseous Substrates (C1 Gas)

The overexploitation of ecosystems for human advancement has significantly impacted global pollution. With the growing human population, there is a high demand for natural resources, which is diminishing oil reserves and at the same time creating a problem of greenhouse gas (GHG) emissions responsible for climate change (Teixeira et al., 2018).

The manufacturing of high-value, useful products by capturing GHG can be an effective solution for minimizing the issue of air pollution. Some GHGs contain gases with only one carbon in their structure, commonly known as C1 gases. Carbon dioxide (CO_2), carbon monoxide (CO), and methane gas (CH_4) are the main C1 gases responsible for climate change problems which can be redirected to be useful as a substrate in the generation of biodiesel. Mostly, the industries release syngas as exhaust gas, which is a mix of CO (major gas), CO_2, CH_4, and H_2 gases. This syngas is converted into biofuels by certain microorganisms in specially designed fermenters (Hu et al., 2016; Fernández-Blanco et al., 2023).

Acetogen bacteria and archaea like methanotrophic bacteria are the most common microorganisms involved in the production of biofuels. Acetogen bacteria follow the reductive acetyl-CoA pathway (also known as the linear Wood–Ljungdahl pathway) to assimilate CO/syngas. In these microbes, CO is either directly converted to acetyl-CoA or firstly CO is oxidized to CO_2 and then forms formate. This formate is then converted to acetyl-CoA. Finally, generated acetyl-CoA is utilized in various metabolic processes to produce biofuels. Similarly, in methanotrophic bacteria, CH_4 is converted into methanol by the enzymatic action of oxygen-dependent, methane monooxygenases enzyme. Formed methanol is further converted into formaldehyde and then to CO_2 (Teixeira et al., 2018).

Several companies all over the globe have demonstrated the use of C1 gases in the production of a variety of products, viz., ethanol, methanol, butanediol, biofuels, amino acids, proteins, lactic acids, polyhydroxyalkanoates, and self-healing concrete. Newlight Technologies, a company in the USA, has demonstrated commercial production of biopolymers, polyhydroxyalkanoate, and poly-β-hydroxybutyrate from CH_4 gas. Precigen, an American biotechnology company formerly known as Intrexon, has obtained a patent for the production of biodiesel (FAME) along with fatty alcohols, n-Butanol, and 2,3-butanediol from *Methylococcus capsulatus* bacteria (Teixeira et al., 2018) (Table 3.2).

3.4 METHODS APPLIED FOR ENHANCING THE YIELD OF MICROBIAL BIODIESEL (MEDIA OPTIMIZATION/ ENGINEERING)

Microbial biodiesel production involves the use of OMs, which hold around 20% of lipid on a dry cell weight basis and can accumulate 70% of lipid content under stress conditions. Some bacteria, fungi, yeast cells, and microalgae are considered OMs due to their capacity to produce SCO. Among these eukaryotic yeasts, microalgae and molds are well characterized by the process of synthesis of triacylglycerides. The higher the lipid content in microbes, the more is the yield of biodiesel (Jeevan Kumar et al., 2020). To upsurge biodiesel production, lipid content in microbial cells is increased by (i) application of environmental shocks by culturing cells on nutrient-deficient media, (ii) genetic engineering of cells by RNAi and CRISPR/Cas9 system, and (iii) co-culturing cells with another partner; both partners contribute to lipid accumulation (Levering et al., 2015; Zhang et al., 2021).

Parameters like carbon source, nitrogen source, and C/N ratio along with pH, temperature, and rate of agitation strongly affect lipid production. Glucose is the ideal carbon source for the production of biodiesel, but it is expensive. So, it is preferred to use cheap feedstocks. Accumulation of lipids in microbial cells is increased by harvesting cells in nutrient-limiting media. Culture media with limiting nitrogen sources showed more deposition of lipids in species like *Yarrowia lipolytica*.

TABLE 3.2
Various Pretreatment Methods and Their Advantages and Disadvantages

Pretreatment	Technique	Advantage	Disadvantage
Mechanical • Chipping • Milling • Grinding	Vibratory ball milling, ball milling, hammer milling, colloid milling, wet disk milling	No chemical used Scalable	Poor sugar conversion Highly energy-intensive process
Ultrasonication	Ultrasound was applied to microbial cells in suitable solvents under pressure	Less processing times Low solvent consumption Greater penetration into hard cell walls	High power consumption
Microwave-assisted size reduction	High-pressure microwave pretreatment is operated in closed reactors within the temperature range from 150°C to 250°C	Cost-efficient Short reaction time Degrades lignin and hemicellulose Applied to acids and alkali and steam Eco-friendly	Increased formation of inhibitors
Acidic	Sulfuric acid (0.2–2.5 wt%) is mixed with biomass and the pretreatment is carried out either at 180°C or above for 5 min or 120°C for 30–90 min	Can be used for a wide range of materials Produce hydrolyzed xylose during pretreatment	Need to use an expensive Hastelloy reactor Difficult to control reaction conditions Toxic degradation products Expensive to remove salt while recycling water
Alkaline	Lignocellulosic biomass is soaked in the solution containing hydroxides of calcium/sodium, potassium/ammonia	Lesser degradation products formed Low reaction temperature	A large amount of water is required Longer residence time Energy-intensive Not efficient for hardwood Catalyst recovery is expensive
Steam explosion/hot water treatment	Exposure of hot steam or hot water up to 160–260°C, 0.69–4.38 MPa pressure for few minutes	Reduces the size of particles Effective for solubilizing hemicelluloses as oligomers Does not require catalysts, chemical products, or corrosion-resistant materials Lignin removed >73% Up to 95% assimilable sugars are obtained after enzymatic hydrolysis Inhibitor compounds in low concentrations compared to acid treatments	Inhibitors can be produced Requires a high demand for water and energy

AFEX (ammonia fiber explosion)	Biomass (1 kg) is treated with hot liquid ammonia (1–2 kg) under high pressure at 90°C for 30 min, followed by sudden release of the pressure	Partial disruption of the fibers leaves short cellulose chains and disrupts lignin Solid 99% recoveredLow inhibitor concentration Removes acetyl groups by deacetylation Herbaceous biomass and agricultural residues are highly susceptible Lignin removed >85%Up to 95% assimilable sugars are obtained after enzymatic hydrolysis	Hardwood low susceptible Not suitable for softwoods The dosage of liquid ammonia is 1–2 kg of ammonia/kg of dry biomass, which increases the costs Residual lignin generates unspecific bonds in the enzymatic hydrolysis
CO_2 explosion	At conditions above the critical point of 31°C and 7.39 MPa, CO_2 behaves as a supercritical fluid that has access to pores in biomass. CO_2 reacts with moisture to form carbonic acid, which further catalyzes biomass degradation	Rapid analysis CO_2 gas is non-toxic and has low critical values Reduced solvent volumes required	Energy-intensive due to the use of high-pressure CO_2 not enough polar for the separation
Organosolv	It operates at 90–120°C (for grasses) and 155–220°C (for wood), with 25–100 min incubation time Solvent and catalyst concentration varies with the type of feedstock	Solvent combination successfully disrupts cell membranes Use of polar solvents decreases the interference of nonlipid compounds Efficiently extracts polar lipids Nonlipid compounds do not interfere, due to the strong polar solvents used A buffered solvent system prevents ionic salt adsorption effects	Requires high volumes of solvents Carcinogenicity Polar lipids have been found in the aqueous phase Time-consuming Difficulty in extracting organic layer without contamination occurring
Biological	Microbes like fungus or bacteria (deconstruction of lignin structure in the cell wall using microbes and/or enzymes under ambient conditions)	Mild pretreatment condition Low energy consumption No chemicals are needed	Slow process and slow throughput Sugar conversion after pretreatment is not high Larger space requirements Need continuous monitoring

Source: Adopted from Zuccaro et al., 2020; Shaheen et al., 2023.

In *Rhodosporidium toruloides*, lipid formation and accumulation are increased by growing cells on nitrogen-rich but sulfate-deficient media. Moreover, it is reported that the absence of elements like Fe, Zn, P, and Mg in media also influences lipid accumulation (Annamalai et al., 2018; Zhang et al., 2021).

Physicochemical parameters to maximize lipid accumulation may vary from species to species. Generally, for a good amount of lipid deposition, the optimum pH is in the range of 5.5–9, the optimum temperature is in the range of 25–37°C, and the agitation rate is between 100 and 400 rpm. Accumulation of the lipids in microbial cells is influenced by regulating the genes, which are involved in fatty acid metabolism. Manipulation of the following genes helps to attain maximum lipid production (Zhang et al., 2021).

3.4.1 Acetyl-CoA Carboxylase

The first step in fatty acid synthesis is carboxylation, carried out by an enzyme acetyl-CoA carboxylase (produced by gene ACC1). Acetyl-CoA carboxylase is a cytosolic enzyme and requires Mg^{2+} and biotin as cofactors. This enzyme performs carboxylation of acetyl-CoA and produces malonyl-CoA, the precursor of fatty acids.

Overexpression of ACC1 showed a twofold increase in lipid content which is reported in *Yarrowia lipolytica*. Also, in *R. toruloids*, ACC1 overexpression along with glyceraldehyde 3-phosphate dehydrogenase (GAPDH) promoter hiked the lipid content up to 61.1% over the control. It is reported that co-expression of *E. coli* ACC1 and thioesterase I (product of tes A gene) demonstrates a sixfold increase in lipid level (Jeevan Kumar et al., 2020; Chawla et al., 2022).

3.4.2 Diacylglycerol Acyl Transferase

Diacylglycerol acyl transferase (DGA) catalyzes the conversion of diacylglycerol (DAG) to triacylglycerol (TAG). DGA is a trans-membrane protein of the ER membrane and performs the very last step of TAG biosynthesis, where the produced TAGs are stored in lipid bodies (LB). It is reported that overexpression of specific *dga* genes (responsible for the formation of proteins; DGA1, DGA2) in *Y. lipolytica* causes the change in size and position of LBs. The DGA2 protein is reported to carry out the accumulation of the lipids while the DGA1 protein controls the flux of fatty acids inside LBs. Higher expression of DGA1 causes the formation of innumerable LBs, which are small in size, while higher expression of DGA2 forms LBs of large sizes. Overexpression of gene *dga1* with the effect of ATP citrate lyase (ACLY) promoter led to a 43.4% rise in the lipid profile of *R. toruloids* as compared to the control. Some mutant strains of yeast cells cannot accumulate lipids and form LBs. This can be altered by the expression of gene; diacylglycerol acyltransferase (DGAT) helps in the accumulation of TAGs. Overexpression of DGAT leads to the development of LBs in yeast cells and thus shows a high TAG deposition in LBs. The transformation of yeast cells with DGAT from *Arabidopsis* showed a 3–9-fold increase in TAG formation and thus represents a direct proportionality between DGAT activity and TAG synthesis (Jeevan Kumar et al., 2020; Chawla et al., 2022).

3.4.3 Δ9-Fatty Acid Desaturase

Δ9-fatty acid desaturase is a protein encoded by gene *ole1* which converts saturated fatty acid into monounsaturated fatty acids. OLE1 is an endoplasmic reticulum membrane protein that adds a double bond at the Δ9 position of fatty acyl-CoAs by utilizing electrons from reduced cytochrome b5 complex. Δ9-fatty acid desaturase converts substrates like palmitic acid (C16:0) and stearic acid (C18;0) into palmitoleic acid (C16:1) and oleic acid (C18;1), respectively. In yeast cells, it is reported that the transcription level of the *ole1* gene gradually increases at low oxygen and temperature conditions. Overexpression of *ole1* in *Y. lipolytica* increases the degree of unsaturation (fatty acids to unsaturated fatty acids) with an 89.4 g/L rise in total lipid content in comparison to control. The same is reported in *R. toruloides*; deposition of lipids is increased by 1.4 × times as compared to the original cells (Chawla et al., 2022).

3.4.4 Glycerol 3-Phosphate Dehydrogenase

Glycerol 3-phosphate dehydrogenase is a cytosolic enzyme that converts dihydroxyacetone phosphate (DHAP) to glycerol 3-phosphate (G3P). This enzyme requires NAD^+ for catalysis and is produced by gene *gpd1*. G3P is utilized in the formation of TAGs. A prior study showed that protein GPD1 is expressed under the influence of osmotic stress. The overexpression of gene *gpd1* in *Y. lipolytica* led to a 40% increase in lipid content when grown on culture media containing glucose. Manipulation of such genes can alter the lipid profiles of microbes and thus help maximize the biodiesel yield (Chawla et al., 2022).

3.4.5 Triacylglycerol Lipase

There are certain genes whose downregulation will help in the synthesis and accumulation of lipids in microbial cells. The gene *tgl* produces the enzyme triacylglycerol lipase, which performs hydrolysis of TAG into monoacylglycerol (MAG) and DAG and thus degrades TAG into FFAs. In *Y. lipolytica*, it is reported that two intracellular lipase genes *tgl3* and *tgl4* are responsible for the hydrolysis of TAG and showed an increase in lipid synthesis and accumulation when these genes were deleted. Thus, the downregulation of *tgl* by genetic engineering showed direct proportionality to high lipid deposition (Chawla et al., 2022).

3.4.6 Phospholipases A2

Cytoplasmic enzyme phospholipase A2 is encoded by the gene *pla2*, which plays diverse roles including signal transduction, lipid metabolism, and oxidative stress tolerance. PLA2 catalyzes hydrolysis reaction on phospholipids and thus destabilizes phospholipids into fatty acid chains and glycerol. The exact mode of action of PLA2 on phospholipid is still under discovery, but it is reported that knocking out the *pla2* gene showed a significant rise in total lipid accumulation in OMs which will benefit the formation of more biodiesel manufacturing (Khan and Hariprasad, 2020).

Another dynamic research revealed symbiotic relationships between microorganisms can benefit to accelerate lipid production when cultured together. Culturing of microbes like microalgae, yeast, and fungi with phototrophic or heterotrophic bacteria showed enhancement in total biomass along with the rise in total lipid content. Both the partners in the co-culture method contribute to each other in metabolic pathways and thus facilitate high lipid biosynthesis and accumulation. The selection of proper partners will ultimately influence the fatty acid composition. The co-cultured batch of organisms *Trichosporonoides spathulata* and *Chlorella vulgaris* unveils deposition of more saturated fatty acids than culturing individually (Levering et al., 2015; Chawla et al., 2022).

3.5 OLEAGINOUS MICROBE BIODIESEL SYNTHESIS PATHWAYS

OMs have recently caught large attention due to their remarkable property to synthesize and accumulate lipids and thus are of great industrial importance in the production of biodiesel. OMs have a unique feature to convert glucose and fermentable sugars (C5 and C6 sugars) into pyruvate by the glycolysis process, which is further utilized for the biosynthesis of lipids. The metabolic lipid synthesis pathway of OMs is different than other microorganisms and various factors contribute to the accumulation of lipids (Jeevan Kumar et al., 2020; Chintagunta et al., 2021).

It is reported that starvation of certain OMs on a nitrogen-deficient medium initiates the synthesis and accumulation of fatty acids or triacylglycerides. The nitrogen starvation condition causes a reduction in cytoplasmic adenosine monophosphate (AMP) concentration which triggers inhibition of the enzyme isocitrate dehydrogenase (ICDH). ICDH is responsible for the downregulation of the Kerbs cycle and causes high deposition of citrate in mitochondria. Also, the pyruvate produced from glycolysis follows Kerbs cycle and accumulates citrate (Chawla et al., 2022). Formed citrate is an essential substrate for the formation of acetyl-CoA. The cytoplasm is the principal site for the formation of lipids by conversion of acetyl-CoA to long-chain FFAs. The citrate translocase system on the mitochondrial membrane assists in the transfer of citrate from mitochondria to the cytoplasm. The cytosolic enzyme ATP-citrate lyase (ACL) is responsible for the cleavage of citrate into acetyl-CoA and oxaloacetate, and this feature is unique to OMs (Kumar et al., 2017).

Further enzyme malate dehydrogenase converts oxaloacetate into malate, which is again translocated to mitochondria via protein malate translocase. Existing cytosolic acetyl-CoA carboxylase (ACC) causes carboxylation of acetyl-CoA to malonyl-CoA under the influence of biotin as a cofactor. Acetyl-CoA, malonyl-CoA, and glycerol-3-phosphate are the three precursors that play a vital role in the biosynthesis of lipids (Jeevan Kumar et al., 2020; Chintagunta et al., 2021). In the presence of the enzyme fatty acid synthase 1 (FAS1) complex, lipid biosynthesis initiates. FAS1 is responsible for the extension of acetyl-CoA and malonyl-CoA carbon chains by performing four sequential biochemical reactions, that is, condensation, reduction, dehydration, and reduction (Chawla et al., 2022). For efficient working, FAS1 requires NADPH. Malase enzyme (ME) is the sole player responsible for the production of NADPH and its presence is only limited to OMs. In FAS1, initial condensation of acetyl-CoA produces malonyl-CoA, which is further reduced at the β-keto group to form alcohol. Generated alcohol is dehydrated by the removal of water molecules and again reduced to form the subsequent saturated fatty acid (SFA). As a result of FAS1 action, after each cycle in FAS1, carbon chain length is elongated by two carbons. The cycle is carried out until there is a formation of acids containing C16, C18, C20, C22, and C30 long carbon chains (Kumar et al., 2017; Jeevan Kumar et al., 2020).

In oleaginous microbes, the elongation of fatty acids is similar to that of palmitic acid biosynthesis, but there is further introduction of double bonds in fatty acid chains that is carried out by the enzyme fatty acyl–CoA desaturase. Introduction of double bonds at the Δ9 position of palmitic acid (C16:0) and stearic acid (C18;0) causes the formation of palmitoleic acid (C16:1) and oleic acid (C18;1), respectively, which is catalyzed by the enzyme Δ9-fatty acid desaturase (product of OLE1 gene) (Jeevan Kumar et al., 2020). The overall biochemical reaction of the palmitate synthesis is as follows:

$$\begin{aligned}\text{Acetyl}-\text{CoA}+7\,\text{malonyl}-\text{CoA}+14\,\text{NADPH}+20\,\text{H}^{+} &\longrightarrow \text{Palmitate}\\ +14\,\text{NADP}^{+}+8\,\text{CoASH}+7\,\text{CO}_2+6\,\text{H}_2\text{O}\end{aligned} \tag{3.1}$$

The biosynthesis of TAG follows the Kennedy pathway in the endoplasmic reticulum. Inside ER, glycerol 3-phosphate dehydrogenase 1 enzyme converts DHAP into G-3P. In the cycle, G-3P undergoes two subsequent acylation reactions. Initially, to the backbone of G-3P, first acylation is carried out to form compound lysophosphatidate (LPA), which undergoes second acylation in the presence of acylglycerol-phosphate acyltransferases (AGPAT) to form phosphatidate (PA). Enzyme phosphatidate phosphate (PAH1) further catalyzes the dephosphorylation of PA into DAG. In the end, the addition of a third acyl group is carried out by DGAT which converts DAG into TAGs. Synthesized TAGs are stored in LB which remain attached to endoplasmic reticulum. The extraction of such accumulated loads of TAGs causes OMs to qualify as the best source for manufacturing biodiesel (Jeevan Kumar et al., 2020; Chawla et al., 2022; Chintagunta et al., 2021) (Figure 3.3).

3.6 MICROBIAL OIL EXTRACTION AND BIODIESEL PRODUCTION

3.6.1 Extraction of Microbial Lipids

After the harvesting step of oleaginous microbes, one of the crucial steps in biodiesel production is the extraction of lipids from microbes which alone costs around 90% of the total biodiesel production cost (Chintagunta et al., 2021). The direct extraction process is one of the most common processes of cell disruption which is generally preferred in small-scale biodiesel fermentations. Simultaneous extraction and transesterification of biomass in one system but different chambers directly produce biodiesel. The biomass is subjected to a continuous flow of fresh solvent, facilitating the lipid transfer which further gets converted to biodiesel in the presence of a catalyst (Tarigan et al., 2019). However, the direct process has a major drawback of low recovery lipids value due to the presence of a rigid cell wall. To maximize lipid

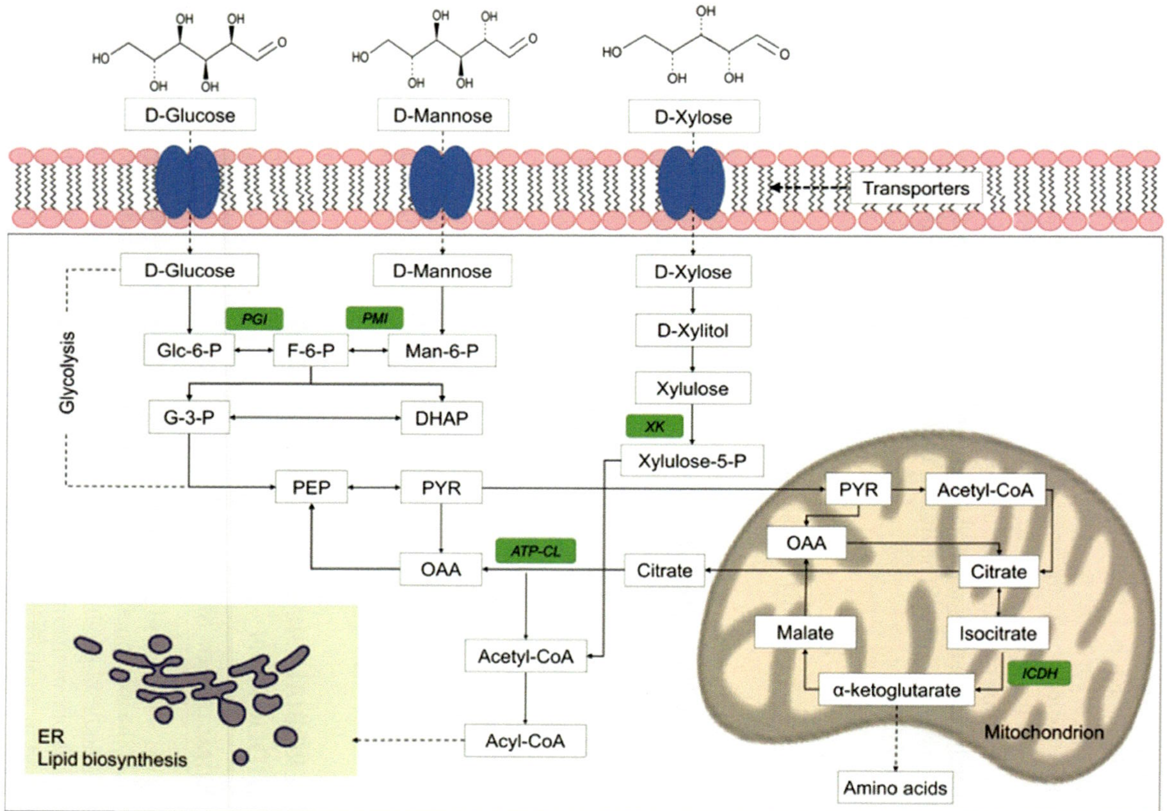

FIGURE 3.3 Cellular pathways, enzymes, and organelles involved in the carbohydrate to lipid conversion. (Figure available under Creative Commons CC-BY license was adopted from Valdés et al., 2020).

recovery from the direct extraction, various pretreatments are needed to perform on microbial cells. Pretreatments involve the use of physical, chemical, and biological approaches to increase the chances of easy cell disruption (Zhang et al., 2021). A variety of cell disruption techniques such as electroporation, bead beating, osmotic shock, supercritical CO_2, microwave, and ultrasonication have been implemented for the efficient rupturing of microbial cells. Each of these cell disruption techniques has certain advantages and limitations. After proper disruption, the cytoplasmic content of cells is subjected to extraction of lipids by the utilization of different solvents. Most commonly, organic solvents methanol and chloroform are preferred as they interact with polar lipids and other biopolymers in the cell and disrupt their integrity causing the release of lipids from other cell waste materials. Non-lipid components are removed in the washing step by the addition of small quantities of water in the ratio of methanol:chloroform:water as 4:8:3 (Khot et al., 2020; Zuccaro et al., 2020).

3.6.2 Transesterification of Microbial Lipids

After the successful extraction of lipids from oleaginous microbes, lipids are converted to diesel by a process called transesterification (Bhatia et al., 2019a). In the process, complex lipids are converted to less viscous, highly flammable fuel, which shares quite similar properties to that of conventional biodiesel. The transesterification reaction of lipids (triacylglycerides) is carried out in the presence of alcohols, such as methanol or ethanol, and a catalyst to give alkyl esters (biodiesel) and glycerol as a by-product. The transesterification reaction involves the lysis of TGA into FFA and further conversion into fatty acid ethyl esters. Depending on the alcohol used in the process, produced biodiesel is categorized into two types, FAME with methanol and fatty acid ethyl ester with ethanol (FAEE) (Singh et al., 2019). Based on the type of mechanism that is followed by a catalyst, the transesterification process is divided into three categories, (1) chemical-catalyzed reaction, (2) non-catalyzed reaction, and (3) enzyme-catalyzed reaction.

The chemical-catalyzed transesterification reaction is subdivided into two classes, homogeneous and heterogeneous-catalyzed reaction (Bhatia et al., 2020). The homogeneous reaction involves the use of both acid and base catalysts in the liquid state. Sulfonic, hydrochloric, phosphoric, and sulfuric acids are examples of homogeneous acid catalysts. Examples of homogeneous base catalysts are sodium hydroxide and potassium hydroxide. In a heterogeneous-catalyzed reaction, the physical state of the acid and base catalyst is solid. As a heterogeneous acid-catalyst, heteropoly acids (HPAs), tungstated zirconia, and sulfated zirconia are used while alkaline earth metal oxides and mixed metal oxides (CaO–MgO) along with hydrotalcite are utilized as a heterogeneous base-catalyst. There are several disadvantages related to utilizing chemical-catalyzed reactions, which are the high operational cost required for the purification of biodiesel and glycerol from a catalyst, large energy consumption, and wastewater management (Norjannah et al., 2016).

On the other hand, in the non-catalyzed reaction of biodiesel production, transesterification is carried out without the use of a catalyst. Supercritical conditions are maintained accurately to carry out maximum biodiesel production without the requirement of a catalyst. Due to controlled conditions, the process is completed in a

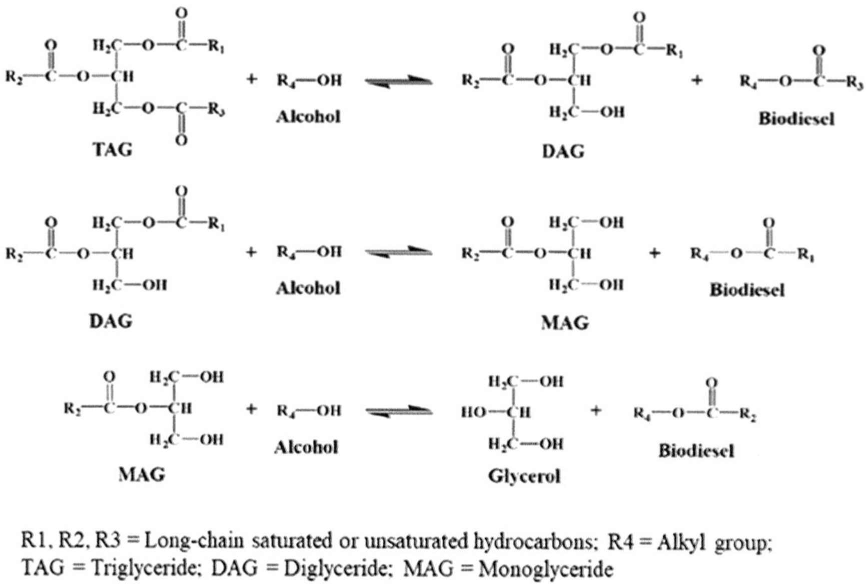

FIGURE 3.4 Schematic of the generalized transesterification reaction mechanism to produce biodiesel (Figure freely available to use was adopted from Leng et al., 2020).

short time (generally 2 min) under a high temperature (250°C–400°C) and high pressure (10–30 MPa). A major limitation of non-catalyzed reactions is the cost of production, as maintenance of physical conditions requires a large amount of energy. To overcome these problems, an enzymatic catalysis approach is utilized. The use of an enzyme called lipase as a catalyst has provided many advantages over other catalyzed processes, such as complete catalysis of fatty acids with a moderate temperature range, high specificity of the substrate, and low wastewater generation with high product quality. Apart from advantages, the enzyme-catalyzed reaction requires more time for enzyme catalysis, and thus the production rate is low. There is a chance that enzymes lose their activity due to the inactivation of enzymes.

At the end of transesterification, formed glycerol settles down in the fermenter because of its high weight while biodiesel floats above glycerol. The biodiesel separation must be carried out fast as there is a chance that the transesterification reaction will reverse. Further produced biodiesel is subjected to purification and the glycerol (by-product) produced is utilized as an important component of the pharma industry and cosmetic industry (D. Singh et al., 2019) (Figure 3.4).

3.7 COMPOSITION OF MICROBIAL OIL

Lipids synthesized by microbes exist in the form of triacylglycerol. These triacylglycerols resemble the same fatty acid profiles as those of edible and non-edible oils which are conventionally used in biodiesel synthesis. Triacylglycerides are composed of one glycerol molecule with three saturated or unsaturated fatty acid chains. These triacylglycerides on hydrolysis release FFA chains. Most oleaginous microbes utilize C5 and C6 sugars as a substrate to accumulate lipid content (Ramos et al., 2019). The oleic, linoleic, stearic, and palmitic acids are the most commonly found fatty acids that are present in microbial lipids. When oleaginous microbes are grown on a LCB-containing medium, lipids synthesized show 50% mono- and di-saturated fatty acids (C18:1 and C18:2) along with 40% lipid mass of saturated fatty acids (C16:0 and C18:0) (Valdés et al., 2020). The oleaginous microbes cultivated on pretreated lignin and its derivatives produce different lipid profiles. In oleaginous yeast, the composition of fatty acids is found in order, oleic acid > palmitic acids > linoleic acids > stearic acids. Depending on the strain of microbe used and the type of substrate utilized, the composition of fatty acid varies (Norjannah et al., 2016; Patel et al., 2017).

3.8 PROPERTIES OF BIODIESEL

Biodiesel is an amalgam of ethyl or methyl esters of long-chain fatty acids, extruded by the method of transesterification of vegetable, animal, or microbial oil with short-chain alcohol, either ethanol or methanol, in the presence of a catalyst. Fuel properties of the biodiesel play a crucial role in combustion. With time, biodiesel in its neat or blended form has emerged as a favorable alternative to conventional diesel, satisfying the physiochemical standards of diesel. The property of biodiesel is important in order to design a compression ignition engine (CIE) fuel system for the neat or blended forms.

3.8.1 C/N Ratio

The yield of biodiesel production is directly proportional to the lipid content stored in OMs. In the lipid accumulation process, the carbon-to-nitrogen ratio acts as a yield-deciding factor (Cho et al., 2011). Culturing the microbial cells under the high carbohydrate concentration and low nitrogen concentration showed high lipid accumulation profiles. Various microbes show different accumulation ranges of lipids on the usage of nitrogen-containing compounds. Nitrogen obtained from inorganic nitrogen compounds leads to the microbial cells retarded lipid accumulation and promotes cell division and the opposite happens when cells are cultured on organic nitrogen compounds (viz., urea, nitrate, nitrite, and ammonia) causing higher lipid accumulation profile and inhibition of cell division. Low nitrogen supply produces starvation-like conditions in culture media that downregulate Krebs cycle, causing a reduction in cell division and thus promoting lipid cumulation in cells. In conclusion, to acquire more lipid content from OMs, the C/N ratio in nutrient media should be set high (Jeevan Kumar et al., 2020; Zuccaro et al., 2020).

3.8.2 Cetane Number

The cetane number (CN) is used to measure the fuel ignition quality, as it indicates the speed at which a fuel auto-ignites while entering the engine. The cetane number affects the fuel combustion and engine performance. Several other properties of fuels such as heating value and density are cetane number dependent. Based on the fatty acid profile of the feedstock utilized for the microbial biodiesel, the US Standard, ASTM 6751, and the European standard, EN 14214, have the lowest acceptable values at 47 and 51, respectively (Hawrot-paw et al., 2020). Several diverse formulas are derived to predict the CN number.

$$CN = -0.1209 \times DU + 65.0958 \tag{3.2}$$

Here, DU is the degree of unsaturation.

The following formula gave a predicted CN value of 51.77 for biodiesel extracted from microalgae *S. obliqus*. The CN number for biodiesel extracted from other microalgae by in situ transesterification was determined at 48 by standard ASTM 976 (Raut et al., 2019). The cetane number is calculated by creating a blend of n-Cetane (100CN) and hepta methyl Cetane (15CN). Studies have found that the predicted CN value of biodiesel using models is similar to that of the actual CN value, thus deducing that thermal properties can be utilized to predict the CN number. The cetane number is determined by the percentage volume of normal cetane in a blend of normal cetane and α-methyl naphthalene, resulting in equivalent ignition characteristics to the fuel under examination when burned in a standardized engine.

3.8.3 Viscosity

In order to measure the internal friction of biodiesels produced, the dynamic or kinematic viscosity is measured at different temperatures (Hawrot-paw et al., 2020). It affects the atomization of the fuel and varies according to the feedstock selected for the production of biodiesel. At 40°C, the viscosity of a biodiesel should vary 1.9–6.0 and 3.5–5.5 mm^2/s according to the ASTM 6571 and EN 14214 norms, respectively. Biodiesel produced by microalgae, *N. gaditana*, was found to have a kinematic viscosity of 3.9 mm^2/s at 40°C as per the ASTM 445 standards. At 40°C the experimentally determined viscosity of *Rhodococcus opacus* PD630 was found to be 4.1 mm^2/s, *M. circinelloides* URM 4182 as 4.45 mm^2/s, *A. terreus* as 4.5 mm^2/s, *Candida* sp. LEB-M3 as 4.28 mm^2/s, and *Y. lipolytica* NCIM 3589 as 4.65 mm^2/s.

The viscosity of *A. terreus* was determined using the equation

$$\ln \eta i = -12.503 + 2.496 \times \ln(Mi) - 0.178 \times N \tag{3.3}$$

The kinematic viscosity of the ith FAME is denoted by ηi, Mi is the molecular weight of the *i*th FAME, and N indicates the number of double bonds.

The kinetic viscosity of biodiesel obtained from *Candida* sp. LEB-M3 was calculated by using

$$\ln \nu = \Sigma \ln(\nu\text{me})\%\text{ME}/100 \tag{3.4}$$

where ν indicates the kinematic viscosity of the sample, me indicates individual methyl ester, and %ME indicates the percentage present of each methyl ester in the mixture of biodiesel (Raut et al., 2019).

3.8.4 Flash Point and Fire Point

Flash point is the lowest temperature at which a fuel will ignite upon providing an ignition source. The volatility of the fuel is inversely proportional to the flash point. Minimum flash point temperature ensures proper safety and handling of the fuel. Fire point measures the lowest temperature at which the fuel will sustain burning for 5 s. Flash point and fire point assist in determining whether the fuel may become hazardous and release inflammable fumes (Jeas et al., 2012).

3.8.5 Iodine Value

The iodine number or iodine value of biodiesel evaluates its stability toward getting oxidized and it's the grams of iodine required to react with 100 g of biodiesel. More unsaturation content will result in higher iodine value; it's the quantification of total unsaturation of the fatty acids. Iodine numbers vary depending on the ester composition and the character of the feedstocks for biodiesel production. A higher iodine value indicates easier oxidation in the air (Yoosaf et al., 2018).

$$\text{Iodinenumber} = (B - S) \times N \times 129 / W \tag{3.5}$$

B – Volume for sodium thiosulfate for blank (mL)
S – Volume for sodium thiosulfate for sample (mL)
W – Weight of the sample (g)

3.8.6 Cold Flow Properties

Cold flow properties of a fuel refer to its behavior at low temperatures, particularly in cold weather conditions. It indicates the efficiency at which the fuel can atomize at low temperatures and is essential in determining the functioning of engines and machines. Key cold flow properties include cloud point, pour point, cold filter plugging point (CFPP), and wax content. Biodiesel displays substandard cold flow behavior. It demonstrates high cloud point (CP) and pour point (PP), high viscosity, low vitality content, and high nitrogen oxide (NO_x) discharge (Monirul et al., 2015; Leng et al., 2020).

3.8.7 Calorific Value

The amount of heat energy released by a unit mass of biofuel during combustion is its calorific value. The calorific values of fuel can be higher and lower depending upon the form in which H_2O is released. After combustion, the amount of H_2O released as heat is condensed to liquid form and called a higher calorific value. The vapor form of H_2O released is known as the lower calorific value. The calorific value is determined using a bomb calorimeter according to the ASTM D240 standard method. The calorific value measures the energy content in a fuel which helps in determining the suitability of the fuel as an alternative to conventional diesel (Yoosaf et al., 2018) (Table 3.3).

TABLE 3.3
Comparison of Properties of Biodiesel Obtained from Bacteria *S. coeliocolor* and Algae with Different Standards (Tyagi et al., 2010; Bhatia et al., 2017)

Properties	S. coelicolor	Algae	IS 15607	ASTM
Viscosity (mm^2/sec), 40°C	1.03	27–26	2.5–6.0	1.9–6
Flash point °C	–	115	120	93
Cloud point °C	3.73	–	–	–
Pour point °C	–	−12	–	–
Cetane number	64.42	41	48	47
Higher heating value (MJ/kg)	36.56	41	–	–
Density (kg/L)	0.82	0.86	–	–

3.9 ECONOMICS OF BIODIESEL

Recent times have shown a significant increase in global energy consumption, which is expected to expand between 2012 and 2040 by 40% (IEO2016). Population and economic growth are the chief reasons for the expansion. Currently, the major sources of energy are crude oil, coal, and gas (Gebremariam and Marchetti, 2018). The ever-increasing fuel prices, resource scarcity, depleted supply, and negative impact on the environment have created an undeniable need for renewable and sustainable resources of energy such as biodiesel. Geopolitical conflicts between the fuel/oil producing and consuming countries have led to an inconsistent supply chain creating global differences (Anwar, 2021). Biodiesel can be extracted from a diverse range of feedstock sources, which can also include wastewater or sludge from various factories, waste animal fat, and waste cooking oil remains. Biodiesel display the properties of biodegradability, lack of sulfur and other aromatics, non-toxic, and emission of lesser air pollutants and GHGs giving it an advantage over conventional fuels. However, they are comparatively low in fuel content and upon burning release nitrogen oxide compounds. With all its ecological benefits, biodiesel is still not extensively used, mainly due to the high production cost (Gebremariam and Marchetti, 2018).

Recent studies aim to resolve the financial burden and reduce the production cost by evolving production technologies for higher yield and productivity, lowering the capital cost, and finding suitable inexpensive feedstocks with high yields (Gebremariam and Marchetti, 2018). Economic analysis of various possible production pathways is a need, along with the different catalysts, feedstock, and purification techniques. To produce biodiesel, the total investment cost includes production technology, plant size or scale of production, raw materials source type, and market value, along with others. Fixed Capital Coast and Operating or Working Capital Costs are the main types of Investment Capital Costs. Fixed Investment Capital Cost includes all the investment required for the installed equipment along with all the additional charges for its maintenance, whereas Working Capital includes the cost of services or products such as raw materials or feedstocks, labor charges, facility-dependant costs, and other irregular expenses required for biodiesel production (Gebremariam and Marchetti, 2018).

Direct plant setup costs and equipment costs are the major categories under capital investment costs. Installation of equipment and instruments, electrical connections, piping, and additional facilities are some of the direct costs for the plant setup. Before the calculation of capital cost, proper designing of the process flow is required along with selecting the type and size of equipment required (Gebremariam and Marchetti, 2018). Operating cost or working capital is required for the varying costs of raw materials, labor, maintenance, renovation, restoration, or repairs. The cost of raw materials holds the major share of the operating cost, including the feedstock source, catalyst, alcohol, and washing water (Gebremariam and Marchetti, 2018; Mizik and Gyarmati, 2021).

The major barrier to the commercial use of biodiesel as an alternative to petroleum diesel has always been the high production cost. As the feedstock capital owns the major share of the capital invested, using cheaper raw materials would reduce the production cost. Inexpensive waste oils or non-edible food crops contain high FFA and water content, reducing the quality and yield of the biodiesel releasing unwanted by-products during production. Microbial biodiesel has emerged as a better alternative, as it requires less land than plant feedstocks (Anwar, 2021).

3.10 FUTURE PERSPECTIVES

Concerns about the environmental impact of fossil fuels and petroleum supplies have led to the hunt for sustainable and renewable methods of producing biofuels. The two biggest issues facing humanity today are the present shortage of petro-diesel and the increased expense of producing biodiesel made from vegetable oils and animal fat. For the industry to manufacture substantial amounts of biodiesel, it is advised that OMs be used. Given the novel approaches to LCB pretreatment and the enhancement of OM's efficiency via genetic modification and adaptive lab evolution, the transformation of LCB into SCO may prove to be a promising technology. After enzymatic hydrolysis, chemical pretreatments including acid, alkaline, and acid/alkaline have shown to be viable approaches for getting large yields of assimilable sugars. Microbial lipids can be a source for the responsible synthesis of renewable biofuels and a potential remedy for the drawbacks associated with the declining supply of conventional fuels. By transesterifying fatty acids with short-chain alcohols, mono-alkyl esters of long-chain fatty acids are generated, which is how biodiesel is formed. The establishment of OMs that produce the fatty acids and short-chain alcohols needed for transesterification and acyltransferases with increased reactivity for short-chain alcohols depends on a knowledge of this process. Moreover, to fight against the usage of fossil fuels, it is imperative to produce microorganisms with both increased efficacy and the ability to use inexpensive feedstock. A creative farming technique incorporating all the starchy or carbon materials – comprising household and industrial wastes – will be crucial to reducing expenses. Our economic research, however, showed that developing microbes that can yield greater lipid productivity in minimum media and reducing the cost of medium components are necessary. To achieve sustainable biodiesel production, pretreatment techniques, process design, government subsidies, policy support, and other technological improvements for lucrative co-product creation must be developed.

3.11 CONCLUSION

The production of biodiesel necessitates significant amounts of transportation fuels owing to its numerous advantages. The industrialization of this technology relies on the development of a cost-effective biodiesel manufacturing process. For biodiesel synthesis, various feedstocks are accessible, including plant oils, OMs, food crops, non-food/energy crops, tree-based oils, and wastes from factory processes. Since LCBs are widely available, have a high cellulose and hemicellulose content, and provide considerable potential for growing OMs, using LCBs as a carbon substrate is the most recommended alternative among these substrates. Pretreatment, selecting of the appropriate OM, the saccharification process for fermentable sugars, the fermentation of hexoses and pentoses, lipid extraction, methanolysis, and other aspects are critical and must be optimized to build a sustainable process.

REFERENCES

Annamalai, N., Sivakumar, N., Oleskowicz-Popiel, P., 2018. Enhanced production of microbial lipids from waste office paper by the oleaginous yeast Cryptococcus curvatus. *Fuel*, 217, 420–426.

Anwar, M., 2021. Biodiesel feedstocks selection strategies based on economic, technical, sustainable aspects. *Fuel*, 283, 119204.

Babatunde, E.O., Gurav, R., Hwang, S. 2023. Recent advances in invasive aquatic plant biomass pretreatments for value addition. *Waste and Biomass Valorization*, 14, 1–25.

Bateni, H., Saraeian, A., Able, C., 2017. A comprehensive review on biodiesel purification upgrading. *Biofuel Research Journal*, *4*(3), 668–690.

Bhatia, S.K., Bhatia, R.K., Yang, Y.H., 2017. An overview of microdiesel—A sustainable future source of renewable energy. *Renewable and Sustainable Energy Reviews*, 79, 1078–1090.

Bhatia, S.K., Gurav, R., Cho, D.-H., Kim, B., Jung, H.J., Kim, S.H., Choi, T.-R., Kim, H.-J., Yang, Y.-H. 2023. Algal biochar mediated detoxification of plant biomass hydrolysate: mechanism study and valorization into polyhydroxyalkanoates. *Bioresource Technology*, 370, 128571.

Bhatia, S.K., Gurav, R., Choi, T.-R., Han, Y.H., Park, Y.-L., Park, J.Y., Jung, H.-R., Yang, S.-Y., Song, H.-S., Kim, S.-H., Choi, K.-Y., Yang, Y.-H. 2019a. Bioconversion of barley straw lignin into biodiesel using Rhodococcus sp. YHY01. *Bioresource Technology*, 289, 121704.

Bhatia, S.K., Gurav, R., Choi, T.-R., Jung, H.-R., Yang, S.-Y., Song, H.-S., Kim, Y.-G., Yoon, J.-J., Yang, Y.-H. 2019b. Effect of synthetic and food waste-derived volatile fatty acids on lipid accumulation in Rhodococcus sp. YHY01 and the properties of produced biodiesel. *Energy Conversion and Management*, 192, 385–395.

Bhatia, S.K., Gurav, R., Choi, T.-R., Kim, H.J., Yang, S.-Y., Song, H.-S., Park, J.Y., Park, Y.-L., Han, Y.-H., Choi, Y.-K., Kim, S.-H., Yoon, J.-J., Yang, Y.-H. 2020. Conversion of waste cooking oil into biodiesel using heterogenous catalyst derived from cork biochar. *Bioresource Technology*, 302.

Bhatia, S.K., Gurav, R., Choi, Y.-K., Lee, H.-J., Kim, S.H., Suh, M.J., Cho, J.Y., Ham, S., Lee, S.H., Choi, K.-Y., Yang, Y.-H. 2021. Rhodococcus sp. YHY01 a microbial cell factory for the valorization of waste cooking oil into lipids a feedstock for biodiesel production. *Fuel*, 301, 121070.

Bhuvaneshwari, S., Hettiarachchi, H., Meegoda, J.N., 2019. Crop residue burning in India: policy challenges potential solutions. *International Journal of Environmental Research Public Health*, 16(5), 832.

Chawla, K., Kaur, S., Kaur, R., Bhunia, R.K., 2022. Metabolic engineering of oleaginous yeasts to enhance single cell oil production. *Journal of Food Process Engineering*, 45(7), 13634.

Chintagunta, A.D., Zuccaro, G., Kumar, M., Kumar, S.P., Garlapati, V.K., Postemsky, P.D., Chel, A.K., Simal-Gara, J., 2021. Biodiesel production from lignocellulosic biomass using oleaginous microbes: prospects for integrated biofuel production. *Frontiers in Microbiology*, 12, 658284.

Cho, S.J., Lee, D.H., Luong, T.T., Park, S.R., Oh, Y.K., Lee, T.H., 2011. Effects of carbon nitrogen sources on fatty acid contents composition in the green microalga, Chlorella sp. 227. *Journal of Microbiology Biotechnology*, 21(10), 1073–1080.

Cho, H.U., Park, J.M., 2018. Biodiesel production by various oleaginous microorganisms from organic wastes. *Bioresource Technology*, 256, 502–508.

Dourou, M., Aggeli, D., Papanikolaou, S., Aggelis, G., 2018. Critical steps in carbon metabolism affecting lipid accumulation their regulation in oleaginous microorganisms. *Applied Microbiology Biotechnology*, 102, 2509–2523.

Fernández-Blanco, C., Robles-Iglesias, R., Naveira-Pazos, C., Veiga, M.C., Kennes, C., 2023. Production of biofuels from C1-gases with Clostridium related bacteria—Recent advances. *Microbial Biotechnology*, 16(4), 726–741.

Ferreira, A., Ribeiro, B., Ferreira, A. F., Tavares, M. L., Vladic, J., Vidović, S., Gouveia, L., 2019. Scenedesmus obliquus microalga-based biorefinery–from brewery effluent to bioactive compounds, biofuels biofertilizers–aiming at a circular bioeconomy. *Biofuels, Bioproducts Biorefining*, 13(5), 1169–1186.

Garg, R., Sabouni, R., Ahmadipour, M., 2023. From waste to fuel: challenging aspects in sustainable biodiesel production from lignocellulosic biomass feedstock's role of metal organic framework as innovative heterogeneous catalysts. *Industrial Crops Products*, 206, 117554.

Gebremariam, S.N., Marchetti, J.M., 2018. Economics of biodiesel production. *Energy Conversion Management*, 168, 74–84.

Gurav, R., Bhatia, S.K., Choi, T.-R., Hyun Cho, D., Chan Kim, B., Hyun Kim, S., Ju Jung, H., Joong Kim, H., Jeon, J.-M., Yoon, J.-J., Yun, J., Yang, Y.-H. 2022. Lignocellulosic hydrolysate based biorefinery for marine exopolysaccharide production and application of the produced biopolymer in environmental clean-up. *Bioresource Technology*, 359, 127499.

Gurav, R., Bhatia, S.K., Choi, T.-R., Kim, H.J., Song, H.-S., Park, S.-L., Lee, S.-M., Lee, H.-S., Kim, S.-H., Yoon, J.-J., Yang, Y.-H. 2020. Utilization of different lignocellulosic hydrolysates as carbon source for electricity generation using novel Shewanella marisflavi BBL25. *Journal of Cleaner Production*, 277, 124084.

Hajilary, N., Rezakazemi, M., Shirazian, S., 2019. Biofuel types membrane separation. *Environmental Chemistry Letters*, 17, 1–18.

Hawrot-Paw, M., Koniuszy, A., Sędłak, P., Seń, D., 2020. Functional properties microbiological stability of fatty acid methyl esters (FAME) under different storage conditions. *Energies*, 13(21), 5632.

Hu, P., Chakraborty, S., Kumar, A., Woolston, B., Liu, H., Emerson, D., Stephanopoulos, G., 2016. Integrated bioprocess for conversion of gaseous substrates to liquids. *Proceedings of the National Academy of Sciences*, 113(14), 3773–3778.

Iragavarapu, G.P., Imam, S.S., Sarkar, O., Mohan, S.V., Chang, Y.C., Reddy, M.V., Kim, S.H., Amradi, N.K., 2023. Bioprocessing of waste for renewable chemicals fuels to promote bioeconomy. *Energies*, 16(9), 3873.

Jeevan Kumar, S.P., Avanthi, A., Chintagunta, A.D., Gupta, A., Banerjee, R., 2020. Oleaginous lipid: a drive to synthesize utilize as biodiesel. In *Practices Perspectives in Sustainable Bioenergy: A Systems Thinking Approach* (pp. 5–129). doi:10.1007/978-81-322-3965-9_6

Kavitha, V., Geetha, V., Jacqueline, P.J., 2019. Production of biodiesel from dairy waste scum using eggshell waste. *Process Safety Environmental Protection*, 125, 279–287.

Khan, M.I., Hariprasad, G., 2020. Human secretary phospholipase A2 mutations their clinical implications. *Journal of Inflammation Research*, 551–561.

Khot, M., Raut, G., Ghosh, D., Alarcón-Vivero, M., Contreras, D., Ravikumar, A., 2020. Lipid recovery from oleaginous yeasts: perspectives challenges for industrial applications. *Fuel*, 259, 116292.

Kim, S.H., Da Yi, Y., Kim, H.J., Bhatia, S.K., Gurav, R., Jeon, J.-M., Yoon, J.-J., Kim, S.-H., Park, J.-H., Yang, Y.-H. 2022. Hyper biohydrogen production from xylose and xylose-based hemicellulose biomass by the novel strain Clostridium sp. YD09. *Biochemical Engineering Journal*, 187, 108624.

Kumar, D., Singh, B., Korstad, J., 2017. Utilization of lignocellulosic biomass by oleaginous yeast bacteria for production of biodiesel renewable diesel. *Renewable Sustainable Energy Reviews*. 73, 654–671.

Lee, H.S., Lee, H.-J., Kim, S.H., Cho, J.Y., Suh, M.J., Ham, S., Bhatia, S.K., Gurav, R., Kim, Y.-G., Lee, E.Y., Yang, Y.-H. 2021a. Novel phasins from the Arctic Pseudomonas sp. B14-6 enhance the production of polyhydroxybutyrate and increase inhibitor tolerance. *International Journal of Biological Macromolecules*, 190, 722–729.

Lee, S.M., Lee, H.-J., Kim, S.H., Suh, M.J., Cho, J.Y., Ham, S., Song, H.-S., Bhatia, S.K., Gurav, R., Jeon, J.-M., Yoon, J.-J., Choi, K.-Y., Kim, J.-S., Lee, S.H., Yang, Y.-H. 2021b. Engineering of Shewanella marisflavi BBL25 for biomass-based polyhydroxybutyrate production and evaluation of its performance in electricity production. *International Journal of Biological Macromolecules*, 183, 1669–1675.

Leng, L., Li, W., Li, H., Jiang, S., Zhou, W., 2020. Cold flow properties of biodiesel the improvement methods: a review. *Energy & Fuels*, 34(9), 10364–10383.

Levering, J., Broddrick, J., Zengler, K., 2015. Engineering of oleaginous organisms for lipid production. *Current Opinion in Biotechnology*, 36, 32–39.

Mizik, T., Gyarmati, G., 2021. Economic sustainability of biodiesel production—a systematic literature review. *Clean Technologies*, 3(1), 19–36.

Monirul, I.M., Masjuki, H.H., Kalam, M.A., Zulkifli, N.W.M., Rashedul, H.K., Rashed, M.M., Imdadul, H.K., Mosarof, M.H., 2015. A comprehensive review on biodiesel cold flow properties oxidation stability along with their improvement processes. *RSC advances*. 5(105), 86631–86655.

Nair, L.G., Agrawal, K., Verma, P., 2022. An overview of sustainable approaches for bioenergy production from agro-industrial wastes. *Energy Nexus*, 6, 100086.

Norjannah, B., Ong, H.C., Masjuki, H.H., Juan, J.C., Chong, W.T., 2016. Enzymatic transesterification for biodiesel production: a comprehensive review. *RSC Advances*, 6(65), 60034–60055.

Patel, A., Karageorgou, D., Rova, E., Katapodis, P., Rova, U., Christakopoulos, P., Matsakas, L., 2020. An overview of potential oleaginous microorganisms their role in biodiesel omega-3 fatty acid-based industries. *Microorganisms*, 8(3), 434.

Patel, A., Mahboubi, A., Taherzadeh, M.J., Rova, U., Matsakas, L., 2021. Volatile fatty acids (VFAs) generated by anaerobic digestion serve as feedstock for freshwater marine oleaginous microorganisms to produce biodiesel added-value compounds. *Frontiers in Microbiology*, 12, 614612.

Patel, A., Arora, N., Mehtani, J., Pruthi, V., and Pruthi, P.A., 2017. Assessment of fuel properties on the basis of fatty acid profiles of oleaginous yeast for potential biodiesel production. *Renewable and Sustainable Energy Reviews*, 77, 604–16.

Patel, A., Mikes, F., Matsakas, L., 2018. An overview of current pretreatment methods used to improve lipid extraction from oleaginous microorganisms. *Molecules*, 23(7), 1562.

Preethi, G. M., Kumar, G., Karthikeyan O. P., Varjani S., Banu, R. J. 2021. Lignocellulosic biomass as an optimistic feedstock for the production of biofuels as valuable energy source: techno-economic analysis, environmental impact analysis, breakthrough and perspectives. *Environmental Technology & Innovation*, 24, 102080.

Ramos, M., Dias, A.P.S., Puna, J.F., Gomes, J., Bordado, J.C., 2019. Biodiesel production processes sustainable raw materials. *Energies*, 12(23), 4408.
Raut, G., Kamat, S., RaviKumar, A., 2019. Trends in production fuel properties of biodiesel from heterotrophic microbes. In *Advances in Biological Science Research* (pp. 247–273). Academic Press.
Rezania, S., Kamboh, M.A., Arian, S.S., Al-Dhabi, N.A., Arasu, M.V., Esmail, G.A., Yadav, K.K., 2021. Conversion of waste frying oil into biodiesel using recoverable nanocatalyst based on magnetic graphene oxide supported ternary mixed metal oxide nanoparticles. *Bioresource Technology*, 323, 124561.
Sasmal, S., Mohanty, K., 2018. Pretreatment of lignocellulosic biomass toward biofuel production. Biorefining of biomass to biofuels: opportunities perception. 203–221.
Shaheen, F., Ravinder, P., Jadhav, R., Valekar, N., Hwang, S., Gurav, R., Jadhav, J., 2023. Oleaginous microbes for biodiesel production using lignocellulosic biomass as feedstock. In *Green Approach to Alternative Fuel for a Sustainable Future* (pp. 271–296). Elsevier.
Singh, D., Sharma, D., Soni, S.L., Sharma, S., Kumari, D., 2019. Chemical compositions, properties, stards for different generation biodiesels: a review. *Fuel*, 253, 60–71.
Tarigan, J. B., Ginting, M., Mubarokah, S. N., Sebayang, F., Karo-Karo, J., Nguyen, T. T., Ginting, J., Sitepu, E. K., 2019. Direct biodiesel production from wet spent coffee grounds. *RSC Advances*, 9(60), 35109–35116.
Teixeira, L.V., Moutinho, L.F., Romão-Dumaresq, A.S., 2018. Gas fermentation of C1 feedstocks: commercialization status future prospects. *Biofuels, Bioproducts Biorefining*, 12(6), 1103–1117.
Tyagi, O.S., Atray, N., Kumar, B., Datta, A., 2010. Production, characterization and development of standards for biodiesel—A review. *Mapan*, 25(3), 197–218.
Uthi, S., Kaliyaperumal, A., Srinivasan, N., Thangavelu, K., Muniraj, I.K., Zhan, X., Gathergood, N., Gupta, V.K., 2022. Microbial biodiesel production from lignocellulosic biomass: new insights future challenges. *Critical Reviews in Environmental Science Technology*, 52(12), 2197–2225.
Valdés, G., Mendonça, R.T., Aggelis, G., 2020. Lignocellulosic biomass as a substrate for oleaginous microorganisms: a review. *Applied Sciences*, 10(21), 7698.
Yellapu, S. K., Kaur, R., Kumar, L. R., Tiwari, B., Zhang, X., & Tyagi, R. D. (2018). Recent developments of downstream processing for microbial lipids and conversion to biodiesel. *Bioresource Technology*, *256*, 515–528.
Yoosaf, P.P., Gopu, C., Kumar, C.S., 2018. Properties of biodiesel blends: an investigative comparative study. In *Emerging Trends in Engineering, Science Technology or Society, Energy Environment* (pp. 483–492). CRC Press.
Zabermawi, N. M., Alsulaimany, F. A., El-Saadony, M. T., El-Tarabily, K. A., 2022. New eco-friendly trends to produce biofuel bioenergy from microorganisms: an updated review. *Saudi Journal of Biological Sciences*.
Zhang, L., Loh, K. C., Kuroki, A., Dai, Y., Tong, Y. W., 2021. Microbial biodiesel production from industrial organic wastes by oleaginous microorganisms: current status prospects. *Journal of Hazardous Materials*. 402, 123543.
Zuccaro, G., Pirozzi, D., Yousuf, A., 2020. Lignocellulosic biomass to biodiesel. In *Lignocellulosic biomass to liquid biofuels* (pp. 127–167). Academic Press.

4 Biohydrogen Production

Harnessing Microbes for Clean Energy Generation

Rohit Topkar, Anmol Patil, Priyanka Magadum, Sarita Wadmare, Ranjit Gurav, Sangchul Hwang, Jyoti Jadhav, and Rahul Jadhav

4.1 INTRODUCTION

The nation's energy reserves, or fuel reserves, have long been seen as an accurate indication of its power and progress economically. Approximately 87% of the energy used today is derived from fossil fuels worldwide (Kamaraj et al., 2020). Nonetheless, the growing world population demands a high level of fossil fuel consumption. In addition to having a detrimental impact on every living thing on earth, our reliance on fossil fuels to provide the world's energy needs is a primary contributor to climate change (Usman et al., 2019). The majority of fossil fuels are non-renewable and are extracted from petroleum origins. Excessive use of petroleum fuels contributes significantly to air pollution, and burning fuels derived from petroleum releases a variety of harmful gases, among them carbon monoxide, sulfur dioxide, nitrogen dioxide, and volatile organic compounds (VOCs). These gases are the cause of greenhouse gas emissions and global warming, which pose a serious threat to humankind. Therefore, intending to combat climate change, sustainable energy sources must be developed that are not reliant on non-renewable resources (Usman et al., 2019). Climate change may be mitigated by utilizing renewable energy sources, such as hydropower, wind power, and solar power. Additionally, using renewable biofuels like biodiesel, bio-H_2, and bioethanol may replace fossil fuels with an energy source that is safe for the environment (Osman et al., 2020).

Since hydrogen (H_2) gas has several benefits over traditional fossil fuels, it is recognized as a contemporary, sustainable, and alternative energy source. Since there are no emissions of contaminants, it is regarded as clean energy. Furthermore, in comparison to natural fossil fuels, H_2 gas has a low heating value and a high energy yield. According to research published by the International Renewable Energy Agency (IRENA), over 95% of the world's H_2 generation is derived from the physio-chemical breakdown of fossil fuels, which poses a serious environmental risk since it produces pollutants. An environmentally friendly method of producing H_2 is via electrically breaking down water molecules. In the method, water is broken down into H_2 and oxygen (O_2) atoms during electrolysis. Since splitting H_2O requires a

 DOI: 10.1201/9781003585398-4

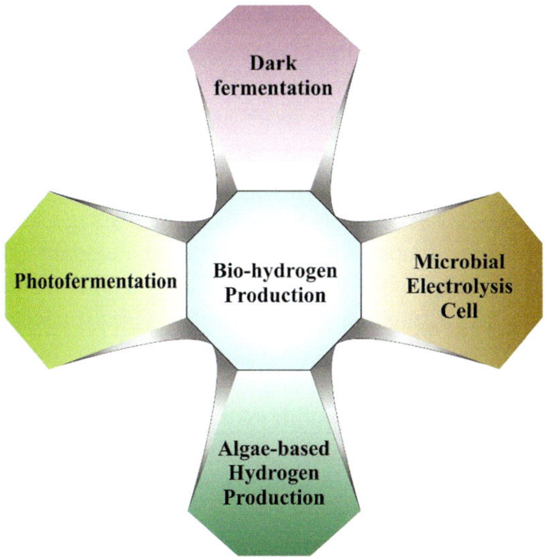

FIGURE 4.1 Different methods for bio-H_2 production.

significant amount of energy, which is provided by the use of petroleum-based fuels, this method is not ecologically friendly (Ferraren-De Cagalitan and Abundo, 2021). The traditional methods of producing H_2 are frequently linked to the creation of environmental risks. A biological strategy has been devised to give a contemporary, more resilient, and readily replicable solution for the bioproduction of H_2 in order to address this issue. Microbial pathways, which are significantly cleaner for the environment and may prove to be a more practical long-term renewable energy source, are used in the biological process of H_2 generation (Yun et al., 2018). Initially, in 1939, bio-H_2 production was discovered by German physiologist Hans Gaffron in green microalgae. However, recently several methods have been investigated for bio-H_2 production, including dark fermentation, photofermentation, microbial electrolysis cells (MEC), and algae-based H_2 products (Sharma et al., 2021), which are illustrated in Figure 4.1.

The capacity of several microorganisms to create bio-H_2 is investigated. Chief producers of H_2 include several archaeans, some eubacteria, and microalgae. There are several advantages to using living organisms for H_2 bioproduction, including high rates of H_2 generation with quick development, low energy needs, less influence on the environment, and less use of water and land. The use of two common methods, namely, dark fermentation and photofermentation, has been established as effective in the bioproduction of H_2. Dark fermentation involves the cultivation of microbes without light to convert complex carbohydrates into H_2 and other by-products. Likewise, photofermentation involves the production of H_2 in the presence of sunlight by a specific group of microorganisms (Ahmed et al., 2021).

An innovative way of producing H_2 has been put into practice by using bacteria's metabolic capacity in conjunction with electrochemistry. This process is sometimes referred to as a "microbial electrolysis cell," which is an amazing ability of some microorganisms to break down organic materials into protons, electrons, and carbon dioxide. Furthermore, the combination of protons and electrons to generate H_2 is opening the door to a future that is far more sustainable (Ferraren-De Cagalitan and Abundo, 2021). Photosynthetic organisms are used in another method of producing H_2. Photosynthetic organisms, including cyanobacteria and green algae, possess an amazing biochemical capacity to use sunlight to convert water molecules into H_2 gas via a process called photolysis. This process has a distinct advantage over other techniques of producing H_2 because it employs water, which is an abundant and cheap substrate. It also requires less space and costs less money (Lazaro and Hallenbeck, 2019).

Of the processes, dark fermentation is the one that is most advised for creating H_2. Dark fermentation produces H_2 gas with minimal energy consumption and may be done on a variety of substrates. Dark fermentation is done in a continuous stirred tank reactor (CSTR), which is specifically built to offer the right temperature and pressure management for the development of H_2-producing organisms, to maximize H_2 bioproduction (Yun et al., 2018). One of the most important variables in the fermentation process of H_2 bioproduction is pH. Just as in microbial cells, pH controls several metabolic processes to provide the best possible output. For dark fermentation, a pH range of 5.5–6.0 is optimum (Sharma et al., 2021). In the process, these microbial cells are helped in the bioproduction of H_2 by a class of enzymes known as nitrogenases and hydrogenases. The majority of bacteria involved in dark fermentation adhere to one of two metabolic pathways: the acetic acid pathway or the butyric acid pathway. The fermentation process's sugar content and the microorganisms' chosen metabolic pathway determine the presumed H_2 production (Ferraren-De Cagalitan and Abundo, 2021).

Currently, the pace of H_2 generation worldwide is not keeping up with the entire amount of demand. Thus, this chapter covers the following topics: (1) different processes of H_2 bioproduction; (2) substrates used for production; (3) factors affecting the production rate; (4) the application of H_2 utilization; and (5) the economic implications involved in H_2 production.

4.2 MICROBIAL PATHWAYS FOR H_2 PRODUCTION

4.2.1 Dark Fermentation

4.2.1.1 Introduction to Dark Fermentation

Anything that can be exploited to the utmost is the best alternative source for mankind's benefit; when it comes to the production of H_2, various fermentation techniques have been practiced. But dark fermentation is the technique that has been in practice for the last few epochs. This technique's main benefit is that it doesn't need a light source because, as its name implies, it operates in the dark (Sarangi and Nanda, 2020).

Dark fermentation involves the utilization of nutrients in absolutely no light with a single source of nutrients which makes it even more demanding for today's industries. In this process, certain microorganisms metabolize organic compounds to yield H_2 gas as a metabolic by-product. The initial phases of this fermentation involve hydrolysis and acidogenesis. During hydrolysis, complex compounds such as carbohydrates, proteins, lipids, etc. are catabolized into simpler compounds and this is possible with the help of certain hydrolytic enzymes produced by bacteria. This is then followed by acidogenesis which transforms previous compounds into VFAs, i.e., volatile fatty acids, alcohols, and hydrogen (Jain et al., 2022).

Anaerobic digestion is significant because it plays a critical part in the organic matter cycle and has applications in waste treatment, bioenergy generation, and environmental remediation.

4.2.1.2 Microorganisms

The inoculum for the production of H_2 can be a pure culture as well as a slightly mixed culture of bacteria (Tapia-Venegas et al., 2015). Bacteria that are involved are *Clostridium butyricum*, *Thermoanaerobacterium* spp., *Bacillus coagulans*, and *Enterobacter aerogenes*. There is less literature and practice for using mixed cultures. Mostly, *Clostridium* and *Thermoanaerobacterium* spp. are preferred for mixed culture inoculum. The major reason for avoiding the mixed culture is to perform scrutiny of several parameters during fermentation, for instance, studying temperature, pH, acidity, basicity, and other environmental factors of each strain used in inoculum (Jain et al., 2022). The active sources for sample collections are river sediments, sludges, and cow excrement. H_2 serves as an essential substrate for various microorganisms in anaerobic environments, and thus H_2 consumers are divided into several categories such as the following.

4.2.1.2.1 Hydrogenotrophic Archaea

These microorganisms make use of H_2 gas as an electron donor in their metabolic pathways. These are commonly found in anaerobic environments and play important roles in the cycling of H_2 and carbon compounds (Tapia-Venegas et al., 2015).

4.2.1.2.2 Homoacetogenic Bacteria

Homoacetogens are a group of bacteria that produce acetate (acetic acid) as their primary metabolic end product from the reduction of carbon dioxide (CO_2) using H_2 as the electron donor. They are important H_2 consumers, especially in environments where other electron acceptors like nitrate or sulfate are absent or present at low concentrations (Sarangi and Nanda, 2020).

4.2.1.2.3 Nitrate- and Sulfate-Reducing Bacteria

Some bacteria are capable of utilizing H_2 to reduce nitrate (NO_3^-) or sulfate (SO_4^{2-}) as alternative electron acceptors in anaerobic respiration. These microorganisms play significant roles in nitrogen and sulfur cycling in anaerobic environments where these compounds are present (Jain et al., 2022).

4.2.1.2.4 Methanogenic Archaea

Methanogens are a group of archaea that produce methane (CH_4) as a metabolic end product from the reduction of CO_2 or other simple carbon compounds using hydrogen as electron donors. Methanogens are significant H_2 consumers in anaerobic environments and play key roles in methane production during anaerobic digestion processes (Tapia-Venegas et al., 2015).

4.2.1.3 Substrates Used in Fermentation

Simple sugars like lactose, sucrose, and glucose are the main substrates utilized in dark fermentation to produce H_2. Due to the extensive research conducted on their metabolic route and their ability to facilitate synthesis, simple sugars are utilized extensively and copiously (Sarangi and Nanda, 2020). Another major source is nitrogen and its quantity used is considered crucial because a little less or more amount can cause ammonification which is not at all favorable for the pure production of H_2. Phosphorus is required for buffering purposes and also acts as a nutrient up to some extent. There are no fixed or mostly followed proportions of all these substrates. Thus, the ratio of C-N-P is formulated with the needs (Jain et al., 2022).

Lignocellulosic materials are mostly opted since they have a high concentration of cellulose, hemicellulose, and lignin that act as the core source of nutrients. Additionally, it is easily available from wood, agricultural residue, etc. (Tapia-Venegas et al., 2015). The only drawback of using lignocellulosic biomass is that it has to be given a pretreatment where the complex compounds are made more easily accessible to the microbes for degradation. The enzymes that are naturally produced by the bacteria are externally added for easy hydrolysis of the compounds. Various methods such as gas sparging, membrane separation, or pressure swing adsorption can be employed for H_2 recovery. The challenge lies in creating cost-efficient ways to treat lignocellulosic biomass and efficient recovery of bio-H_2. Other feedstocks used for dark fermentation are food waste, paper mill effluent, microalgal biomass, palm oil refinery effluent, spent coffee grounds, etc. The discards from livestock are also equivalently used for dark fermentation of H_2, for instance, solid waste, liquified manure, silage juices, etc. (Jain et al., 2022). (Lin et al. 2017)

4.2.1.4 Advantages

The advantages of dark fermentation for the production of H_2 are highlighted over any other methods used for the production of H_2.

4.2.1.4.1 Renewable Feedstock

A wide range of substrates makes the availability of nutrients. Any alternative source of sugars and nitrogen can be used without causing complexes in the general metabolic pathways of bio-H_2 production (Tapia-Venegas et al., 2015).

4.2.1.4.2 Versatility

Large groups of bacteria can be used in the form of mixed inoculum or even a single strain can be used. The microorganisms used are thermophilic and also mesophilic which gives a vast ground for the optimization of procedures and various physio-chemical factors (Balachandar et al., 2013).

4.2.1.4.3 Syngas Production

The by-products produced are of diverse forms including VFAs, alcohols, acids, etc. with H_2. These other by-products can also add to the parallel production for the industry (Jain et al., 2022).

4.2.1.4.4 *Waste Utilization and Cost*

The waste generated after dark fermentation is nearly negligible and waste from other industries can be utilized as the source for the production of H_2. The cost efficiency is very high since the production of H_2 requires fermenters of basic designs such as membrane bioreactors (MBRs), fixed bed reactors, CSTRs, tank reactors, etc. (Sarangi and Nanda, 2020; Balachandar et al., 2013).

4.2.1.5 Limitations

The limitation of using the dark fermentation process can be lost if the sole purpose of production is H_2, since the other metabolized by-products such as alcohols and acids are formed. Overproduction of these by-products can cause the least recovery of H_2 (Jain et al., 2022). These difficulties result from the process's intricate and delicate nature, which calls for the meticulous regulation and optimization of a number of variables, including temperature, pH, alkalinity, hydraulic retention time (HRT), H_2, and partial pressure of carbon dioxide. The yield of H_2 production can be negatively impacted by even the smallest variation (Tapia-Venegas et al., 2015).

4.2.2 Photofermentation

4.2.2.1 Introduction to Photofermentation

Photosynthetic bacteria convert acetic acid and some organic compounds into H_2 and carbon dioxide with the help of sunlight; this process is termed photofermentation. Photofermentation occurs in conditions where there is an abundance of luminescence and anaerobic conditions. The convenience of overall production is carried out by gathering various enzyme systems, carbon flow (TCA cycle), photosynthetic membrane, etc. The key enzymes involved in the production of H_2 are nitrogenases and hydrogenases; nitrogenase enzymes are said to be differing genetically, for instance, *nif* and *vnf*, and they are based on the metal ion they retain (Mus et al., 2018).

The most common bacteria used for photofermentation are PNS (photosynthetic non-sulfur) bacteria. The role of light is to provide the external energy required during the process since the reaction is not spontaneous. Photofermentation requires undivided attention for the maintenance of the parameters involved in fermentation, such as temperature, alkalinity, etc. (Song et al., 2022). The instance for the reaction of production of H_2 using acetate as the organic source is shown below.

$$2CH_3COOH + 2H_2O \longrightarrow 4H_2 + 2CO_2,\ \Delta G0 = +104\,\text{kJ} \tag{4.1}$$

Mostly the bioreactors used for photofermentation are continuous stirred tank bioreactors (CSTBR) and cylindrical bioreactors (Melitos et al., 2021). These days the industry is seeing a rise in the use of sophisticated modern bioreactors, such as transparent low-pressure airlift fermenters shown in Figure 4.2. Airlift fermenters combine mass transfer and heat processes into one vehicle for the effective production of H_2. H_2 was first produced from mixed acids, and in the process the most common and efficient bacterium *E. coli* was used. The production was possible due to

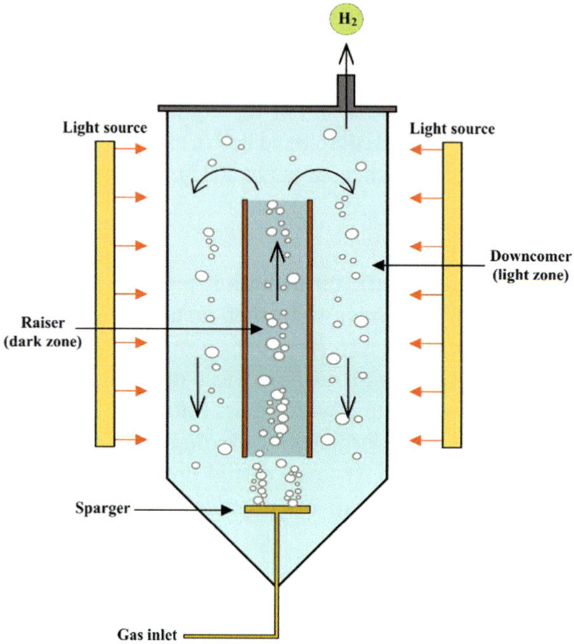

FIGURE 4.2 Transparent airlift fermenter design.

hydrogenase enzymes present in the bacteria encoded by multiple operons. This enzyme is considered to tackle adverse environmental effects and regulate bacteria's efficiency (Trchounian, 2015).

4.2.2.2 Microorganisms

The very common bacteria involved in photofermentation are PNS (photosynthetic non-sulfur), also called purple, and they retain the ability to perform conversion of VFAs to H_2 and CO_2 in anaerobic conditions. Precisely *Rhodobacter sphaeroides*, *R. capsulatus*, *R. sulfidophilus*, *Rhodopseudomonas palustris*, and *Rhodospirillum rubrum* are the widely used ones (Gupta et al., 2024). The metabolism of these bacteria plays a crucial role. They can produce H_2 in anaerobic conditions. Organic acids, alcohols, and even H_2 work as electron donors. They utilize light sources for various pathways in their metabolism. They survive at low oxygen levels such as in sediments, ponds, etc. They play a vital role in nitrogen and carbon cycling. These bacteria have hydrogenase enzyme that acts as the core for producing H_2, and the optimum temperature required for them ranges between 30°C and 35°C. Malate, succinate, pyruvate, and lactate are a few substrates on which the bacteria have shown the utmost activity for the production of H_2 (Schuchmann et al., 2018).

The ideal pH range is 6.8–7.5, while the ideal light level is 6–6000 lux. Nitrogenase is an enzyme that produces hydrogen and adenosine diphosphate (ADP) from ATP (adenosine triphosphate) (Melitos et al., 2021).

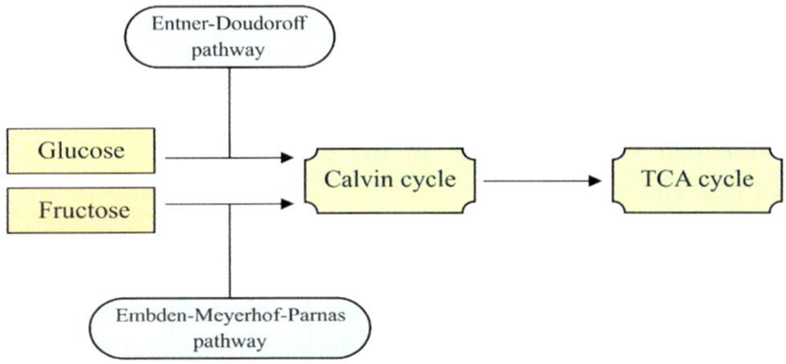

FIGURE 4.3 The overall scheme of carbon metabolism in PNS bacteria.

4.2.2.3 Substrates Used in Fermentation

PNS employs a large variety of substrates, although only a few substrates are supported for high yield. Different multiple routes are employed by PNS depending on the species used in the fermentation. When taking into account the direct substrate-to-yield ratio, organic acids including lactic acid, succinic acid, pyruvic acid, and malic acid exhibit excellent performance. Sugar is the most promising source of nutrition and is also an often-utilized substrate. Microbes often use the Embden–Meyerhof–Parnas (EMP) route when using fructose as a substrate for energy, but they generally follow the Entner–Doudoroff pathway when using glucose. The TCA cycle is eventually reached by both routes converging into the Calvin cycle (Gupta et al., 2024). Figure 4.3 illustrates the general carbon metabolism strategy used by PNS bacteria.

The production of H_2 is not only on the substrate but also on the metal ion present, for example, Ni or Mg ions. The production can be enhanced if the glucose is replaced by other sugars and a significant amount of glycerol in the complex medium.

4.2.2.4 Advantages

Several advantages of photofermentation are as follows (Das and Basak, 2021; Reungsang et al., 2018).

4.2.2.4.1 Utilization of Renewable Resources

Reusing substrates can create hope for sustainable development where every component used is reused for several other purposes.

4.2.2.4.2 Production of Value-Added Products

Significant products such as H_2 gas (H_2), organic acids (e.g., acetic acid, butyric acid), and alcohols (e.g., ethanol) can act as by-products and add extra benefits to the industry.

4.2.2.4.3 Versatility and Flexibility

This can be performed using a variety of microorganisms, including purple non-sulfur bacteria like *Rhodobacter* and *Rhodopseudomonas*. These bacteria are versatile and strong enough to tackle and sustain in harsh conditions, which will ultimately reduce condition maintenance efforts.

environment. Microalgae help to convert light into chemical energy by separating protons and oxygen molecules, further assistance from photosystems is used (Khetkorn et al., 2017).

4.2.4.2.1 H_2 Production by Nitrogen-Fixing Microalgae

Plectonema boryanum was first used to produce H_2. This technique was adapted to reduce the steps involved in the traditional way of H_2 production. Due to nitrogenase activity, H_2 was produced in the absence of nitrogen. Another type of cyanobacteria involved was filamentous cyanobacteria with heterocyst cells. These bacteria can convert organic substrates into H_2 in a light-driven, by stoichiometric manner. A study conducted in Japan used a procedure that involved a separate dark fermentation to show that they could recycle algal colonies, and a hollow fiber ultrafiltration unit is used to feed the products of the algal fermentation to the bacterial photobioreactors that are photosynthetic (Benemann, 2000).

4.2.4.2.2 H_2 Production by *C. reinhardtii*

When DCMU suppresses PSII, *C. reinhardtii* is reported to be able to photoproduce H_2. However, when 2,5-dibromo-3-methyl,6-isopropyl-p-benzoquinone is added, it prevents the cytochrome b6f complex from functioning, and no H_2 evolution takes place (Stuart and Gaffron, 1972).

4.2.4.3 Substrates Used in Process

Rich in proteins and carbohydrates, the lipid-extracted microalgal biomass is also a suitable substrate for fermentative H_2 generation. Research on microalgal biomass ought to focus on the following: pretreatment, co-fermentation, HPB domestication, and continuous operation. The utilization of additional macroalgal species as substrates, such as *Ulva lactuca*, to boost H_2 generation and increase the viability of large-scale applications (Wang et al., 2021). Both photosystems I (PSI) and II (PSII) are active when employing light as the substrate. In PSII, light energy is extracted and water is split into electrons and oxygen. These electrons are triggered in PSI and used to create H_2. Although the ability to directly convert light energy into H_2 makes direct biophotolysis an appealing method of producing bio-H_2, its inherent H_2 production rate is very low due to the high reaction free energy (+237 kJ/mol H_2) and the by-product O_2's inhibitory effects on the involved enzymes (Xuan et al., 2023).

Microalgae depend on nitrogen as a vital nutrient for growth. It is an essential component of proteins, nucleic acids, and chlorophyll, all of which are required for photosynthesis and cellular processes. Thus, for microalgae cultures to continue producing biomass at the optimum range, an adequate amount of nitrogen is essential (Yaakob et al., 2021). If the nitrogen is limited, then microalgae can face physiological changes, which will enhance their yield of H_2. Microalgae may redirect resources from biomass synthesis toward metabolic processes that produce H_2, like the development of hydrogenase enzymes, when nitrogen is scarce. This resource redistribution aids in maintaining cellular energy balance and redox state, which in turn encourages the production of H_2 (Goswami et al., 2021).

Numerous studies have looked into the connection between microalgae's ability to produce H_2 and their availability of nitrogen. Research by Grechanik et al. (2021)

showed that, in contrast to nitrogen-replete circumstances, nitrogen deprivation in the green alga *Chlamydomonas reinhardtii* enhances H_2 production rates. Similarly, Kosourov et al. (2021) demonstrated that in the same species, hydrogenase activity and H_2 generation were induced by nitrogen starvation. Nitrogen control techniques are frequently used to maximize H_2 production from microalgae. These tactics help supplement or deprive the growing medium of nitrogen to regulate its availability. Protocols involving nitrogen deprivation or famine are frequently employed to stimulate the synthesis of hydrogen in microalgae cells (Grechanik et al., 2021).

4.2.4.4 Advantages

4.2.4.4.1 Renewable and Sustainable

Since H_2 is produced as a by-product of the metabolism of microalgae, which need carbon dioxide and light to thrive, microalgae-based H_2 generation is regarded as a renewable and sustainable method. This method can lessen greenhouse gas emissions and is not dependent on finite fossil fuel supplies (Razu et al., 2019).

4.2.4.4.2 Versatile Cultivation

Open ponds, closed photobioreactors, wastewater treatment plants, and other systems can all be used to develop microalgae. They are adapted to a wide range of topological backgrounds since they can flourish in a variety of environmental circumstances and do not require specific areas.

4.2.4.4.3 CO_2 Mitigation

Industrial flue gases and agricultural runoff are just two of the sources of carbon dioxide that microalgae can absorb and store. Microalgae-based systems can produce H_2 as a clean energy carrier and help reduce CO_2 emissions by using CO_2 for photosynthesis (Ahmed et al., 2021).

4.2.4.5 Limitations

4.2.4.5.1 Expensive Production Costs

When considering traditional fossil fuels and alternative renewable energy sources, the costs associated with growing microalgae and producing H_2 are still comparatively high. The infrastructure capital, the energy inputs needed for cultivation, and the downstream processing needed for H_2 extraction are quite challenging (Ahmed et al., 2021).

4.2.4.5.2 Low H_2 Yield

Although microalgae can produce H_2 as a metabolic by-product, there is still not much H_2 produced overall per unit of biomass. To increase the efficiency of H_2 production, strain selection, growth conditions, and genetic engineering techniques are obliged to optimize for maximum yield. Oxygen Sensitivity: the amount of oxygen can directly prevent microalgae from producing H_2. The presence of oxygen inhibits the activity of hydrogenase, which lowers the rates at which H_2 is produced. To get around this restriction, techniques such as creating oxygen-tolerant hydrogenases or anaerobic environments are being investigated (Ahmed et al., 2021).

4.2.4.5.3 Technological Difficulties

Scaling up microalgae-based H_2 generation systems from lab-scale research to industrial applications requires critical planning and action. It will need attention to problems including reactor design, harvesting and dewatering techniques, and scalability to make microalgae-based H_2 production commercially feasible (Limongi et al., 2021).

4.3 MICROBIAL STRAINS FOR BIO-H_2 PRODUCTION THROUGH DARK FERMENTATION

Both pure and slightly mixed bacterial cultures can be used as the inoculum to produce H_2 (Tapia-Venegas et al., 2015). *Clostridium butyricum*, *Thermoanaerobacterium* spp., *Bacillus coagulans*, and *Enterobacter aerogenes* are the bacteria that are involved. For mixed culture inoculum, *Clostridium* and *Thermoanaerobacterium* spp. are typically preferred. Several characteristics should be closely examined during fermentation, such as temperature, pH, acidity, basicity, and other environmental aspects of each strain employed in the inoculum. This is the main justification for not using a mixed culture (Jain et al., 2022). According to Tapia-Venegas et al. (2015), river sediments, sludges, and cow dung are the active sources for sample collection. In anaerobic settings, H_2 is a necessary substrate for a variety of microbes.

4.4 BIOREACTOR FOR BIO-H_2 PRODUCTION

The construction and mode of operation of bioreactors strongly influence the efficiency of substrate conversion and the rate of bio-H_2 formation. An ideal bioreactor should function optimally at a low HRT for maximum H_2 bioproduction without raising the cost of production. Different bioreactors have been designed for various pathways of bio-H_2 production (Usman et al., 2019). For instance, in dark fermentation and MEC, the by-products (like CO_2) get produced in the process, and it can interfere with the purity of H_2 produced. An ideal bioreactor should operate for the separation of by-products from produced H_2. In the case of light-dependent fermenters, a bioreactor should be designed to pass light for the optimal production of H_2 (Ferraren-De Cagalitan and Abundo, 2021). The use of appropriate bioreactors made with different manufacturing materials strongly affects the overall cost of production and thus the economy of H_2 bio-generation (Kamaraj et al., 2020). Batch fermentation is mostly preferred in lab-scale studies, which has recently gained popularity due to its ease of operation and versatility. However, the reason batch bioreactors are not used for industrial-level production of H_2 is because of fluctuations in the production of bio-H_2, as the initial rate of production decreases upon utilization of substrates. Also, batch operations are often prone to the deposition of inhibitory products, causing interference in the total yield of H_2. On the other hand, continuous fermentation is utilized for large-scale H_2 production, for which the process of a CSTR is selected which is represented in Figure 4.5. CSTR involves the use of the stirrer for unceasing production of H_2 by stirring the culture continuously with constant mixing speed, to obtain the maximum rate of production of H_2. One of the drawbacks of using CSTR is that the H_2 yield decreases sometimes due to the overloading of biomass in the fermenter. Also, as the HRT of organic matter in CSTR increases, the acidogenic effect of microbes shifts to the methanogenic effect, which causes the increase in methanogenic organisms in fermenters, ultimately ceasing the H_2 bioproduction (Usman et al., 2019).

In batch fermentation, the end product yield is often calculated at the end of the fermentation process which depends on the amount of substrate utilized (represented as mol H_2/mol hexose). Theoretically, 4 mol H_2/mol hexose is the maximum yield that can be attained in batch fermentation, but in reality the total yield does not exceed more than 3 mol H_2/mol hexose (Lalman et al., 2013). In continuous fermentation, the microbial count is maintained at a high concentration for the constant fermentation process and rapid bioproduction of H_2. The overall yield of continuous fermentation is calculated as volumetric H_2 production rate (VHPR), instead of mol H_2/mol hexose. Previously recorded data showed the highest H_2 yield in continuous fermentation by utilizing sucrose-containing wastewater is 15 L H_2/L/h (Yun et al., 2018). Along with

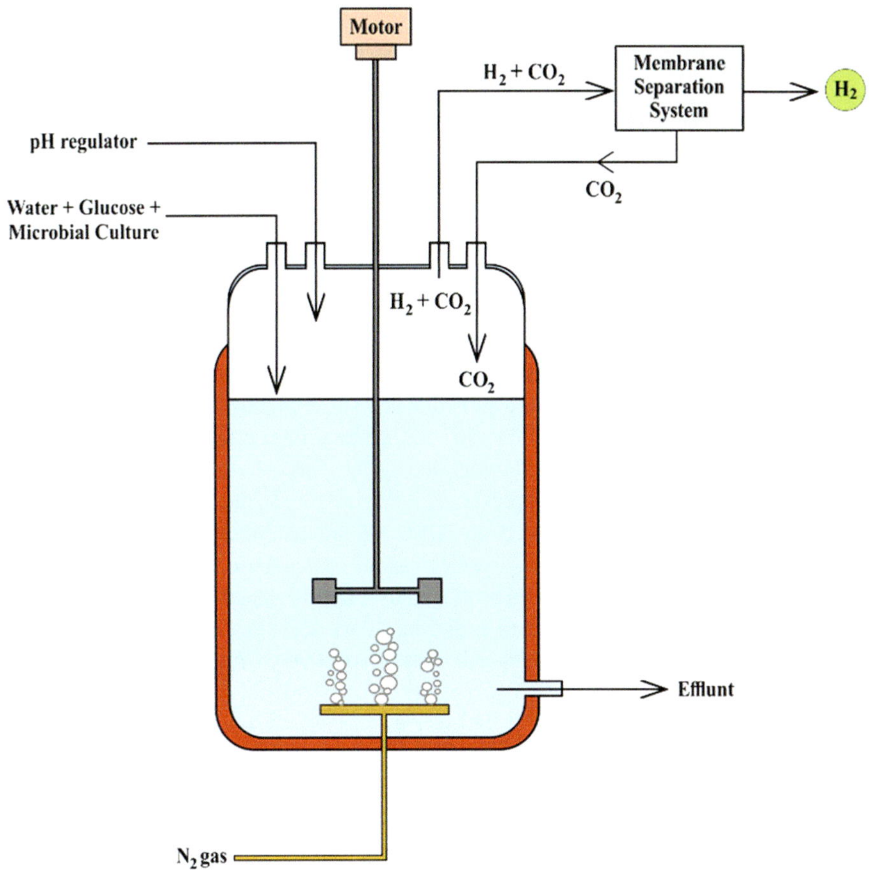

FIGURE 4.5 Design of continuous stirred tank reactor (CSTR) for dark fermentation.

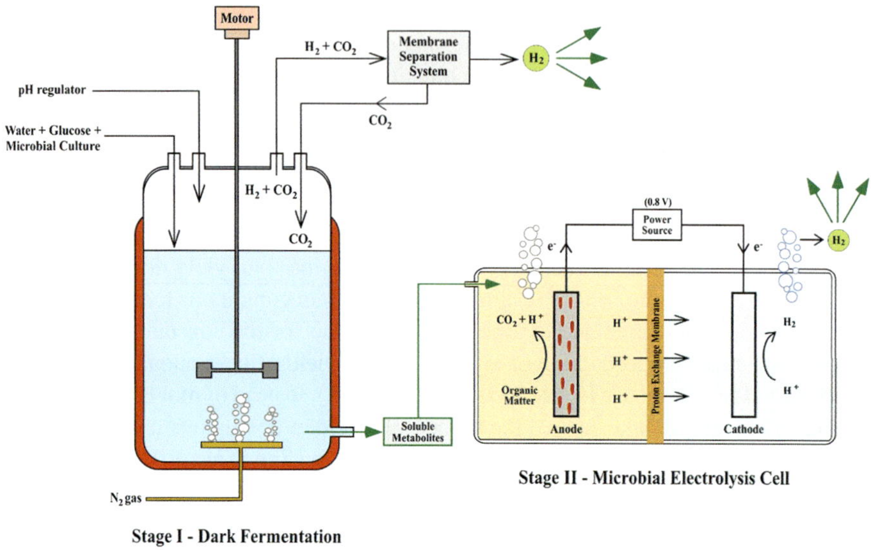

FIGURE 4.6 Design of an integrated hybrid system between CSTR and MEC.

CSTR, various other fermenters have also been developed, which include upflow anaerobic sludge bed (UASB), anaerobic fluidized bed reactor (AFBR), anaerobic biofilm reactor (ABR), etc. (Qu et al., 2022).

To improve the overall H_2 yield, a hybrid H_2 production system is introduced that includes the use of two or more technologies for maximizing the overall yield of H_2. One integrated system involves the use of both systems, dark fermentation and photo-fermentation. In this fused system, during dark fermentation, light-independent organisms transform organic substrates into H_2 and VFAs. Produced VFAs are utilized by certain photosynthetic microbes as a substrate and subsequently convert VFAs into H_2, thus helping in maximizing the total yield of H_2 generation (Ferraren-De Cagalitan and Abundo, 2021). Another developed integrated system involves the use of dark fermentation along with MEC which is represented in Figure 4.6. In the process, VFAs generated in dark fermentation are fed to MEC anode chambers as a substrate for exoelectrogenic microbes. In MEC, exoelectrogens utilize these substrates for the production of H_2. The total bio-H_2 yield produced by a hybrid system of dark fermentation and MEC is 57% in comparison to individual dark fermentation system yield and MEC system yield, which were 35% and 34%, respectively (Yun et al., 2018).

4.4.1 Factors Affecting H_2 Production

H_2-producing organism differs in metabolic pathways for the synthesis of H_2. Various physical and chemical factors are reported to strongly influence these metabolic processes causing a reduction in total H_2 yield by microbes. Control of these parameters can significantly increase H_2 production yield (Qu et al., 2022). These factors include pH, temperature, nutrients and substrate concentration, species of microbe used, and the special properties of fermenters, like HRT and organic loading rate of the substrate (Usman et al., 2019). The influencing factors responsible for affecting H_2 yield are as follows.

4.4.1.1 pH

Enzymes play a crucial role in the development of organisms and their metabolic activities. In the case of H_2-producing organisms, certain H_2-producing enzymes are strongly affected by fluctuations in the pH. Changes in pH not only affect total H_2 yield but can also alter the fermentation pathway causing the production of unwanted end products (Qu et al., 2022). It is reported that optimal H_2 production occurs at higher pH levels. Enzyme hydrogenase loses its activity at a low pH range causing a reduction in H_2 generation and thus causing a decline in viable microbial count, while at alkaline conditions, many of the H_2-producing organisms sporulate and actively produce H_2 due to the optimal working of the hydrogenase enzyme. A higher pH range also eliminates the growth of contaminating methanogenic microbes. Previously reported data suggests that to obtain an ideal yield of H_2, the pH range should be set between 5.5 and 6 (Sharma et al., 2021).

4.4.1.2 Temperature

Like pH, temperature greatly impacts the production rate of H_2, as temperature influences (1) the enzymatic activity of H_2-producing enzymes, (2) the growth rate of H_2-producing organisms, and (3) the substrate degradation to form a product. It has been reported that various species of microbes produce optimal H_2 at varying temperature ranges. For mesophilic organisms, the temperature range is maintained at 25–40°C, while for thermophilic organisms, 40–65°C temperature is maintained. The majority of the literature shows that most research on bio-H_2 production is conducted using mesophilic microbes (Sharma et al., 2021). In anaerobic fermentation, the utilization of mesophilic organisms favors the highest H_2 yield at moderate temperatures because of effective substrate conversion, but there is no efficient recovery of H_2. Also, surpassing the threshold range of temperature causes a decrease in the H_2 production rate. In industries, mesophilic organisms like *Clostridium* and *Enterobacter* sp. are widely preferred for higher H_2 production rates (Usman et al., 2019). Recent studies have demonstrated that thermophilic organisms (like *Thermobacterium* sp.) could be beneficial for fermentation at high temperatures for maximum H_2 yield. Fermentation at high temperatures (around 60°C) accelerates the rapid conversion of the raw or waste material as a substrate for easy uptake to thermophilic organisms. Fermentation at high temperature favors low solubility of H_2 in the liquid phase, and thus the recovery of H_2 is effortless (Qu et al., 2022). Also, the production of contaminating methanogenic microbes is inhibited at high temperatures. However, the major limitation of fermentation at high temperatures is the energy cost, required for heating and the maintenance of the process (Jain et al., 2022).

4.4.1.3 Nutrients and Substrate Concentration

In the various processes of H_2 generation, the provided substrate for the cultivation of microorganisms greatly impacts the overall production of H_2 gas. Substrates that are rich in carbon sources like glucose, sucrose, or starch have been demonstrated to

TABLE 4.1
The Concentrations of Different Substrates and How They Affect H_2 Production

Substrate	Concentration	pH	Temperature	Yield	Reference
Glucose	1.3–1.7 g VSS/g-glucose	5.5	37°C	18.8 g H_2/kg	Zhang et al., 2008
Glucose	34–37 g VSS/L	5.5	37°C	17.7 g H_2/kg	Zhang et al., 2008
Sucrose	33.3 g VSS/L	6.4	35°C	23.5 g H_2/kg	Chen and Lin, 2003
Sucrose	5.1 g VSS/L	6.7	35°C	27 g H_2/kg	Chang and Lin, 2004
Sucrose	6.3 g VSS/L	4	35°C	15.6 g H_2/kg	Wang et al., 2007
Molasses	6.8 g VSS/L	5.5	35°C	131.2 g H_2/kg	Lay et al., 2010
Cheese whey	4.5 g VSS/L	5.9	37°C	14.6 g H_2/kg	Davila-Vazquez et al., 2009
Cheese whey	9 g VSS/L	5	30°C	0.2 g H_2/kg	Castelló E et al., 2009
Brewery wastewater	2 g COD/L	5.5	37°C	1.5 mol H_2/mol fructose	Pachiega et al., 2019)
Alcohol industry wastewater	45g COD/L	5.5	37°C	125.1 mL H_2/g COD removed	Poontaweegeratigarn et al., 2012
Beverage wastewater	20 g COD/L	5.5	37°C	3.76 mol H_2/ mol-sucrose utilized	Sivagurunathan and Lin, 2020
Dairy industry wastewater	8.12–15.44 g COD/L	3.7–4.3	24–30°C	2.56 mol H_2/mol carbohydrate	da Silva et al., 2019

gain maximum H_2 yield. However, the utilization of pure sugars for the fermentations is not economically feasible. In the context of economic feasibility, wastewater from several industries (food industry, olive-processing mills, dairy industry, and brewery industry) having carbohydrate-rich contents is utilized as a cheap substrate for the cultivation of H_2-producing organisms (Jain et al., 2022). Along with the constituents of substrate, the concentration of substrates also affects the H_2 production rate. Data reported by Nagarajan et al. (2019) and Usman et al. (2019) suggests that the maximum yield of H_2 can be obtained at 10 g/L by utilizing glucose as a substrate. However, the H_2 yield gradually decreases as the concentration of glucose exceeds 10 g/L (Kamaraj et al. 2020). The concentrations of various substrates are mentioned in Table 4.1.

Along with carbon sources, the presence of nutrients and trace elements influences the rate of generation of H_2. The phosphorous and nitrogen content in the substrate affects the growth of H_2-producing organisms, as these are critical elements that are present in the proteins, enzymes, and nucleic acids, yet a higher concentration of these elements increases the pH level inside microbial cells causing induction of an unsuitable condition for the production of H_2 (Jain et al., 2022). Like nitrogen and phosphorous, other trace metal ions are also critical to the development of microbes in fermentation. Trace elements like Fe, Zn, K, I, Na, Ni, Co, Mn, Cu, and Ca act as cofactors for enzymes, thus assisting in the efficient conversion of the substrate into the product. Fe ion plays a crucial role as a part of enzyme hydrogenase and is an important factor of the enzyme ferredoxin, which influences the synthesis of H_2. However, the excess amount of trace elements causes inhibition of H_2-producing organisms (Qu et al., 2022).

4.4.1.4 Microbial Species

Many microorganisms have been investigated for their ability to produce bio-H_2 on a variety of substrates. Out of all the microbial species, two that produce H_2—*Enterococcus* and *Clostridium*—have drawn the most interest from various investigations. *Enterobacter* species are rod-shaped, Gram-negative microbes that are facultative anaerobes with spore-forming ability. On the other hand, *Clostridium* is an anaerobic, spore-forming microbe with a Gram-positive nature and having a rod shape. Most of the previously published research is conducted by selecting a single type of H_2-generating species in a batch culture (Kim et al., 2022a; Kim et al., 2022b). Data recorded by Lin et al. (2007) showed that H_2 generated by batch cultivation of *Clostridium tyrobutyricum* FYa102 on glucose and peptone is 1.47 mol H_2 per mol glucose at 35°C and pH 7.2 (Usman et al., 2019). Generation of H_2 using mixed cultures can be beneficial as these organisms survive on a variety of substrates and also demonstrate tolerance to pH and temperature. However, the H_2 production rate has been reported to decrease due to the presence of H_2 and sulfur-consuming microbes in a mixed culture (Khan et al., 2018).

Another way to manufacture sustainable H_2 is to use green algae, as the majority of green algae have the inherent ability to make H_2. *Chlorella*, *Haematococcus*, *Dunaliella*, and *Tetraselmis* are the four genera of green algae that have dominated the market potential for bio-H_2 production. *Chlorella* is the most commonly used microalgae in this genus and is a dietary supplement for humans and animals. *C. sorokiniana* is the species of the *Chlorella* genus that yields the most H_2 due to its fast rate of development and enormous capacity for nitrogen absorption. In addition, *C. sorokiniana* yields additional beneficial by-products such as carotenoids, fatty acids, and antioxidants. They also demonstrate tolerance to temperature and other environmental stressors (Jimenez-Llanos et al., 2020).

4.4.1.5 Hydraulic Retention Time

The quantity of material that must stay in contact with bacteria in the bioreactor for a specific period of time in order to produce the intended end product is known as the hydraulic retention time or HRT. Since HRT is the primary element influencing the pace at which bacteria produce bio-H_2, it is important to consider when assessing the effectiveness of H_2 produced in a bioreactor (Usman et al., 2019). HRT in a CSTR is maintained by regulating the washing rate in a bioreactor, which resists the metabolic shift of H_2-generating microorganisms from an acidogenic to a methanogenic state. Short HRT inhibits the growth of slow-growing methanogenic bacteria, and thus during the initial time of bioreaction an optimal amount of H_2 is generated in bioreactors. Data recorded in various literatures suggests that generally short HRT of 2–8 hours is excellent for obtaining the optimal yield of H_2 without production of contaminating methane by methanogenic microbes. It is reported that the maximum H_2 yield, 2.45 ± 0.24 mol H_2/mol glucose, is generated by maintaining HRT for 6 hours (Usman et al., 2019). It is possible to increase HRT to 72 hours by harnessing mixed cultures of H_2-generating microbes, without the production of methanogenic microbes in bioreactor. Many factors contribute to HRT such as the substrate composition used for cultivation, the strain of microbe used, and the organic load in the bioreactor (Sharma et al., 2021).

4.4.1.6 Organic Loading Rate

The concentration of organic matter that may be added to the fermenter's unit volume to produce the highest amount of H_2 is defined by the phrase "organic loading rate (OLR)" which is taken into account based on the design of the fermenter and the microbe being utilized (Usman et al., 2019). The overflow in the fermenter is represented by the OLR, which is typically inversely proportional to the total amount of carbohydrates given in as a substrate. Initially, at moderate OLR of substrate, the H_2 yield gradually rises but because of the overcrowding impact of nutrients on microorganisms, high OLR results in poor H_2 generation in the fermenter (Usman et al., 2019). Based on previously reported data, it is possible to generate significant H_2 from mixed cultures by adjusting the OLR to 320 g COD/L/d for HRT of 3 hours, setting the pH to around 6, and maintaining a temperature of 35°C (Prabakar et al., 2018).

4.4.2 Advance Technologies to Enhance the Yield of H_2

Along with the previously mentioned characteristics, core advancement in H_2 generation may be attained by choosing inexpensive reactors, high H_2-producing strains, and less expensive substrates with the right pretreatment. In the current situation, it is necessary to look for sophisticated ways to employ cutting-edge technology to provide high H_2 production. Recent years have seen significant advancements in the production of H_2, including the application of genetic engineering and use of modern facilities with manufacturing technologies to increase H_2 output (Usman et al., 2019). A few of these techniques are listed below.

4.4.2.1 Pretreatment Technologies for Improvement of H_2 Production

Several feedstocks and raw materials act as complex substrates that cause difficulty to uptake as a nutrient source for the development of H_2-generating species, thus causing low production of H_2 in the bioreactors. Several pretreatment procedures were evaluated to reduce complicated substrates to simpler ones, depending on the kinds of substrates that were available, to convert complex materials into fermentable substrates. These techniques include mechanical pretreatment, enzymatic pretreatment, acid hydrolysis, alkaline hydrolysis, sonication, and many more procedures (Chandrasekhar et al., 2015).

Several new alternative pretreatment strategies have lately been employed in addition to conventional ones for increased H_2 generation. One such instance was the modification of red mud, a solid by-product from the bauxite processing industry. A high yield of H_2 generation of 198.62 mL/g-VS was obtained by pretreating brewer waste grain with calcinated red mud at a concentration of 10 g/L. Nanoparticles are now used in the creation of renewable energy. In a previously recorded study, a higher increase in H_2 generation was observed when batch fermentation of distillery effluent was carried out with the addition of nickel oxide (NiO) and hematite (Fe_2O_3) nanoparticles (NPs). The H_2 production was found to be increased by an order of 1.2–4.5 greater when Fe_2O_3 and NiO NP were added together than when NP was added separately (Prabakar et al., 2018).

4.4.2.2 Bioaugmentation

The method of "bioaugmentation" involves adding particular microbial strains or consortia to pre-existing microbial communities to boost the efficiency of H_2 production. Higher substrate concentrations in the dark fermentation reactor are said to induce VFAs to build up, which gradually lowers the pH of the system and prevents H_2 from forming. A natural microbial community can overcome inhibition and increase substrate conversion efficiency by introducing the right microbial strains (Usman et al., 2019). Because of the retention duration utilized in a CSTR treating sugarcane juice, it was shown that coculturing the augmented stringent anaerobe bacteria *Clostridium* with native facultative anaerobe *Klebsiella pneumoniae* increased bio-H_2 production (Chandrasekhar et al., 2015).

4.4.2.3 Gas Extractive Membrane Systems

Bioreactors with continuous cultures are aided by gas sparging/flushing systems to improve H_2 production. Even though gas sparging and flushing have been demonstrated to enhance process performance, continuous H_2 extraction would need a significant amount of energy if implemented on a large scale since gas compression is required and the produced end gas is highly diluted (Ramírez-Morales et al., 2019). As an alternative, membrane technology is a strong method for continually extracting H_2. In an attempt to develop a H_2 extractive membrane reactor and enhance the recovery of produced H_2, researchers have coupled membrane modules with reactors. A membrane is a thin physical barrier that allows chemicals to be transported selectively. Pressure or concentration gradients across the membrane serve as the driving force, causing two phases to separate. According to their morphology, the membranes are generally divided into three types: asymmetric, porous, and dense membranes. In the process, the equilibrium of the reaction is shifted to the product side as a result of the constant removal of products. The negative impact of H_2 pressure buildup in the bioreactor can be mitigated by the use of membrane technology for fermentative H_2 separation (Ramírez-Morales et al., 2015).

These days, membrane systems based on silicon and palladium (Pd) are thought to be viable options for separating the H_2/CO_2 gas mixture created by thermo-catalytic processes that are conducted at high temperatures (200–700°C). Nevertheless, biological processes do not produce H_2 at these temperatures, and thus other membranes must be employed. Polymeric membranes, on the other hand, would be more suitable because their working temperatures are normally lower than 100°C. For this reason, the generation of nonporous polymeric membranes with artificial mixes that mimic bio-H_2 mixtures is the subject of the current study (Ramírez-Morales et al., 2019).

4.4.2.4 Genetic Engineering

Several strategies may be used to increase the production of bio-H_2, and many of them rely on the use of metabolic or protein engineering to accomplish their objectives. The knowledge of present metabolic processes and metabolomics forms the basis for exploring new routes or modifying the major ones that already exist. Metabolic engineering can tackle limiting factors in bio-H_2 production by boosting substrate consumption, increasing electron flow to H_2-producing pathways, and oxygen-resistant H_2-evolving enzymes. Several metabolic engineering techniques have been reported for raising bio-H_2 production in both phototrophic and dark fermentative environments (Lazaro and Hallenbeck, 2019).

Enhancing H_2 production requires genetic engineering of important genes that encode enzymes involved in H_2 metabolism. The enzymes nitrogenase and hydrogenase cooperate to derive H_2 from reducing agents as a substrate. The primary obstacle to synthesizing adequate H_2 over time is hydrogenase's sensitivity to oxygen, which limits productivity. Ferredoxin (Fd) gets electrons from the electron transport chain and transfers them to hydrogenase. Even under optimum enzyme expression situations, competition from alternative assimilatory routes for reducing agents limits the overall H_2 yield (Anwar et al., 2019). To get above this obstacle, fusion proteins have been developed. An in vivo culture of *Chlamydomonas reinhardtii* was used to evaluate a fusion complex between Fd and hydrogenase. The complex displayed higher rates of synthesis and better resistance to oxygen than the single enzyme exhibited. Data on site-targeted mutagenesis in *Chlamydomonas reinhardtii* revealed that knockout mutants for the flavodiiron protein (FDP) photoproduced higher amounts of H_2 than the wild type. This shows that electrons are transported preferentially toward the production of H_2 by blocking an opposing pathway (like the FDP-mediated O_2 photoreduction pathway) (Limongi et al., 2021).

Modern revolutionized research accomplished the use of non-coding microRNAs (miRNAs) to significantly impact the synthesis of H_2. miRNAs are a group of small non-coding RNAs that perform an important role in the modulation of several cellular mechanisms inside eukaryotes through negative regulation of gene expression. Overexpression of miRNAs has the potential to impact the target gene's mRNA levels. *Chlamydomonas reinhardtii* miRNAs altered expression patterns and caused significant modifications in the expression of genes and metabolism, which was highly associated with the generation of H_2 through photosynthesis. Therefore, regulatory miRNAs can be used to enable prolonged H_2 generation in microalgae. miRNAs may be one of the possible answers to problems with sustainable H_2 generation (Anwar et al. 2019).

4.5 Waste Sources for Bio-H_2 Production

Developments in the field of renewable energy findings have demonstrated that cheaper biomasses and certain environmentally hazardous waste streams can be used to generate H_2, a renewable gas that can be used for energy. Biomass has been defined as any naturally occurring organic material that is ecologically friendly, including agronomical by-products, plants and trees, terrestrial and aquatic plants, lumber and timber debris, grasses, urban wastes, animal leftovers (such as slurry or manure), etc.

Since utilizing biomass has virtually no CO_2 consequences when compared with using traditional feedstocks, it is an acceptable choice (Kamaraj et al. 2020).

Agricultural residual waste is one of the primary raw sources used in today's industry to manufacture H_2. Regarding the creation of agricultural waste, the utilization of lignocellulosic biomass has created a new paradigm for the low-cost production of H_2. LB is a sustainable and developing source of bioenergy that converts CO_2 and sun energy into chemical energy. The main ingredients of LB are cellulose, hemicellulose, and lignin; these components are widely available in both farming and non-agriculture soils and take relatively little time to produce (Singh et al. 2023). To optimize the LB's gasification performance, certain treatments are needed. To further improve biomass's suitability for thermochemical bio-H_2 generation, several treatments, including physical, chemical, and thermal pretreatment techniques, can be used. Physical pretreatments, or mechanical treatments, are applied to LB in order to improve the surface area and decrease particle size (Ma et al. 2020).

The Food and Agriculture Organization (FAO) estimates that 1.6 billion tons of food waste are created annually worldwide. Foods have a high organic content, which gives the microbial population favorable energy for growing. Rice bran, cheese whey, oil mill waste, bread waste, and many other culinary wastes are among the most common types of waste that are produced. This trash is regularly disposed of in landfills, which typically results in unpleasant environmental effects such as groundwater pollution, smells, and greenhouse gas (CH_4) emissions. Converting these waste materials into energy materials can bring the benefit of creating "best from waste" (Jain et al., 2022). The aforementioned waste materials have shown to have great potential as feedstock for a dark fermentative system that produces bio-H_2. The pace of bio-H_2 generation and output is dependent on several important factors, including temperature, ideal pH, critical partial pressure, and the pre-operation processing of food rejects. Large-scale bio-H_2 harvesting requires consideration of other factors such as moisture load, nutritional content, and biodegradability (Kamaraj et al., 2020).

Another sustainable source for fermentative H_2 generation is sewage sludge (SS). Municipal sewage sludge is a valuable resource for producing H_2 due to its high glycerol and glucose content. It has been demonstrated that treating the sludge will boost H_2 production. The majority of research has focused on physical pretreatments, including heat, ultrasound, microwave, UV radiation, and sterilizing techniques. Sludge's main application limits, however, are caused by its low carbon-to-nitrogen ratio and thus poorer yield production (Prabakar et al., 2018). Also, the formation of municipal solid wastes (MSW), whose volume is growing daily, is a problem that large cities are facing today. These wastes include cheap and abundant material, which gives them the ability to produce a variety of microbes. Furthermore, MSW includes macro- and micronutrients such as proteins, lipids, carbs, minerals, and vitamins. It is possible to think of this nutrient-rich MSW as a good source for H_2 fermentation (Kamaraj et al., 2020).

The selection and accessibility and the concentration of biomass, the price, and the yield, or volume of product produced, all influence the appropriate waste selection for H_2 generation. Data from research published by Sivagurunathan and Lin (2020) showed the combination of enriched mixed culture and brewery wastewater at a pH of 5.5, a temperature of 38.3°C, and an OLR of 16.6 g/L results in a greater generation of H_2, demonstrating the significant influence of substrate concentrations on the final output. Data reported by Ahmed et al. (2021) have demonstrated the optimal conditions for H_2 production were found when the activated sludge was pretreated for 15 minutes, the initial pH was 5.0, the concentration of corn starch powder was 22.34 mg/mL, and the fermentation temperature was 40°C.

Using textile desizing water (TDW) with coagulation pretreatment, Lin et al. (2020) have demonstrated the production of fermentative bio-H_2. Its result indicates that large concentrations of coagulation pretreatment reduce the high concentration of harmful chemicals, which suppresses the H_2-producing microbes. At a substrate concentration of 15 g total sugar/L, the formation of H_2 is specifically reduced by coagulation pretreatment. The rise in H_2 generation capacity to 120% indicates the efficacy of the coagulation pretreatment. The coagulation pretreated TDW produced an HPR of 3.9 L/(L-d) and a hydrogen yield (HY) of 1.52 mol/mol hexose at a concentration of 15 g total sugar/L (Kothari et al., 2017).

In the analysis by Okolie et al.. (2020), it was found that soybean straw at 500°C, 1:10 load feed concentration, particle size of 0.13 mm, and 45 minutes of residence time produced the highest amounts of bio-H_2 generation and syngas at 6.62 mmol/g and 14.91 mmol/g, respectively. Employing a two-step method of dark fermentation and microbial electrolysis, Intanoo et al. (2016) generated bio-H_2 from a variety of agro-industrial wastes and by-products, including dairy products, fruit juice, sugar, paper products, the processing of fruit, and liquor. The combined effect of these waste materials yielded a maximum COD removal rate of 78.5 ± 5.7% and a high generation of H_2 of 1608.6 ± 266.2 mL H_2/g COD used.

There are several different approaches to improving the generation of bio-H_2, according to recent research. These strategies can be categorized into multiple areas:

A. The first step in improving the biodegradability of substrates is pretreatment. Pretreatment techniques involve the use of biocatalysts, acid-alkali treatment, and thermo-aero-size reduction. Utilization of such techniques has proved to generate higher production of H_2 (Rawat et al., 2023)
B. By expressing heterologous proteins, metabolic engineering is a cutting-edge technology that helps organisms overcome thermodynamic constraints, redirect metabolic pathways, and enhance substrate consumption. One other area of attention to increase the efficiency of H_2 generation is electron flux for proton reduction performance which gene editing can also make feasible.
C. It is feasible to increase the process yield and boost the effectiveness of substrate consumption by combining the methods of dark fermentation or microbial electrolysis with photofermentation. In the process of producing biofuels and other bio-based goods, this integration process might contribute to a more sustainable approach (Chandrasekhar et al., 2015)
D. The yield of bio-H_2 generation may be increased by using H_2-selective polymeric membranes. This also helps to reduce the expenditures of the purification process (Sazali et al., 2020)

E. The application of nanotechnology has received a lot of attention lately. It has been demonstrated that using substrates supplemented with nanoparticles derived from inorganic elements like Fe and Zn increases the H_2 fermentation's efficiency in the dark. (Sun et al. 2019)

4.6 Applications of Bio-H_2

In the present climate, using bio-H_2 can be extremely important. There are many different uses for H_2 gas utilization across all industries. A few of these are discussed next.

4.6.1 Green H_2

All economic sectors—transportation, electricity, industry, residential, commercial, and construction—can benefit from H_2. Waste materials have the potential to be utilized in the production of bio-H_2 since they are reasonably less costly feedstock and easily accessible. The resulting reduction of air pollution and the solving of waste management issues through the creation of H_2 from recycled rubbish are two of the largest environmental benefits of using waste (Kumar and Agrawal, 2020). Because it employs less fossil fuel, using ecologically friendly H_2 offers a sustainable and beneficial choice by potentially mitigating the consequences of climate change by lowering greenhouse gas emissions. Via efficient utilization of resources, allegedly hazardous components can potentially be transformed into valuable energy sources through the H_2 generating process (Ahmed et al., 2021; Kamaraj et al., 2020).

4.6.2 Industrial Significance

Industries that have historically been challenging to decarbonize, including the manufacture of steel and cement, stand to benefit greatly from the introduction of green H_2. Green H_2 may dramatically cut carbon emissions in several areas by taking the place of fossil fuels. In the case of steel production, for example, coal may be replaced by green H_2 when it is utilized as a reducing agent to make steel from iron ore. Also, H_2 has the potential to decarbonize the rail industry in addition to the transportation sector. Before employing H_2 to power fuel cells, a few technological obstacles need to be investigated, including bioreactor efficiency, biomass feedstock preconditioning, waste processing, and purification (Osman et al., 2022). Utilizing fuel cells in cars, ships, and trains is another way that bio-H_2 may be utilized for transportation. Numerous businesses can use bio-H_2 as a feedstock. The International Energy Agency (IEA) estimates that by 2020 oil refineries will require 38 million tons of H_2 annually, while other sectors will require 51 million tons (Nazir et al., 2020). By 2022, the generation of H_2 for industrial usage is expected to reach a global market size of $154.1 billion.

4.6.3 Energy Storage

The capacity of green H_2 to store energy is among its most promising uses. Green H_2 may be produced via utilizing microbial systems or in other cases electrolysis using excess of renewable energy, such as that of solar or wind power. A steady supply of energy may then be ensured by storing the H_2 and converting it back to electricity as needed. H_2 is easy to store and transport by truck or pipeline as per need. When combined, H_2 and MFCs generate power without releasing any harmful by-products. In the absence of sulfur, a *C. sorokiniana* strain produces bio-H_2. 8.9 mA of current was created when 27.09 mL of H_2 was introduced into the polymer electrolyte membrane fuel cells (PEMFC), which is where the produced H_2 is transported and turned into electricity (Chader et al., 2011).

4.6.4 Household Use

Green H_2 can also be integrated into existing gas pipelines to power residential appliances. This allows households to use renewable energy for heating, cooking, and other daily activities, reducing the reliance on natural gas.

4.7 Environmental and Economic Implications

Clean energy resources are increasingly crucial due to their environmental benefits, as fossil fuels currently meet our energy needs, but technical, political, and economic factors may make future supply unpredictable (Kamaraj et al., 2020). Alternative fuel sources include natural gas, biofuels, H_2, and synthetic gas. Of these, H_2 is the greatest since it produces a lot of energy, is renewable, emits no greenhouse emissions, and can be used to create power in fuel cells. H_2 production produces just water as a waste, making it the cleanest renewable energy source. The adoption of H_2 as an energy carrier is hampered by infrastructure needs, safety issues, and manufacturing costs. The future of H_2 is dependent on technological breakthroughs and governmental changes. However, doing this will cost a lot of money in terms of more eco-friendly production techniques, effective storage solutions, and related infrastructure (Hassan et al., 2023).

One of the numerous advantages of H_2 over fossil fuels is its high heat of combustion (142.9 kJ/g), which exceeds that of coal and gasoline combined. At between 30% and 60%, H_2 has a higher thermal efficiency than the majority of conventional fuels. Hydrogen is also sold as an upgrading (enhancing) agent for petroleum distillates and products made from crude oil by hydrocracking, hydroprocessing, and other methods. Hydrocracking process involves breaking of heavier hydrocarbons with the use of H_2 into lighter derivatives and increase the hydrogen to carbon ratio. (Osman et al., 2022). In addition, we are dealing with waste management issues as a result of our rapidly developing and urbanizing society. The primary cause is a lack of proper infrastructure for collection and disposal. By 2050, 3.4 billion tons of MSW are expected to be produced, according to UN estimates. Waste management may be made more sustainable by using biological H_2 generation. High H_2-yielding microorganisms should be employed to improve bio-H_2 generation. Concerning sustainable waste management and process cost reduction, many types of waste, including food waste, industrial waste, and agricultural waste combined with H_2 generation, provide a bright future (Kumar et al., 2022).

The commercial usage of bio-H_2 includes the following: ammonia is regarded as the best fertilizer in the agriculture business. The Haber process catalyzes the reaction of H_2 and nitrogen elements, hence advancing the ammonia synthesis process. Ammonia is also provided to a variety of industries, including fuel cells, polymer manufacture, explosives, refrigerants, medications, and gas sensors. Moreover, H_2 is frequently utilized as a reducing agent to recover metals from ores and to initiate oxy-H_2 furnaces in industrial metallurgical processes. During the exothermic reaction process, H_2 and oxygen may mix at extremely high temperatures (3000°C), producing oxy-H_2 flames that can be utilized for cutting and welding on non-ferrous metals (Osman et al., 2022).

Production of H_2 from organic waste is an economical way for renewable biofuels. After that, the working conditions need to be decided and based on that production cost and revenue could be calculated annually. Bio-H_2 synthesis is part of the waste treatment process, and thus no additional treatment is necessary to meet effluent standards. The present net worth of H_2, internal cost of return, and payback period all contribute to the process's profitability. It has several advantages, including increased income from agricultural food crops as a result of their energy conversion, lower waste management costs, the substitution of non-renewable sources, and the use of sustainable biomass (Li et al., 2012). Approximately two-thirds of the organic materials included in solid waste from municipalities may be used to produce bio-H_2. In addition to solid waste, other potential sources for H_2 generation include food waste, wastewater, animal waste, and agricultural residual waste. Numerous research is conducted to increase the economic viability of producing H_2; however, there are a few major issues that need to be resolved. The anticipated expense of resolving the engineering and technical problems is \$1.42 million (Ahmed et al., 2021). Even though producing H_2 by dark fermentation has numerous advantages, it is not economically feasible to produce H_2 on a big scale since every challenge must first be overcome (Yun et al., 2018).

4.8 Challenges and Future Perspectives

The global advancement of H_2 energy technology is gaining momentum, with its usage, distribution, and utility becoming increasingly important in research, planning, and policy-making. H_2 derived from fossil fuels has a more positive carbon footprint impact (Burmistrz et al., 2016) The generation and consumption of energy are major contributors to the first two of these major environmental concerns: greenhouse gas emissions, overuse of natural resources, and inefficient waste management. Relying less on fossil fuels and more on bio-H_2 would only have a 0.6% influence on the climate (Balat and Kırtay, 2010). H_2 may be produced utilizing a variety of methods and basic energy sources. Advancements in the field have identified limiting elements in microbial processes for renewable H_2 production, providing a strong foundation for designing enhanced catalysts based on molecular processes and specifics. Biophotolysis, a process of converting solar energy to H_2, faces challenges like low conversion efficiencies and hydrogenase sensitivity to oxygen, but alternative methods could offer potential benefits. It is necessary to make modifications to the current bio-H_2 production procedures to improve fermentation and open up new opportunities for the generation of bio-H_2 from sustainable biomass. To overcome the challenges of converting organic waste into bio-H_2, such as poor yields and a slow rate of H_2 generation, large working reactor volumes, appropriate testing conditions, cutting-edge technology, and various storage and transportation facilities are needed. Integrative pretreatment techniques need to be assessed economically even though they demonstrate greater rates of bio-H_2 production to establish a solid foothold in the energy-generating industry. Furthermore, the use of agricultural waste for bioenergy generation is more common than that of industrial effluents (Martino et al., 2021). (Usman et al. 2021)

4.9 CONCLUSION

H_2 is an alternative fuel that is receiving a lot of interest from researchers to store energy and build prototype automobiles. Biological H_2 production promises to be a sustainable supply of H_2 for the future H_2 economy. While methods for producing bio-H_2 have several advantages, they can only be applied on a limited level. Although fermentation technologies are environmentally benign, they are inefficient in producing bio-H_2 and converting sunlight. By fermenting H_2 from waste materials including industrial effluent, residues, and agricultural waste, including food waste, the H_2 economy may alleviate environmental challenges. Metagenomics and genetic alterations are two recent technological developments that seek to make microbial-assisted H_2 production commercially, operationally, and economically viable in the near future. A sustainable H_2-driven economy depends on large-scale bio-H_2 production techniques, including pretreatment technologies that turn waste into useful energy resources. H_2 may be produced by organisms from waste streams or carbohydrates; however, poor yields and ineffective waste treatment are still problems. Methane digestion, photofermentation, or methane-electrode catalysis (MECs) in hybrid two-stage systems or metabolic engineering might be used to improve the situation. Increasing current densities and locating inexpensive cathode materials are challenges. Within the next few decades, a breakthrough could allow for practical implementation.

ACKNOWLEDGMENT

Mr. Rohit R. Topkar is thankful to the Department of Biotechnology, Willingdon College, Sangli and the Department of Biotechnology, Shivaji University, Kolhapur for providing resources and facilities. Ms. Anmol Patil is thankful to the Department of Biotechnology, Smt. Kasturbai Walchand College, Sangli for providing resources and facilities. Ms. Priyanka Magadum is thankful to the Department of Environmental Science, College of Non-Conventional Vocational Courses for Women, Kolhapur for providing resources and facilities. Mrs. Sarita M. Wadmare is thankful to the Department of Biotechnology and the Department of Microbiology, Willingdon College, Sangli. Prof. Jyoti P. Jadhav thanks the Department of Biotechnology, Government of India for the Interdisciplinary Program for Life Sciences (IPLS) (Reference no. BT/PR4572/ INF/22/147/2012) for giving research facilities. Mr. Rahul R. Jadhav is thankful to the Chhatrapati Shahu Maharaj Research, Training and Human Development Institute (SARTHI), Pune for providing the CSMNRF fellowship award and the Department of Biotechnology, Willingdon College, Sangli and the Department of Biotechnology, Shivaji University, Kolhapur for providing resources and facilities.

REFERENCES

Ahmed, S.F., Rafa, N., Mofijur, M., Badruddin, I.A., Inayat, A., Ali, M.S., Farrok, O. and Yunus Khan, T.M., 2021. Biohydrogen production from biomass sources: metabolic pathways and economic analysis. *Frontiers in Energy Research*, 9, p.753878.

Almatouq, A., Babatunde, A.O., Khajah, M., Webster, G. and Alfodari, M., 2020. Microbial community structure of anode electrodes in microbial fuel cells and microbial electrolysis cells. *Journal of Water Process Engineering*, 34, p.101140.

Anwar, M., Lou, S., Chen, L., Li, H. and Hu, Z., 2019. Recent advancement and strategy on bio-hydrogen production from photosynthetic microalgae. *Bioresource Technology*, 292, p.121972.

Balachandar, G., Khanna, N. and Das, D., 2013. Biohydrogen production from organic wastes by dark fermentation. In *Biohydrogen* (pp. 103–144). Elsevier.

Balat, H. and Kırtay, E., 2010. Hydrogen from biomass–present scenario and future prospects. *International Journal Of Hydrogen Energy*, 35(14), pp.7416–7426.

Benemann, J.R., 2000. Hydrogen production by microalgae. *Journal of Applied Phycology*, 12, pp.291–300.

Boesen, T. and Nielsen, L.P., 2013. Molecular dissection of bacterial nanowires. *MBio*, 4(3), pp.10–1128.

Burmistrz, P., Chmielniak, T., Czepirski, L. and Gazda-Grzywacz, M., 2016. Carbon footprint of the hydrogen production process utilizing subbituminous coal and lignite gasification. *Journal of Cleaner Production*, 139, pp.858–865.

Castelló, E., y Santos, C.G., Iglesias, T., Paolino, G., Wenzel, J., Borzacconi, L. and Etchebehere, C., 2009. Feasibility of biohydrogen production from cheese whey using a UASB reactor: links between microbial community and reactor performance. *International Journal of Hydrogen Energy*, 34(14), pp.5674–5682.

Chader, S., Mahmah, B., Chetehouna, K., Amrouche, F. and Abdeladim, K., 2011. Biohydrogen production using green microalgae as an approach to operate a small proton exchange membrane fuel cell. *International Journal Of Hydrogen Energy*, 36(6), pp.4089–4093.

Chandrasekhar, K., Lee, Y.J. and Lee, D.W., 2015. Biohydrogen production: strategies to improve process efficiency through microbial routes. *International Journal of Molecular Sciences*, 16(4), pp.8266–8293.

Chang, F.Y. and Lin, C.Y., 2004. Biohydrogen production using an up-flow anaerobic sludge blanket reactor. *International Journal of Hydrogen Energy*, 29(1), pp.33–39.

Chen, C.C. and Lin, C.Y., 2003. Using sucrose as a substrate in an anaerobic hydrogen-producing reactor. *Advances in Environmental Research*, 7(3), pp.695–699.

da Silva, A.N., Macêdo, W.V., Sakamoto, I.K., Pereyra, D.D.L.A.D., Mendes, C.O., Maintinguer, S.I., Caffaro Filho, R.A., Damianovic, M.H.Z., Varesche, M.B.A. and de Amorim, E.L.C., 2019. Biohydrogen production from dairy industry wastewater in an anaerobic fluidized-bed reactor. *Biomass and Bioenergy*, 120, pp.257–264.

Das, S.R. and Basak, N., 2021. Molecular biohydrogen production by dark and photo fermentation from wastes containing starch: recent advancement and future perspective. *Bioprocess and Biosystems Engineering*, 44, pp.1–25.

Davila-Vazquez, G., Cota-Navarro, C.B., Rosales-Colunga, L.M., de León-Rodríguez, A. and Razo-Flores, E., 2009. Continuous biohydrogen production using cheese whey: Improving the hydrogen production rate. *International Journal of Hydrogen Energy*, 34(10), pp.4296–4304.

Ferraren-De Cagalitan, D. D. T., & Abundo, M. L. S., 2021. A review of biohydrogen production technology for application towards hydrogen fuel cells. *Renewable and Sustainable Energy Reviews*, 151, 111413.

Fudge, T., Bulmer, I., Bowman, K., Pathmakanthan, S., Gambier, W., Dehouche, Z., Al-Salem, S.M. and Constantinou, A., 2021. Microbial electrolysis cells for decentralised wastewater treatment: the next steps. *Water*, 13(4), p.445.

Goswami, R.K., Mehariya, S., Obulisamy, P.K. and Verma, P., 2021. Advanced microalgae-based renewable biohydrogen production systems: a review. *Bioresource Technology*, 320, p.124301.

Grechanik, V., Naidov, I., Bolshakov, M. and Tsygankov, A., 2021. Photoautotrophic hydrogen production by nitrogen-deprived Chlamydomonas reinhardtii cultures. *International Journal of Hydrogen Energy*, 46(5), pp.3565–3575.

Gupta, S., Fernandes, A., Lopes, A., Grasa, L. and Salafranca, J., 2024. Photo-Fermentative Bacteria Used for Hydrogen Production. *Applied Sciences*, 14(3), p.1191.

Hassan, Q., Algburi, S., Sameen, A.Z., Jaszczur, M. and Salman, H.M., 2023. Hydrogen as an energy carrier: properties, storage methods, challenges, and future implications. *Environment Systems and Decisions*, pp.1–24.

Intanoo, P., Chaimongkol, P. and Chavadej, S., 2016. Hydrogen and methane production from cassava wastewater using two-stage upflow anaerobic sludge blanket reactors (UASB) with an emphasis on maximum hydrogen production. *International Journal of Hydrogen Energy*, 41(14), pp.6107–6114.

Jafary, T., Daud, W.R.W., Ghasemi, M., Bakar, M.H.A., Sedighi, M., Kim, B.H., Carmona-Martínez, A.A., Jahim, J.M. and Ismail, M., 2019. Clean hydrogen production in a full biological microbial electrolysis cell. *International Journal of Hydrogen Energy*, 44(58), pp.30524–30531.

Jain, R., Panwar, N.L., Jain, S.K., Gupta, T., Agarwal, C. and Meena, S.S., 2022. Bio-hydrogen production through dark fermentation: an overview. *Biomass Conversion and Biorefinery*, pp.1–26.

Jimenez-Llanos, J., Ramirez-Carmona, M., Rendón-Castrillón, L. and Ocampo-López, C., 2020. Sustainable biohydrogen production by Chlorella sp. microalgae: A review. *International Journal of Hydrogen Energy*, 45(15), pp.8310–8328.

Kamaraj, M., Ramachandran, K. K., & Aravind, J., 2020. Biohydrogen production from waste materials: benefits and challenges. *International Journal of Environmental Science and Technology*, 17(1), 559–576.

Khan, M.A., Ngo, H.H., Guo, W., Liu, Y., Zhang, X., Guo, J., Chang, S.W., Nguyen, D.D. and Wang, J., 2018. Biohydrogen production from anaerobic digestion and its potential as renewable energy. *Renewable Energy*, 129, pp.754–768.

Khan, M.Z., Nizami, A.S., Rehan, M., Ouda, O.K.M., Sultana, S., Ismail, I.M. and Shahzad, K., 2017. Microbial electrolysis cells for hydrogen production and urban wastewater treatment: A case study of Saudi Arabia. *Applied Energy*, 185, pp.410–420.

Khetkorn, W., Rastogi, R.P., Incharoensakdi, A., Lindblad, P., Madamwar, D., Pandey, A. and Larroche, C., 2017. Microalgal hydrogen production–A review. *Bioresource Technology*, 243, pp.1194–1206.

Kim, S.H., Da Yi, Y., Kim, H.J., Bhatia, S.K., Gurav, R., Jeon, J.-M., Yoon, J.-J., Kim, S.-H., Park, J.-H., Yang, Y.-H. 2022a. Hyper biohydrogen production from xylose and xylose-based hemicellulose biomass by the novel strain Clostridium sp. YD09. *Biochemical Engineering Journal*, 187, 108624.

Kim, S.H., Kim, H.J., Bhatia, S.K., Gurav, R., Jeon, J.-M., Yoon, J.-J., Kim, S.-H., Ahn, J., Yang, Y.-H. 2022b. Biohydrogen production from glycerol by novel Clostridium sp. SH25 and its application to biohydrogen car operation. *Korean Journal of Chemical Engineering*, 39(8), 2156–2164.

Kosourov, S., Böhm, M., Senger, M., Berggren, G., Stensjö, K., Mamedov, F., Lindblad, P. and Allahverdiyeva, Y., 2021. Photosynthetic hydrogen production: Novel protocols, promising engineering approaches and application of semi-synthetic hydrogenases. *Physiologia Plantarum*, 173(2), pp.555–567.

Kothari, R., Kumar, V., Pathak, V.V. and Tyagi, V.V., 2017. Sequential hydrogen and methane production with simultaneous treatment of dairy industry wastewater: bioenergy profit approach. *International Journal Of Hydrogen Energy*, 42(8), pp.4870–4879.

Kumar, A. and Agrawal, A., 2020. Recent trends in solid waste management status, challenges, and potential for the future Indian cities–A review. *Current Research in Environmental Sustainability*, 2, p.100011.
Lalman, J.A., Chaganti, S.R., Moon, C. and Kim, D.H., 2013. Elucidating acetogenic H2 consumption in dark fermentation using flux balance analysis. *Bioresource Technology*, 146, pp.775–778.
Lay, C.H., Wu, J.H., Hsiao, C.L., Chang, J.J., Chen, C.C. and Lin, C.Y., 2010. Biohydrogen production from soluble condensed molasses fermentation using anaerobic fermentation. *International Journal of Hydrogen Energy*, 35(24), pp.13445–13451.
Lazaro, C.Z. and Hallenbeck, P.C., 2019. Fundamentals of biohydrogen production. In *Biohydrogen* (pp. 25–48). Elsevier.
Li, Y.C., Liu, Y.F., Chu, C.Y., Chang, P.L., Hsu, C.W., Lin, P.J. and Wu, S.Y., 2012. Techno-economic evaluation of biohydrogen production from wastewater and agricultural waste. *International Journal Of Hydrogen Energy*, 37(20), pp.15704–15710.
Lim, S.S., Yu, E.H., Daud, W.R.W., Kim, B.H. and Scott, K., 2017. Bioanode as a limiting factor to biocathode performance in microbial electrolysis cells. *Bioresource Technology*, 238, pp.313–324.
Limongi, A.R., Viviano, E., De Luca, M., Radice, R.P., Bianco, G. and Martelli, G., 2021. Biohydrogen from microalgae: production and applications. *Applied Sciences*, 11(4), p.1616.
Lin, C.Y., Chiang, C.C., Nguyen, M.L.T. and Lay, C.H., 2017. Enhancement of fermentative biohydrogen production from textile desizing wastewater via coagulation-pretreatment. *International Journal Of Hydrogen Energy*, 42(17), pp.12153–12158.
Lin, P.Y., Whang, L.M., Wu, Y.R., Ren, W.J., Hsiao, C.J., Li, S.L. and Chang, J.S., 2007. Biological hydrogen production of the genus Clostridium: metabolic study and mathematical model simulation. *International Journal of Hydrogen Energy*, 32(12), pp.1728–1735.
Ma, Y., Shen, Y. and Liu, Y., 2020. State of the art of straw treatment technology: Challenges and solutions forward. *Bioresource Technology*, 313, p.123656.
Martino, M., Ruocco, C., Meloni, E., Pullumbi, P. and Palma, V., 2021. Main hydrogen production processes: An overview. *Catalysts*, 11(5), p.547.
Melitos, G., Voulkopoulos, X. and Zabaniotou, A., 2021. Waste to sustainable biohydrogen production via photo-fermentation and biophotolysis– A systematic review. *Renewable Energy and Environmental Sustainability*, 6, p.45.
Mona, S., Kumar, S.S., Kumar, V., Parveen, K., Saini, N., Deepak, B. and Pugazhendhi, A., 2020. Green technology for sustainable biohydrogen production (waste to energy): a review. *Science of the Total Environment*, 728, p.138481.
Mus, F., Alleman, A.B., Pence, N., Seefeldt, L.C. and Peters, J.W., 2018. Exploring the alternatives of biological nitrogen fixation. *Metallomics*, 10(4), pp.523–538.
Nagarajan, D., Lee, D.J. and Chang, J.S., 2019. Recent insights into consolidated bioprocessing for lignocellulosic biohydrogen production. *International Journal of Hydrogen Energy*, 44(28), pp.14362–14379.
Nanda, S., Rana, R., Sarangi, P.K., Dalai, A.K. and Kozinski, J.A., 2018. A broad introduction to first-, second-, and third-generation biofuels. *Recent advancements in biofuels and bioenergy utilization*, pp.1–25.
Nazir, H., Louis, C., Jose, S., Prakash, J., Muthuswamy, N., Buan, M.E., Flox, C., Chavan, S., Shi, X., Kauranen, P. and Kallio, T., 2020. Is the H2 economy realizable in the foreseeable future? Part I: H2 production methods. *International Journal of Hydrogen Energy*, *45*(27), pp.13777–13788.
Okolie, J.A., Nanda, S., Dalai, A.K. and Kozinski, J.A., 2020. Optimization and modeling of process parameters during hydrothermal gasification of biomass model compounds to generate hydrogen-rich gas products. *International Journal of Hydrogen Energy*, 45(36), pp.18275–18288.

Osman, A. I., Deka, T. J., Baruah, D. C., & Rooney, D. W., 2020. Critical challenges in biohydrogen production processes from the organic feedstocks. *Biomass Conversion and Biorefinery*, 1–19.
Osman, A.I., Mehta, N., Elgarahy, A.M., Hefny, M., Al-Hinai, A., Al-Muhtaseb, A.A.H. and Rooney, D.W., 2022. Hydrogen production, storage, utilisation and environmental impacts: a review. *Environmental Chemistry Letters*, pp.1–36.
Pachiega, R., Rodrigues, M.F., Rodrigues, C.V., Sakamoto, I.K., Varesche, M.B.A., De Oliveira, J.E. and Maintinguer, S.I., 2019. Hydrogen bioproduction with anaerobic bacteria consortium from brewery wastewater. *International Journal of Hydrogen Energy*, 44(1), pp.155–163.
Poontaweegeratigarn, T., Chavadej, S. and Rangsunvigit, P., 2012. Hydrogen production from alcohol wastewater by upflow anaerobic sludge blanket reactors under mesophilic temperature. *International Scholarly and Scientific Research & Innovation*, 6, pp.293–296.
Prabakar, D., Manimudi, V.T., Sampath, S., Mahapatra, D.M., Rajendran, K. and Pugazhendhi, A., 2018. Advanced biohydrogen production using pretreated industrial waste: outlook and prospects. *Renewable and Sustainable Energy Reviews*, 96, pp.306–324.
Qu, X., Zeng, H., Gao, Y., Mo, T. and Li, Y., 2022. Bio-hydrogen production by dark anaerobic fermentation of organic wastewater. *Frontiers in Chemistry*, 10, p.978907.
Ramírez-Morales, J.E., Tapia-Venegas, E., Campos, J.L. and Ruiz-Filippi, G., 2019. Operational behavior of a hydrogen extractive membrane bioreactor (HEMB) during mixed culture acidogenic fermentation. *International Journal of Hydrogen Energy*, 44(47), pp.25565–25574.
Ramírez-Morales, J.E., Tapia-Venegas, E., Toledo-Alarcón, J. and Ruiz-Filippi, G., 2015. Simultaneous production and separation of biohydrogen in mixed culture systems by continuous dark fermentation. *Water Science and Technology*, 71(9), pp.1271–1285.
Rawat, S., Rautela, A., Yadav, I., Misra, S. and Kumar, S., 2023. A Comprehensive Review on Enhanced Biohydrogen Production: Pretreatment, Applied Strategies, Techno-Economic Assessment, and Future Perspective. *BioEnergy Research*, pp.1–24.
Razu, M.H., Hossain, F. and Khan, M., 2019. Advancement of bio-hydrogen production from microalgae. *Microalgae biotechnology for development of biofuel and wastewater treatment*, pp.423–462.
Reungsang, A., Zhong, N., Yang, Y., Sittijunda, S., Xia, A. and Liao, Q., 2018. Hydrogen from photo fermentation. *Bioreactors for microbial biomass and energy conversion*, pp.221–317.
Sarangi, P.K. and Nanda, S., 2020. Biohydrogen production through dark fermentation. *Chemical Engineering & Technology*, 43(4), pp.601–612.
Saravanan, A., Karishma, S., Kumar, P.S., Yaashikaa, P.R., Jeevanantham, S. and Gayathri, B., 2020. Microbial electrolysis cells and microbial fuel cells for biohydrogen production: Current advances and emerging challenges. *Biomass Conversion and Biorefinery*, pp.1–21.
Sazali, N., Mohamed, M.A. and Salleh, W.N.W., 2020. Membranes for hydrogen separation: A significant review. *The International Journal of Advanced Manufacturing Technology*, 107(3), pp.1859–1881.
Schuchmann, K., Chowdhury, N.P. and Müller, V., 2018. Complex multimeric [FeFe] hydrogenases: biochemistry, physiology and new opportunities for the hydrogen economy. *Frontiers in Microbiology*, 9, p.421914.
Sharma, A.K., Ghodke, P.K., Manna, S. and Chen, W.H., 2021. Emerging technologies for sustainable production of biohydrogen production from microalgae: A state-of-the-art review of upstream and downstream processes. *Bioresource Technology*, 342, p.126057.
Shi, M., Jiang, Y. and Shi, L., 2019. Electromicrobiology and biotechnological applications of the exoelectrogens Geobacter and Shewanella spp. *Science China Technological Sciences*, 62, pp.1670–1678.
Singh, R., Kumar, R., Sarangi, P.K., Kovalev, A.A. and Vivekanand, V., 2023. Effect of physical and thermal pretreatment of lignocellulosic biomass on biohydrogen production by thermochemical route: a critical review. *Bioresource Technology*, 369, p.128458.

Sivagurunathan, P. and Lin, C.Y., 2020. Biohydrogen production from beverage wastewater using selectively enriched mixed culture. *Waste and Biomass Valorization*, 11, pp.1049–1058.

Song, H., Luo, S., Huang, H., Deng, B. and Ye, J., 2022. Solar-driven hydrogen production: Recent advances, challenges, and future perspectives. *ACS Energy Letters*, 7(3), pp.1043–1065.

Stuart, T.S. and Gaffron, H., 1972. The mechanism of hydrogen photoproduction by several algae: II. The contribution of photosystem II. *Planta*, 106(2), pp.101–112.

Sun, Y., He, J., Yang, G., Sun, G. and Sage, V., 2019. A review of the enhancement of biohydrogen generation by chemicals addition. *Catalysts*, 9(4), p.353.

Tapia-Venegas, E., Ramirez-Morales, J.E., Silva-Illanes, F., Toledo-Alarcón, J., Paillet, F., Escudie, R., Lay, C.H., Chu, C.Y., Leu, H.J., Marone, A. and Lin, C.Y., 2015. Biohydrogen production by dark fermentation: scaling-up and technologies integration for a sustainable system. *Reviews in Environmental Science and Bio/Technology*, 14, pp.761–785.

Tartakovsky, B., Manuel, M.F., Wang, H. and Guiot, S.R., 2009. High rate membrane-less microbial electrolysis cell for continuous hydrogen production. *International Journal of Hydrogen Energy*, 34(2), pp.672–677.

Trchounian, A., 2015. Mechanisms for hydrogen production by different bacteria during mixed-acid and photo-fermentation and perspectives of hydrogen production biotechnology. *Critical Reviews in Biotechnology*, 35(1), pp.103–113.

Usman, M., Kavitha, S., Kannah, Y., Yogalakshmi, K. N., Sivashanmugam, P., Bhatnagar, A., & Kumar, G., 2021. A critical review on limitations and enhancement strategies associated with biohydrogen production. *International Journal of Hydrogen Energy*, 46(31), 16565–16590.

Usman, T. M., Banu, J. R., Gunasekaran, M., & Kumar, G, 2019. Biohydrogen production from industrial wastewater: an overview. *Bioresource Technology Reports*, 7, 100287.

Wang, J., Xu, S., Xiao, B., Xu, M., Yang, L., Liu, S., Hu, Z., Guo, D., Hu, M., Ma, C. and Luo, S., 2013. Influence of catalyst and temperature on gasification performance of pig compost for hydrogen-rich gas production. *International Journal of Hydrogen Energy*, 38(33), pp.14200–14207.

Wang, K., Khoo, K.S., Chew, K.W., Selvarajoo, A., Chen, W.H., Chang, J.S. and Show, P.L., 2021. Microalgae: the future supply house of biohydrogen and biogas. *Frontiers in Energy Research*, 9, p.660399.

Wang, Y., Mu, Y. and Yu, H.Q., 2007. Comparative performance of two upflow anaerobic biohydrogen-producing reactors seeded with different sludges. *International Journal of Hydrogen Energy*, 32(8), pp.1086–1094.

Xuan, J., He, L., Wen, W. and Feng, Y., 2023. Hydrogenase and nitrogenase: key catalysts in biohydrogen production. *Molecules*, 28(3), p.1392.

Yaakob, M.A., Mohamed, R.M.S.R., Al-Gheethi, A., Aswathnarayana Gokare, R. and Ambati, R.R., 2021. Influence of nitrogen and phosphorus on microalgal growth, biomass, lipid, and fatty acid production: an overview. *Cells*, 10(2), p.393.

Yang, E., Mohamed, H.O., Park, S.G., Obaid, M., Al-Qaradawi, S.Y., Castaño, P., Chon, K. and Chae, K.J., 2021. A review on self-sustainable microbial electrolysis cells for electro-biohydrogen production via coupling with carbon-neutral renewable energy technologies. *Bioresource Technology*, 320, p.124363.

Yun, Y. M., Lee, M. K., Im, S. W., Marone, A., Trably, E., Shin, S. R., … & Kim, D. H., 2018. Biohydrogen production from food waste: current status, limitations, and future perspectives. *Bioresource Technology*, 248, 79–87.

Zhang, Z.P., Show, K.Y., Tay, J.H., Liang, D.T. and Lee, D.J., 2008. Biohydrogen production with anaerobic fluidized bed reactors—A comparison of biofilm-based and granule-based systems. *International Journal of Hydrogen Energy*, 33(5), pp.1559–1564.

5 Biomethane Production
From Organic Waste to Green Gas

Ying Xu and Chen Zhang

5.1 INTRODUCTION

Organic wastes (e.g., sewage sludge, food wastes, agricultural wastes), which have dual attributes of pollution and resources, are generated in large quantities every year and are expected to intensify with population growth and economic development. For instance, by 2022, annual sludge production of the world's three major economies (China, the United States, and the European Union) has exceeded 180 million tons (calculated with a moisture content of 80%), and global food waste is expected to reach 1.3 billion tons per year by 2025 (Li et al., 2021b; Yao et al., 2018b). Unless treated properly, these organic wastes may lead to severe environmental pollution.

Obtaining biomethane from organic wastes by anaerobic digestion, which can reduce pollution, has been regarded as a promising method to achieve global sustainable development goals, especially those of mitigating the effects of climate change and achieving clean energy (Li et al., 2021a; Yao et al., 2018b). Theoretically, the sludge (180 million tons, calculated with a moisture content of 80% and volatile solid (VS)/total solid (TS) of 0.60) can produce 10.8 billion m^3 of CH_4 through methanogenic fermentation (Xu et al., 2024), generating approximately 108.56 billion kWh of electricity. Assuming an annual per capita consumption of 900 kWh, this would cover the electricity needs of 121 million people all year round. However, in practice, the biomethane from anaerobic digestion of waste activated sludge is usually less than half of the theoretical value. It is still a major challenge to obtain efficient methanogenesis of organic wastes.

5.2 THEORETICAL BASIS FOR BIOMETHANE GENERATION FROM ORGANIC WASTES

5.2.1 Basic Concept

In general, biomethane production from organic waste can be defined as a series of processes by which the methanogenic consortia break down organic substrates into methane under anaerobic conditions. There are three key factors in the process (Agrawal et al., 2023): methanogenic consortia (Ahn et al., 2014); organic substrate (Amin et al., 2021); and the interactions between methanogenic consortia and organic

DOI: 10.1201/9781003585398-5

substrate. Among these three key factors, methanogenic consortia typically depend on the characteristics of the organic substrate, and the microbial community and metabolic activity are usually determined by the interaction with the organic substrate. Unlike the conventional biomethane generation from wastewater, most of the organic substrates are insoluble and the interactions between the methanogenic consortia and solid-state substrates are weak in the biomethane generation from organic wastes, which limit the biomethane generation efficiency of organic wastes.

5.2.2 Stoichiometry and Prediction of Biomethane Production

In organic waste, C, H, O, and N are usually the main elements in organic components, and theoretically, the main organic matter in organic waste can be represented by the apparent molecular formula $C_nH_aO_bN_d$. Therefore, in the case where the main organic matter ($C_nH_aO_bN_d$) is fully biodegradable and would be fully converted by the anaerobic consortia into CH_4, CO_2, and NH_3, the theoretical production of biomethane can be determined by calculations based on the Buswell equation (Van Lier et al., 2008):

$$C_nH_aO_bN_d + \left(n - \frac{a}{4} - \frac{b}{2} + \frac{3d}{4}\right)H_2O \rightarrow \left(\frac{n}{2} + \frac{a}{8} - \frac{b}{4} - \frac{3d}{8}\right)CH_4 + \left(\frac{n}{2} - \frac{a}{8} + \frac{b}{4} + \frac{3d}{8}\right)CO_2 + dNH_3 \tag{5.1}$$

The contents of organic elements C, H, O, and N in a certain organic waste can be determined by experimental methods, and the chemical formula of this organic waste can be calculated based on the specific weight. According to this equation, the theoretical biomethane production can be calculated, assuming that the organic fraction of the organic waste was completely converted into biomethane. McCarty (1964) presented the chemical formula ($C_5H_7O_2N$) for waste activated sludge, and Parkin and Owen (1986) proposed the chemical formula ($C_{10}H_{19}O_3N$) for the organic fraction of primary sludge. Based on these chemical formulas, the theoretical methane production from primary sludge and waste activated sludge is 0.7 m^3 CH_4/kg VS and 0.5 m^3 CH_4/kg VS, respectively, at standard temperature and pressure (STP).

In addition, assuming complete mineralization of the organic substrate, the carbon oxidation state (C-ox. state) in organic substrates also can affect the proportion of theoretical methane in biogas as shown in Figure 5.1, which represents the theoretical composition of the biogas produced as a function of the average oxidation state of carbon in specific substrates.

5.2.3 Important Considerations in Biomethane Generation

The theory and technology of biomethane generation from organic wastes have a long history of development, spanning nearly a century. Research aimed at increasing the generation efficiency of biomethane has been focused on the following three aspects (Agrawal et al., 2023). Identification and physiological analysis of functional microbes: several important functional microbes, such as hydrolysis-acidification bacteria

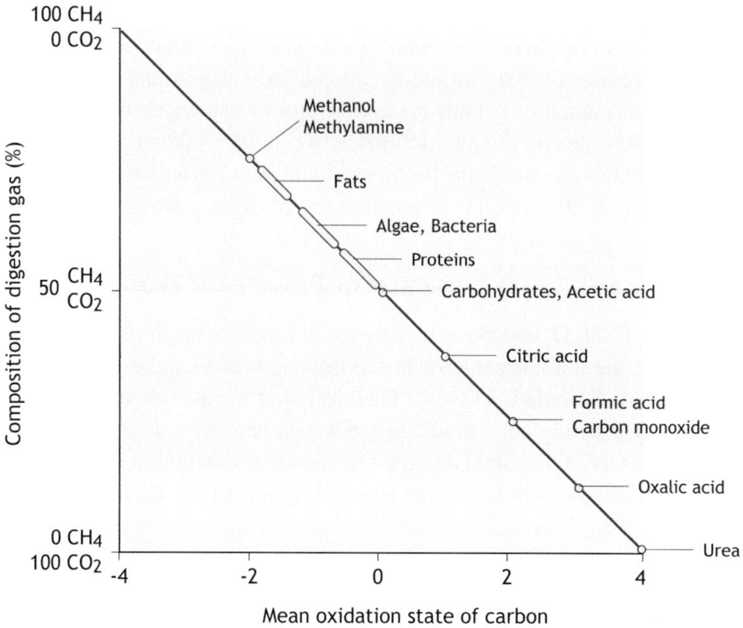

FIGURE 5.1 Theoretical composition of the biogas produced in relation to the mean oxidation state of the carbon in specific substrates (adapted from Van Lier et al., 2008).

(e.g., *Proteobacteria*, *Chloroflexi*, *Actinobacteriota*, *Bacteroidota*, and *Firmicutes*), methanogenic archaea (e.g., *Euryarchaeota*, *Halobacterota*, *Crenarchaeota*, and *Thermoplasmatota*), and microbes (e.g., *Geobacter*) with special metabolic pathways (such as direct interspecies electron transfer) have been identified (Werner et al., 2011; Xu et al., 2020a), and their corresponding physiological characteristics have been explored (Amaresan et al., 2020; Lyu and Whitman 2019; Werner et al., 2011). These studies have helped to establish a theoretical foundation for optimizing biomethane generation by regulating the main parameters (i.e., pH, temperature, and redox potential) and editing functional genes (Yao et al., 2018a). Notably, analyses of the physiological characteristics (e.g., metabolic functions and gene expression) of these microbes have been based primarily on pure culture systems, even though these characteristics may change owing to the interactions among microbes and those between microbes and inert substances in complex environments. Owing to this mismatch, it is challenging to realize the biomethane generation of complex organics solely by identifying the functional microbes (Ahn et al., 2014). Clarification and optimization of biomethane generation processes: in general, biomethane generation frameworks involve successive chemical and biochemical processes. To clarify the biomethane generation mechanics from a macro perspective and provide theoretical bases for identifying the rate-limiting step, the biomethane generation process typically has been artificially decomposed into multiple reaction stages according to the metabolites and main microbes involved in the reaction, such as three-reaction stages (hydrolysis, acidogenesis, and methanogenesis) and four-reaction stages (hydrolysis, acidogenesis, acetogenesis, and methanogenesis). Through the obtained knowledge, several measures for optimizing biomethane generation processes have been recommended, leading to the introduction of multistage biomethane generation, two-phase biomethane generation, high-solid biomethane generation, and co-digestion of different organic wastes (Xu et al., 2020a). Although these recommendations have promoted the engineering application of biomethane generation, the bioconversion efficiency of organic wastes remains to be enhanced. Moreover, the positive effect of these measures is limited by the type of substrate. For example, the multistage biomethane generation performance of food waste is superior to that of sewage sludge (Chatterjee and Mazumder 2019; Amin et al., 2021). Determination of substrate characteristics and their effects on biomethane generation performance: biomethane generation performance depends on the characteristics of complex substrates (Angelidaki and Sanders 2004; Campuzano and González-Martínez 2016). The dominant functional microbes, metabolic pathways, and rate-limiting steps in biomethane generation can be modified by regulating the substrate characteristics through pretreatment (Gonzalez et al., 2018; Xu et al., 2020a). Therefore, to promote biomethane generation, it may be of significance to clarify the relationship between the substrate characteristics and anaerobic bioconversion efficiency of organic matter into biomethane. However, the characteristics of different organic wastes vary, and the common substances that influence these characteristics remain to be identified.

5.3 THE MAIN CHARACTERISTICS OF REPRESENTATIVE ORGANIC WASTES

5.3.1 Sewage Sludge

Sewage sludge, which has complex nature with multi-component, multi-medium, and semi-rigid structure, is one representative of organic wastes. Notably, 30–50% of organic matter, 30–50% of nitrogen, and 95% of phosphorus in wastewater are transferred to sewage sludge. The characteristics of sewage sludge show a remarkable diversity, which is fundamentally due to the different nature of the influent wastewater received and is particularly influenced by the industrial wastewater discharged into the sewage system. Sewage sludge is classified into two main categories: primary sludge and waste activated sludge. Primary sludge, originating from the mechanical and preliminary treatments, is rich in organic matter that is easily degraded by microorganisms, and when treated by the fermentation process it shows a high biogas production potential and good dewatering properties. Waste activated sludge is significantly different from primary sludge, being a complex microbial community, Protozoa and parasites are mixed with organic debris from cell decomposition and a small number of inert materials (silt, clay, carbonates, silicates) (Canziani and Spinosa 2019). Waste activated sludge contains 99% water and the sludge age greatly affects its biodegradability (Xu et al., 2018c). In general, achieving highly efficient fermentation of waste activated sludge means breaking through the bottleneck of sewage sludge fermentation, mainly because waste activated sludge is more complex in structure, composition, and properties.

Table 5.1 summarizes the main characteristics of dewatered waste activated sludge in China. Typically, in China, the organic content in waste activated sludge fluctuates

TABLE 5.1
Main Physical and Chemical Indexes of Dewatered Waste Activated Sludge in China

Index	Number of Samples	Range	Average Value	Standard Deviation
Moisture content (%)	196	50.9–86.2	78.3	5.9
TS (%)	196	13.7–49.1	21.7	5.9
Sand/TS (%)	88	14–56.8	34.3	12.2
Sand/IS (%)	88	30.2–78.9	62.1	14.5
Sand particle size (D50, μm)	88	21.6–57.3	40.0	10.0
Sand particle size (D90, μm)	88	83.6–127	104.1	14.7
pH	196	6.6–8.2	7.1	0.2
Conductivity value (ms/cm)	8	0.62–1.07	0.77	0.17
VS/TS (%)	196	14.2–73.0	42.8	10.4
C (g/kg DS)	5	321.3–355.7	329.7	–
H (g/kg DS)	5	51.5–64.0	55.3	–
N (g/kg DS)	196	7.4–54.9	27.2	11.8
S (g/kg DS)	5	8.5–11.4	9.4	–
P (g/kg DS)	196	2.2–48.3	17.1	7.6
K (g/kg DS)	196	0.8–17.5	4.3	4.5
As (g/kg DS)	196	1.0–156.6	28.3	29
Cd (g/kg DS)	196	0–44.1	3.1	5.4
Cr (g/kg DS)	196	7.9–5,370.0	335.9	655.1
Cu (g/kg DS)	196	8.4–4,598.2	667.1	1,019.3
Hg (g/kg DS)	196	0.2–8.8	1.7	1.2
Ni (g/kg DS)	196	5.7–653.2	62.9	87.2
Pb (g/kg DS)	196	9–4,660	143	429.4
Zn (g/kg DS)	196	37.6–27,300	1,367.3	3,217.4
Antibiotics (μg/kg DS)	25	0.83–38,700	8,390	–
Apeo (μg/kg DS)	14	0–33,810,000	887,000	–

TABLE 5.1 (*Continued*)
Main Physical and Chemical Indexes of Dewatered Waste Activated Sludge in China

Index	Number of Samples	Range	Average Value	Standard Deviation
Bisphenol A substances (μg/kg DS)	11	34.6–127,000	10,500	–
Hormone (μg/kg DS)	12	0–981	178	–
Organochlorine insecticide (μg/kg DS)	7	9.0–3,200	327	–
Perfluorinated compounds (μg/kg DS)	12	0–9,980	796	–
Phthalatic esters (μg/kg DS)	9	680–282,000	48,400	–
Polybrominated diphenyl esters (μg/kg DS)	15	3.46–7,100	1,020	–
Polychlorinated biphenyl (μg/kg DS)	10	3.14–1,400	81	–
Polycyclic aromatic hydrocarbon (μg/kg DS)	24	100–170,000	15,900	–
Microplastic pollutants ($\times 10^3$/kg DS)	79	1.60–56.4	22.7	12.1

Source: Adapted from Dai X.H. (2022).
DS: Dry solid

between 23% and 73%, far lower than in developed countries such as the United Kingdom, Germany, and the United States. Moreover, the inorganic components are mainly composed of SiO_2 and multivalent metals, among which the average particle size of inorganic particles is below 60 μm. These characteristics make the waste activated sludge a more stable semi-rigid structure and make it more difficult for the organic matter in the waste activated sludge to be bio-converted into methane. In order to break through the bottleneck of fermentation of waste activated sludge, based on the semi-rigid structure of sludge, Xu et al. (2017a, 2017b, 2018a, 2018b, 2020a) and Dai et al. (2017a, 2017b) re-identified the key structural components and key organic matter of waste activated sludge, investigated the interactions between key structural components, and explored the impact of this interaction on sludge fermentation, thereby revealing the mechanism of how the key structural components in waste activated sludge change the spatial conformation of easily degradable organic matter in key organic matter through complexation, adsorption, cross-linking, and hydration based on electronic and steric effects, resulting in poor biodegradability of sludge organic matter.

5.3.2 Food Waste

Food waste is another representative of organic wastes. Large amounts of food waste are generated every year. Usually, food waste can be considered as a general term for both catering waste and kitchen waste. Catering waste refers to the leftover food from restaurants, unit canteens, and the processing waste of fruits, vegetables, meat, noodles, etc. from the back kitchen, which contains oil and organic solid-liquid mixtures. Kitchen waste refers to easily perishable organic waste discarded in daily life, such as leftovers and fruit peels. In China, catering waste is the main component of food waste, and its main characteristics can be summarized into two aspects:

1. The physicochemical characteristics: Compared to some other countries, such as the UK and South Korea, food waste in China is high in water (with a moisture content of approximately 75%–85%), salt, and oil and often contains a complex mix of paper, plastic, metal, and glass (Giwa et al., 2019; Negri et al., 2020). The higher moisture content increases the difficulty of garbage removal and leads to a lower calorific value of food waste (2000–3000 kcal/kg). The high content of organic compounds (the ratio of the VS to TS is 85–95%), especially sugars and proteins, leads to the extreme decay and deterioration of food waste. Due to factors such as diet, the oil content is relatively high. The pH value with a mean value of 5.3 is mostly acidic and the nutrients are relatively rich, which is prone to decay and acidification. With high salt content, direct landfill treatment of food waste poses a risk of soil salinization and alkalization (Jin et al., 2021).
2. Space-time differences: The chemical composition of different types of food is different, so the biochemical properties of food waste vary according to the changes in meals throughout the day and different eating habits in different regions. Usually, the food waste generated from lunch and dinner has a higher oil content compared to that generated from breakfast. In addition, comparing the food waste from different regions such as developed countries in Europe with that in China, it is found that there is no significant difference in the polysaccharide and protein content of the food waste in China, but the oil content is generally higher than that in European countries. The food waste is usually rich in natural fibers, carbons, proteins, fats, lipids, vitamins, and trace minerals that are embedded in a biodegradable organic matrix. Figure 5.2 shows the chemical compositions of several foods in food waste.

5.4 BIOMETHANE GENERATION FROM REPRESENTATIVE ORGANIC WASTES

Usually, theoretical biomethane production is affected by two aspects, one is the abiotic factors such as the characteristics of the substrate and the interactions between the substrate and methanogenic consortia. Table 5.2 shows the theoretical biomethane production of different simple organic matters (Angelidaki and Sanders 2004),

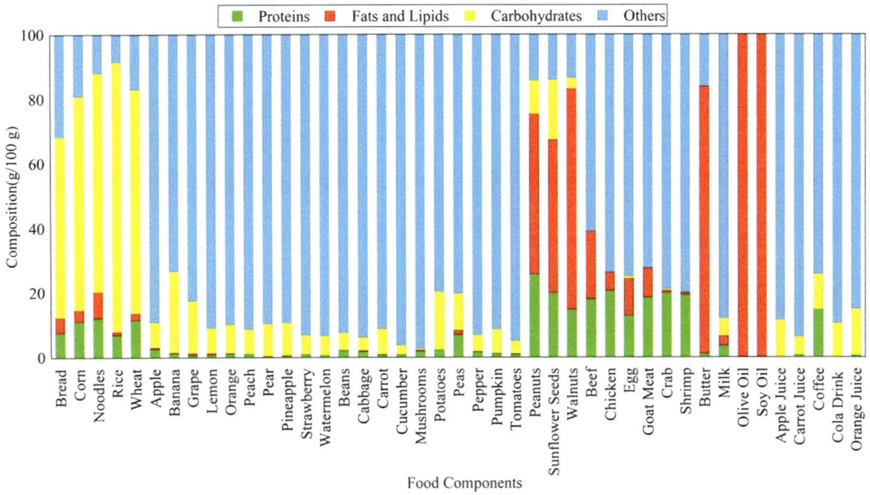

FIGURE 5.2 Chemical composition of individual food components (adapted from An et al., 2024).

TABLE 5.2
Theoretical Characteristics of Typical Substrate Components

Substrate type	**Composition**	**COD/VS (g/g)**	**CH_4 yield (L CH_4/g VS)**	**CH_4 (%)**
Carbohydrate	$(C_6H_{10}O_5)_n$	1.19	0.35	50
Protein[a]	$C_5H_7NO_2$	1.42	0.35	50
Lipids	$C_{57}H_{104}O_6$	2.90	0.35	70
Ethanol	C_2H_6O	2.09	0.35	75
Acetate	$C_2H_4O_2$	1.07	0.35	50
Propionate	$C_3H_6O_2$	1.51	0.35	58

Source: Adapted from Angelidaki and Sanders (2004).
[a] Nitrogen is converted to NH_3

and Figure 5.3 illustrates the proposed key abiotic interactions in our previous study (Xu and Dai 2021). We proposed that in the methanogenic fermentation process of organic wastes, organic wastes itself is a solid-liquid mixture with a large number of solid-liquid interfaces. At the same time, the solid-state organics in organic wastes interact with microbial cells (solid-state), forming a "solid-liquid-solid" mode methanogenic fermentation system. Unlike in the methanogenesis of organic wastewater, in which microbial cells can come into direct contact with the dissolved substrate, in this system, effective contact between the functional microbial cells and the solid-state organics in organic wastes is a prerequisite for achieving highly efficient

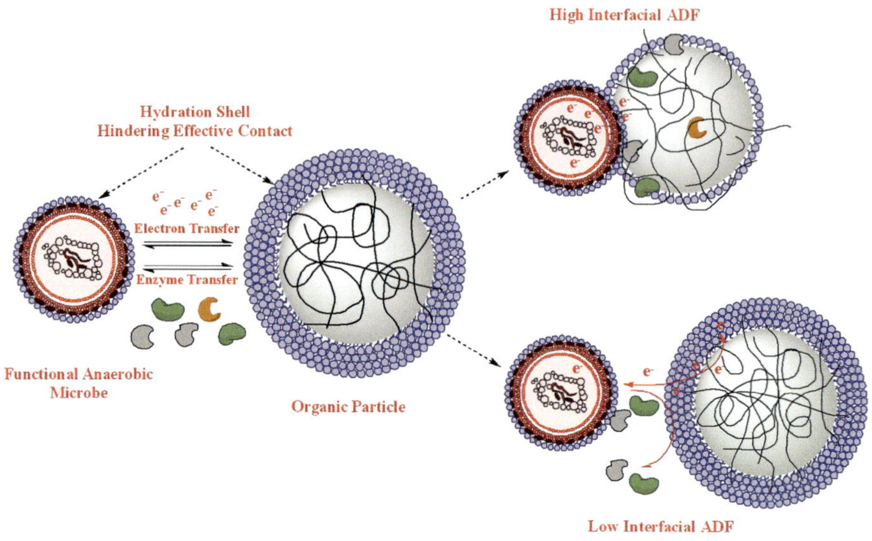

FIGURE 5.3 Diagrammatic sketch of the basic anaerobic biological treatment for organic waste with high and low interfacial abiotic driving forces (ADF: abiotic driving force) (adapted from Xu and Dai 2021).

methanogenesis of organic wastes (Xu et al., 2024). Based on the polar molecules and induced dipole interactions and the theory of non-covalent interactions at the solid-liquid interface, the interfacial abiotic forces, containing the Lifshitz–van der Waals, Lewis acid–base, and electrostatic forces, are proposed to drive the effective contact in the "solid-liquid-solid" mode fermentation system, as shown in Figure 5.3. In fact, organic wastes are composed of various simple organic matters with varying levels of content, which can significantly affect biomethane production. Another one is the biotic factor that can directly affect the activity of methanogenic consortia and key enzymes.

5.4.1 Common Biotic Factors Affecting Biomethane Production

5.4.1.1 Temperature

Temperature is one of the most important parameters affecting biomethane generation, as it not only affects microbial growth and metabolic rate but also affects the biomethane production and the proportion of biomethane in biogas (Appels et al., 2008). Acetotrophic methanogens are one of the most sensitive groups to temperature rise. When the temperature exceeds 70 °C, the degradation of propionic acid and butyric acid is greatly affected. Compared with high-temperature fermentation (55–70 °C), moderate temperature fermentation (37 °C) has lower energy consumption and is less susceptible to impact loads. Kim et al.'s (2017) research has shown that moderate temperature fermentation has a stronger ability to treat high organic load rates. High-temperature fermentation has a higher metabolic rate, higher specific growth rate, and higher pathogen damage rate, while also increasing biogas production (Chew et al., 2021; Zhang et al., 2014). However, excessively high temperatures can affect the fermentation process, leading to a reduction in stability, low quality of effluent, increased toxicity, and sensitivity to environmental impacts. In addition, severe and frequent temperature fluctuations can also have adverse effects on biomethane production, so it is necessary to maintain relatively stable temperature during fermentation.

5.4.1.2 pH

pH plays an important role in biomethane generation. The growth of different methanogenic consortia requires different pH ranges. The optimal pH range for bacteria is 4.0–8.5, and the optimal working range for methanogens is generally 6.5–7.2 (Zhang et al., 2014). If the pH value is below 6.3, it will promote the production of a large amount of organic acids and H_2S and inhibit the growth of methanogens. When pH is above 7.8, it will promote the generation of free ammonia, leading to ammonia inhibition. During two-phase fermentation process, a specific acidogenic microbial population can be selected by adjusting the pH value in the acidification reactor, so as to control the acidogenic pathway and subsequent methanogenesis. There are reports indicating that pH significantly affects the compositions of VFA. Horiuchi et al. (2002) found that the main products change from butyric acid to acetic acid and propionic acid when the pH changes from 5 to 8; this conversion is reversible under acidic and alkaline conditions and is not affected by the dilution rate.

5.4.1.3 Volatile Fatty Acids

Volatile fatty acids (VFAs) are one of the most important intermediate products during fermentation, which are small molecule acids with less than six carbon atoms, mainly including acetic acid, propionic acid, butyric acid, and valeric acid. The accumulation of VFAs can inhibit the activity of methanogens, thereby limiting biomethane generation. During the fermentation process, ammonia nitrogen and VFAs can form a weak buffering system. When the concentration range of both is suitable, more organic acids can be neutralized by higher concentrations of ammonia nitrogen, and the organic load of the fermentation system can be further increased. The overall process is also more stable. Nielsen and Ahring (2007) showed that free ammonia derived from the conversion of N-containing organics combined with VFAs produced by hydrolysis-acidification bacteria established a weak buffering system, thereby enhancing the tolerance of intermittent fermentation systems to ammonium and VFAs. while the establishment of early warning indicators, such as the ratio of propionic acid to acetic acid, the ratio of bicarbonate alkalinity to total alkalinity, and the ratio of VFAs to bicarbonate alkalinity, can be effective in providing timely predictions of potential risks in the system (Ren et al., 2018).

5.4.1.4 Carbon-Nitrogen Ratio

The fermentation performance is significantly affected by the carbon-nitrogen ratio, and the optimal carbon-nitrogen ratio is beneficial for the growth of methanogenic consortia and maintaining a stable environment. An excessive carbon-nitrogen ratio leads to insufficient nitrogen sources for maintaining cell biomass, resulting

chemical, biological pretreatments) before fermentation and the exogenous additives (zero-valent iron, nanomaterials, conductive materials) into bioreactor (Xu et al., 2020a).

5.4.2.1.1 *Sludge Pretreatment*

The sludge pretreatment can convert complex solid-state organics into low molecular weight and easily biodegradable organics. It mainly includes physical pretreatment, chemical pretreatment, and biological pretreatment. However, it is worth noting that these pretreatments mainly aim to increase the initial organic solubilization and hydrolysis. New isoelectric pretreatment methods that do not aim to increase the initial solubility of organic matter or hydrolysis have been widely published in recent years (Xu et al., 2021, 2022, Xu et al., 2018b, Xu et al., 2020b). This pretreatment precisely destroys the key structural component in sludge to disrupt the complex sludge structure and increase the spatial extensibility of sludge organic molecules, enhancing the bioavailability and biodegradability of organic molecules in sludge, thereby improving the efficiency of sludge organic matter in subsequent methanogenesis fermentation without high initial organic solubilization and hydrolysis.

5.4.2.1.1.1 Physical Pretreatment Physical pretreatment mainly includes thermal pretreatment, ultrasonic pretreatment, microwave pretreatment, etc. Thermal pretreatment for sludge can decompose cell membranes and increase organic solubilization. At the same time, thermal pretreatment is beneficial for killing bacteria and pathogens, reducing sludge volume, and enhancing sludge dewatering. Thermal pretreatment can be mainly divided into low-temperature pretreatment (<100 °C) and high-temperature pretreatment (>100 °C). Ferrer et al. (2008) studied that pretreatment at 70 °C for 9 hours before high-temperature fermentation increased biogas production by 57.5%. However, after 72 hours of pretreatment, biogas production was lower, which may be related to the initial inhibition caused by VFA accumulation or the biodegradation of some soluble compounds during the thermal pretreatment process. Bougrier et al. (2006) found that before mesophilic fermentation, treating at 170 °C for 30 minutes increased biogas production by 59%. Although high-temperature pretreatment can increase the solubility of organic matter in sludge, it can also lead to sludge agglomeration, particle size increase, and Maillard reaction above 180 °C, forming difficult to degrade organics (Xu et al., 2020a). Both ultrasonic pretreatment and microwave pretreatment have thermal and non-thermal effects on sludge. The thermal effect is caused by the interaction between the radiated electric field and dipole molecules, resulting in molecular rotation and friction, leading to an increase in temperature. The non-thermal effect is associated with rapid changes in the dipole orientation of the macromolecules that make up the cell membrane, which inhibits intermolecular hydrogen bonds and thus disrupts the structure of the cell membrane. Ultrasonic pretreatment can produce bubbles and high shear forces, leading to the release of intracellular organic matter, thereby achieving solubilization of sludge organics. Specific energy is a parameter that has a significant impact on sludge during ultrasonic pretreatment. Bougrier et al. (2006) found that ultrasonic pretreatment at a specific energy of 9350 kJ/kg before mesophilic fermentation reduced the solubility and particle size of sludge, improved the biodegradability of particle components, and increased biomethane production by 53%.

5.4.2.1.1.2 Chemical Pretreatment Chemical pretreatment mainly includes acid and alkali pretreatment, ozone pretreatment, and Fenton pretreatment. Acid and alkaline pretreatment mainly improves the fermentation efficiency of sludge by destroying sludge flocs, collapsing the structure of extracellular polymeric substances, and accelerating the hydrolysis of sludge organic matter. The type and dosage are key parameters affecting their effectiveness. Maryam et al. (2021) found that after pretreatment with 4 g NaOH/g TS, biomethane production increased by 50.4% after mesophilic fermentation. During this process, NaOH effectively accelerated the swelling of particulate organic matter. Although high doses usually lead to a large amount of organic solubilization in sludge, excessive may even reduce the activity of methanogenic consortia. The amount of acid and alkali added should be kept within a reasonable range to improve the methanogenesis fermentation efficiency of sludge. Ozone, as a strong oxidant, can quickly react with the bacterial cell, converting polysaccharides, proteins, and lipid membranes in bacterial cells into simpler molecules, improving biodegradation rates, and increasing biogas production during fermentation. The impact of ozone pretreatment on methanogenesis fermentation efficiency mainly depends on the dosage of ozone. Le et al. (2019) found that after pretreatment with a dosage of 0.02 g O_3/g TS, the biomethane enhancement rate was 53%, and the COD and VS removal rates were significantly increased by 64.31% and 61.06%, respectively. When the ozone dose increases to 0.15 g O_3/g TS, the organic matter in the sludge will mineralize, leading to low methanogenesis fermentation efficiency. The ferrous ions in the Fenton reagent can catalyze the decomposition of H_2O_2, producing highly active hydroxyl radicals that oxidize difficult to degrade compounds in sludge and convert them into biodegradable compounds. The highly active hydroxyl radicals can also decompose extracellular polymeric substances and organics with supramolecular structure. Erden and Filibeli (2010) found that Fenton pretreatment before high-temperature fermentation resulted in a higher solid reduction and biomethane production, with a 30% increase in biomethane production. They also observed a decrease in sludge resistance to dehydration after Fenton pretreatment.

5.4.2.1.1.3 Biological Pretreatment Biological pretreatment is an environmentally friendly technology that uses mainly aerobic and enzymatic methods to promote and improve the hydrolysis of sludge organics. Aerobic can be carried out by treating sludge with air and aerobic or facultative bacteria before fermentation. Micro-aeration technology mainly injects oxygen into the pretreatment system, which is beneficial for accelerating the hydrolysis of complex organics by improving the hydrolysis activity of endogenous microorganisms. Ahn et al. (2014) achieved a 20% increase in biogas production under the conditions of an aeration rate of 0.05 vvm and pre-aeration time of 24 hours. Adding hydrolytic enzymes to the pretreatment system can hydrolyze microbial cell wall, improve sludge organic solubilization, and at the same time, these enzymes can decompose difficult to biodegrade macromolecular organics into small molecule substances, increasing biogas production in the subsequent methanogenesis fermentation. Odnell et al. (2016) found that the addition of *Bacillus subtilis* protease increased biomethane production by 37%. The addition of enzymes does not require any additional energy input or changes in substrate conditions, but this ability is only effective if the added enzymes have high activity and a sufficiently long lifespan.

5.4.3 Exogenous Additives

Different types of additives increase the biogas production of sludge by promoting microbial growth and metabolism, mainly through nutritional supplements and enhancement of electron transfer. This is generally because the extra nutrients increase buffering capacity and microbial activity, and electron transfer is enhanced, thereby accelerating the rate of biochemical reactions during methanogenic fermentation (Xu et al., 2020a).

5.4.3.1 Zero-Valent Iron

The addition of zero-valent iron (ZVI) reduces the redox potential of the fermentation process, which favors the growth of anaerobic organisms. ZVI can enhance the activity of many important enzymes in the acidification process and increase biomethane generation (Liu et al., 2012). Feng et al. (2014) found that the addition of ZVI effectively enhanced the degradation of proteins and cellulose from sewage sludge, resulting in a 43.5% increase in biogas production. At the same time, fluorescence in situ hybridization analysis showed that the microbial abundance of homotypic acetic acid producing bacteria and hydrogenotrophic methane producing bacteria was higher than the control, which decreased hydrogen partial pressure and created favorable thermodynamic conditions for methanogenesis fermentation of sludge.

5.4.3.2 Nanomaterials

Compared with ZVI, nano-ZVI can more thoroughly and quickly remove a wider range of pollutants, which is conducive to the complete degradation of organic matter and shortens reaction time. Nanomaterials are currently the most important additives for enhancing methanogenesis fermentation performance. Nanomaterials are able to facilitate the process of extracellular electron transfer more strongly due to their significantly increased specific surface area and good electron conductivity, thus effectively enhancing the efficiency of biomethane generation. Nano-Al_2O_3, carbon-based nanomaterials, nano-ZnO, and nano-CuO are common nano additives. Hassanpourmoghadam et al. (2023) found that 120 mg/L of nano-Fe_3O_4 increased biogas production by 70%, and they revealed that nano-Fe_3O_4 can reduce the inhibitory effect of sulfides on biomethane production and increase solid hydrolysis rate or interspecies electron transfer.

5.4.3.3 Conductive Materials

The addition of conductive materials (CMs), including carbon-based CMs and metal-based CMs, can serve as electronic pipelines to facilitate direct interspecies electron transfer (DIET) between anaerobic microorganisms, thereby improving methanogenesis fermentation efficiency. CMs not only act as a conductive medium but also enrich electroactive microorganisms, promote the establishment of the electrotrophic microbial community structure, and greatly improve the synthesis and metabolism of organic matter and biomethane production (De Vrieze et al., 2016). Both carbon-based and metal-based CMs can enhance electron transfer between synthetic microbial communities; however, they show differences in the degree of enhancement and have different application limitations. To enhance the interspecies electron transfer and overcome the limitations, a combination of metal-based CMs and carbon-based CMs in the same fermentation system has been widely studied (Liu et al., 2021) and using multiple CMs together has proven to be a promising strategy for applying CMs to methanogenic fermentation.

According to the characteristics of complex solid-state organics fermentation systems, it can be proposed that proton transfer may become a rate-limiting step for enhancing biomethane production through enhanced electron transfer in complex multiphase fermentation systems. Liu et al. (2022a) utilized metal–organic framework (MOF) materials with high proton transfer ability to enhance the methanogenesis fermentation of sludge and improve the yield of biomethane. Focusing on enhancing proton-coupled electron transfer through exogenous materials may be an important research direction for breaking through the bottleneck of methanogenesis fermentation of complex substrate in multiphase environments in the future.

5.4.3.4 Food Waste

Food waste has a high potential for generating renewable energy during fermentation due to its high biodegradability and rapid hydrolysis, making it suitable for the growth of methanogenic consortia. Although food waste has high biomethane production potential, the organic matter in food waste is rapidly hydrolyzed and acidified, leading to the accumulation of VFAs which inhibit methane production and disrupt the fermentation system.

5.4.3.4.1 Pretreatment

Similar to sludge pretreatment, there are mainly physical, chemical, and biological pretreatment methods for food waste. Physical pretreatment mainly includes mechanical pretreatment and thermal pretreatment. The main purpose of mechanical pretreatment is to decompose or grind solid particles in the food waste and change the organic particle size. Small particle sizes can increase specific surface area and generate more contact with the substrate. Izumi et al. (2010) studied the effect of pretreatment by grinding on methanogenesis fermentation of food waste: reducing particle size from 0.843 to 0.391 mm increased biomethane production by 28%. However, excessively small particle sizes can lead to excessive accumulation of VFAs, thereby reducing biomethane yield. Ariunbaatar et al. (2015) studied the effect of thermal pretreatment on methanogenesis fermentation of food waste and found that low-temperature thermal pretreatment (50 °C or 80 °C) can significantly increase biomethane production, with an increase of 44–46%. Chemical pretreatment mainly includes acid pretreatment, alkali pretreatment, and advanced oxidation pretreatment. Strong acids and strong alkalis can dissolve lipid substances, enhance the bioavailability of organic matter in food waste, and have a positive impact on lignin-rich food waste, mainly by destroying macromolecular structures to promote hydrolysis. However, it is important to note that the appropriate concentration of acid or alkali should be selected during acid or alkali pretreatment. Excessive concentrations may inhibit the activity of methanogens and thus decrease biomethane production. At the same time, excessive acid concentration may cause the production of inhibitors such as phenolic compounds, furans, and carboxylic acids, which may hinder the methanogenesis fermentation process (Zhang et al., 2014). In addition, the free radicals generated by ozone pretreatment can directly or indirectly degrade macromolecular organic matter such as lipids and proteins, which is beneficial for improving the biodegradation rate of food waste. Compared with other chemical pretreatment methods, ozone pretreatment has the advantages of not increasing salt concentration, not having residual oxidants, not producing toxic substances, and eliminating pathogenic bacteria. However, ozone pretreatment may lead to the production of refractory substances. This pretreatment method has high operating costs, and the produced biomethane is not sufficient to compensate for the required energy consumption (Ariunbaatar et al., 2014). Biological pretreatment mainly includes aerobic pretreatment, anaerobic pretreatment, and enzymatic pretreatment. Aerobic pretreatment mainly improves the decomposition of organics in food waste through micro-aeration technology. Anaerobic pretreatment mainly pre-ferments food waste before methanogenesis fermentation, which can improve the degradation degree of food waste; enzymatic pretreatment mainly achieves the efficient hydrolysis of macromolecular organic matter in food waste by adding specific enzymes, such as proteolytic enzymes, peptidases, carbohydrate enzymes, and lipases. Enzymatic pretreatment has the advantages of simple operation and no secondary pollution, but its high cost limits its wide application (Ariunbaatar et al., 2014; Ren et al., 2018).

5.4.3.4.2 Co-fermentation

When food waste is co-fermented with organic wastes such as sludge, animal manure, straw, etc. it leads to increased stability of the fermentation system. This is mainly because the reasonable combination of different organic wastes can enhance the buffering capacity of the system and improve the balance of nutrients. At the same time, it can also dilute the content of toxic substances, thereby improving process stability and increasing biogas production. El-Mashad and Zhang (2010) investigated the co-fermentation of dairy manure (52%) and food waste (48%) under moderate temperature conditions. After 20 days of fermentation, biomethane production increased by 14.4%. They also proposed that this co-fermentation balanced the nutrients in the bioreactor, providing a more stable environment for methanogenic consortia, and found that at a 1:2 ratio between vegetable waste and food waste, biogas production increased by 30%. The co-fermentation of vegetable waste and food waste provides buffering capacity, enhances process stability, balances the availability of nutrients, and provides favorable environmental conditions for the survival of methanogenic consortia (Agrawal et al., 2023).

5.4.3.4.3 Two-Stage Fermentation

During food waste fermentation for methane production, the hydrolytic acidifying bacteria are able to rapidly convert organic matter in food waste to VFAs, but methanogenic bacteria are unable to further convert these VFAs to biomethane at the same rate. To achieve the synergistic effect of hydrolysis-acidification bacteria and methanogens on the metabolic efficiencies toward food waste, a two-stage fermentation strategy is developed for food waste, where hydrolysis-acidogenesis and methanogenesis are carried out in separate reactors (An et al., 2024; Chatterjee and Mazumder 2019). Figure 5.4 illustrates the main steps in the two-stage fermentation of food waste.

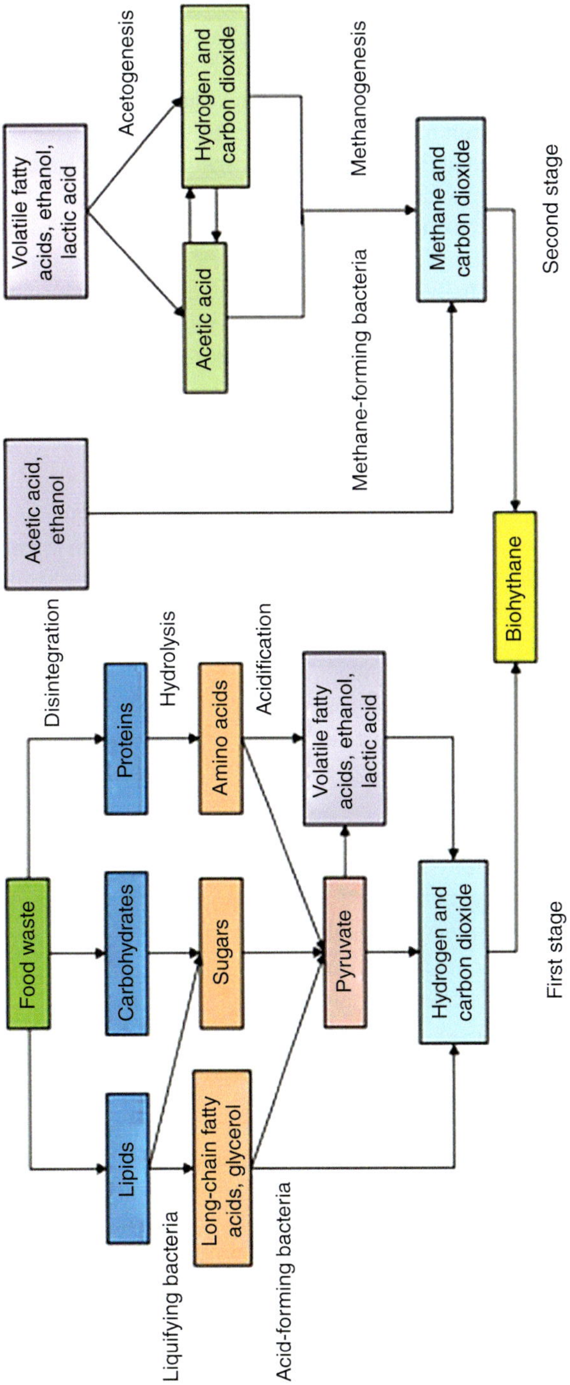

FIGURE 5.4 Schematic of biohythane production from food waste through two-stage fermentation (adapted from An et al., 2024).

In the first stage, the complex organic matter in food waste is first hydrolyzed to low molecular soluble organic matter. The acidogenic bacteria then utilize the hydrolyzed organic matter to produce VFAs such as formic, acetic, propionic, butyric, and valeric acids. The VFAs and other intermediate products are converted by the acetogenic bacteria into acetic acid, ammonia, H_2, and CO_2. In the second stage, the products of the first step are converted to CH_4 and CO_2 by methanogenic bacteria. In general, hydrogenotrophic and acetoclastic methanogens account for about 30% and 70% of biomethanogenesis, respectively (Jiang et al., 2019). Methanol can be also converted to CH_4 and H_2O during methylotrophic methanogenesis. The hydrolysis-acidification process of food waste usually occurs very rapidly, and methanogenic bacteria are very sensitive to the accumulation of VFAs in the previous step. Therefore, methanogenesis can be a rate step in the food waste fermentation process.

5.5 MAIN METHANOGENIC CONSORTIA AND METABOLIC PATHWAYS

Methanogenic fermentation is a critical microbial process where interactive microbial populations transform various complex organic compounds into methane through processes such as hydrolysis, acidogenesis, acetogenesis, and methanogenesis. The anaerobic consortia mainly include hydrolysis-acidification bacteria, methanogenic archaea, and their intricate interspecies symbionts, all of which are central components of methanogenic fermentation metabolism. At the phylum level, bacterial communities in the fermentation process mainly consist of *Firmicutes*, *Proteobacteria*, *Bacteroidetes*, *Chloroflexi*, *Synergistetes*, and *Actinobacteria*, while the archaeal communities mainly contain *Euryarchaeota* (Liu and Whitman 2008).

5.5.1 Main Methanogenic Consortia

5.5.1.1 Main Microorganisms Related to Hydrolysis

There are five typical hydrolyzing bacterial phyla such as *Firmicutes*, *Proteobacteria*, *Bacteroidetes*, *Chloroflexi*, *and Actinobacteria*, which secrete extracellular enzymes such as glucosidases, proteases, lipases, and cellulases (Lu et al., 2023). The fungal groups (e.g., *Mucor*, *Aspergillus*, *Sugiyamaella*) also exhibit hydrolytic activity in the macromolecule hydrolysis. Fungi can produce a variety of lignocellulose-degrading enzymes, which are crucial for biological degradation of lignocellulose, and their mycelium can penetrate cellulose to form pores, increasing the surface area available for enzymatic attack and thus enhancing the hydrophilicity of cellulose. Moreover, fungal populations can produce substantial amounts of H_2 and CO_2, which benefit hydrogenotrophic methanogenic metabolism with high ammonia tolerance, thus improving the buffering capacity of fermentation (Xing et al., 2020). The hydrolysis of organic matter typically involves multiple microorganisms. Depending on the specificity of the secreted hydrolyzers, bacteria with different functions undergo different biodegradation steps to form different intermediates. Therefore, community complexity, functional redundancy, and microbial physiological plasticity significantly determine the hydrolytic capability of microbes during fermentation periods (Ma et al., 2019; Zhang et al., 2024).

5.5.1.2 Main Acidogenic Bacteria

Acidogenic bacteria grow at rates 30–40 times faster than methanogens, which are usually affected by fluctuations in operating conditions including temperature and pH. Acidogenic bacteria are mainly distributed among *Bacteroidetes*, *Firmicutes*, *Proteobacteria*, *Chloroflexi*, *Actinobacteria*, *Acidobacteria*, and *Synergistetes*. Many genera under the acidogenic bacterial phyla (e.g., *Bifidobacterium*, *Lactobacillus*, *Clostridium*, etc.) also possess hydrolytic functions, which can independently promote the degradation of macromolecules (Amin et al., 2021). Acidogenic bacteria are widely present in well-functioning digesters, where they can ferment hydrolysis products into VFAs. Notably, the significant proliferation of acid-producing bacteria reduces the pH in the reactor and disrupts the interactions between bacteria and archaea that sustain methane production.

Pyruvate plays a crucial role in the acid-producing metabolic pathway which can be converted into a variety of products, such as acetic acid, propionic acid, butyric acid, ethanol, propanol, butanol, H_2, and CO_2. Acidogenic metabolic pathways are generally categorized into acetic-acetic alcohol fermentation, propionic fermentation, butyric fermentation, mixed acid fermentation, lactic acid fermentation, and pyruvate fermentation based on the type of major liquid products (Zhou et al., 2018). Specific metabolic processes are illustrated in Figure 5.5. Acetic, propionic, and butyric acids are the main components of VFAs. Increased methane productivity is associated with increased production of VFAs, which provide sufficient feedstock (acetate and H_2/CO_2) for methanogenic bacteria. Accumulation of VFAs may lower pH and affect the activity of methanogenic bacteria, thereby greatly inhibiting the methanogenic process. Therefore, increasing the production of VFAs does not imply

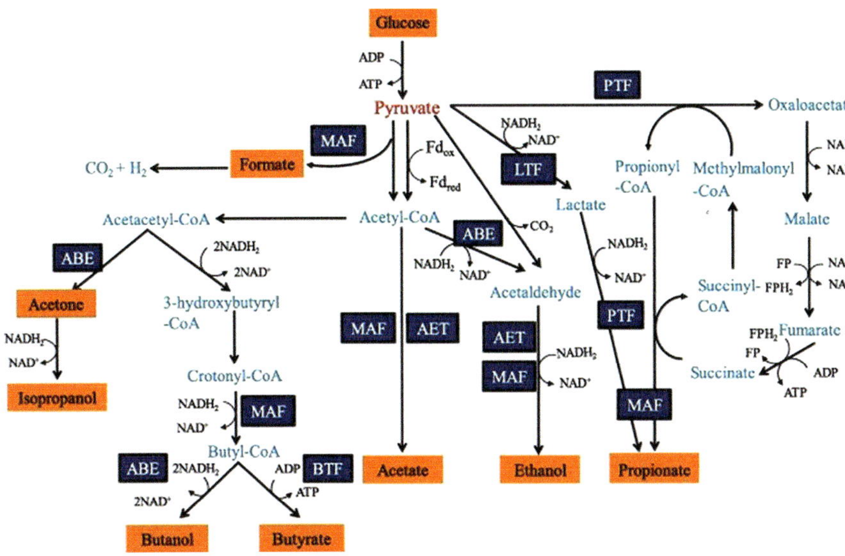

FIGURE 5.5 The main acid-producing metabolic pathways (adapted from Zhou et al., 2018).

an increase in methane production. Acidogenic bacteria greatly influence the concentration and distribution of VFAs, and clarifying the function of acidogenic bacteria is important to optimize the acid production process for efficient product production.

5.5.1.3 Main Acetogenic Bacteria

Homotypic acetate production and reciprocal decarboxylation metabolism (i.e., reciprocal oxidation of acetate, butyrate, and propionate) during acetate production are major factors in the production of methanogenic substrates. Homoacetogenic bacteria and syntrophic acetate-oxidizing bacteria are two representatives of acetogenic bacteria, and Figure 5.6 shows the main metabolic pathways of homoacetogenic fermentation and syntrophic acetate oxidation.

5.5.1.3.1 Homoacetogenic Bacteria

Homoacetogenic bacteria are a group of strictly anaerobic bacteria capable of synthesizing acetate from C1 compounds (e.g., CO_2, CO, and some other organic substances) under anaerobic conditions. Homoacetogenic bacteria include *Acetobacter vinelandii*, *Acetobacter xylosus*, *Clostridium acetobacter*, *Clostridium thermoautotrophicum*, and *Thiobacillus riftani* (Amin et al., 2021).

Homoacetogenic bacteria use H_2 and CO_2 to produce acetic acid via the acetyl coenzyme A pathway. Homoacetogenic bacteria can utilize both inorganic carbon sources (e.g., CO and CO_2) for autotrophic growth and organic matter (e.g., alcohols, sugars, and organic acids) as energy and carbon sources for heterotrophic growth. Conditions such as acidic pH, alkaline pH, low temperatures, and high H_2 concentrations may stimulate homoacetogenic fermentation pathway and contribute to the tendency for microbiomes to be dominated by homoacetogenic bacteria (Zhang et al., 2024). In methanogenic fermentation systems, methanogens and homoacetogens are the primary consumers of H_2, with a competitive relationship where methanogens usually prevail due to their energetic advantage. However, some studies have reported that homoacetogenic bacteria outcompete methanogenic bacteria at low temperatures, which may be due to the fact that homoacetogenic bacteria grow faster than methanogenic bacteria at the same low temperatures (Liu and Conrad 2011).

5.5.1.3.2 Syntrophic Acetate-Oxidizing Bacteria (SAOB)

Improvement of the syntrophic acetate oxidation (SAO) process is important for enhancing biomethane production efficiency and restoring dysfunctional fermentation systems. The SAO process is catalyzed by functional bacteria. SAOB mainly include *Syntrophaceticus schinkii*, *Syntrophomonas wolfei*, *Syntrophus aciditrophicus*, *Syntrophobacter wolinii*, and *Syntrophorhabdus aromaticivoran*. Since very little energy is released during the SAO process, this energy is not sufficient to support the growth of SAOB alone, so they must establish a close energy exchange relationship with the hydrogenotrophic methanogens in what is known as an adaptive rigid symbiotic relationship. This energetic disadvantage in the symbiotic relationship explains the slow growth of SAOB as well as their high dependence on their symbiotic bacteria, which also well explains the difficulty of isolating and stably cultivating SAO isolates, and also this leads to a weaker competition of SAOB with

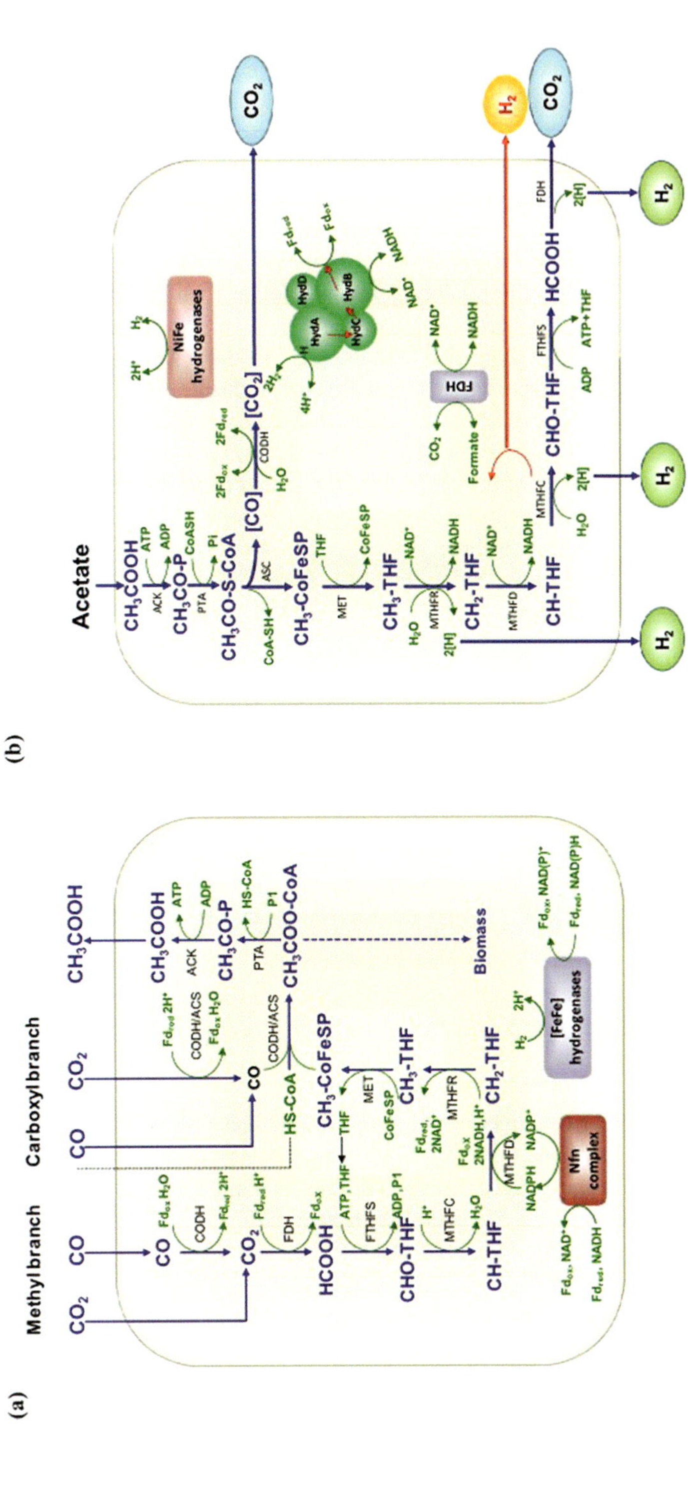

FIGURE 5.6 The main metabolic pathways of homoacetogenic fermentation and syntrophic acetate oxidation: (a) Homoacetogenic fermentation pathway; (b) biochemical pathways of syntrophic acetate oxidation (adapted from Zhang et al., 2024).

acetate-nutrient methanogenic bacteria in normal anaerobic systems (Liu et al., 2022b). Since SAO is thermodynamically and energetically feasible only under low hydrogen partial pressures (10–80 Pa), it needs to be combined with hydrogenotrophic methanogenic bacteria to be useful in fermentation processes. The combination of SAO and the hydrogenotrophic methanogenic pathway is also important for maintaining fermentation stability in stressful environments (Pan et al., 2021), and SAOB may prefer to cooperate with *Methanoculleus* due to their high ammonia tolerance.

5.5.1.4 Main Methanogens

Methanogens are a group of obligately anaerobic archaea which play a pivotal role in natural anaerobic environments by metabolizing a variety of organic and inorganic substances to ultimately yield methane gas. To date, scientific research has identified over 65 distinct species of methanogenic archaea, which are categorized into at least three orders, seven families, and 19 genera within the Archaea domain. *Methanobrevibacter*, *Methanococcus*, *Methanosarcina*, *Methanomicrobium*, *Methanosphaera*, and *Methanospirillum* are considered to be the main microorganisms responsible for methane production. Despite the wide variety of methanogenic bacteria, they can only utilize three major groups of substrates(H_2, CO_2, methyl-group-containing compounds and acetate)to generate biogenic methane. Most organic matter, such as carbohydrates, long-chain fatty acids, and alcohols, cannot be utilized directly by methanogens, but need to be converted by complex microbial interactions into substrates that can be used by methanogenic bacteria. Based on substrate type, traditional euryarchaeotal methanogenic lineages can be classified into four main groups: hydrogenotrophic (H_2 and CO_2), aceticlastic (acetate), methylotrophic ($X\text{-}CH_3$), and H_2-dependent methylotrophic (H_2 and $X\text{-}CH_3$) (Evans et al., 2019). Moreover, the hydrogenotrophic methanogens have a much higher maximum growth rate than the acetoclastic methanogens with doubling times of 4–12 hours.

5.5.2 Main Metabolic Pathways of Methanogenesis

The methanogenic pathway is complex and requires a number of unique coenzymes and membrane-bound enzyme complexes as shown in Figure 5.7, which shows the main metabolic pathways of methanogenesis.

5.5.2.1 Hydrogenotrophic Methanogenesis

Most methanogens are hydrogenotrophic organisms that utilize H_2 as the primary electron donor in the reduction of CO_2 to methane. Hydrotrophic methanogens are highly adaptable and widely distributed, and their presence can be detected in almost all methanogenic orders, such as *Methanocorpusculum*, *Methanosphaera*, *Methanobacterium*, and *Methanoculleus* (Demirel and Scherer 2008; Sun et al., 2015). The hydrogenotrophic methanogenesis pathway consists of three distinct stages. In the first stage, CO_2 binds to MFR and is reduced to the formyl level. In the first reduction step, H_2-reduced ferredoxin (Fd) acts as a direct electron donor. In the second stage, the side group of the formyl group in formyl-methyl-furan is transferred to H_4MPT. This reaction was catalyzed by formyltransferase (Ftr) to form hypomethyl-H_4MPT. In the third stage, hypomethyl-H_4MPT was reduced to

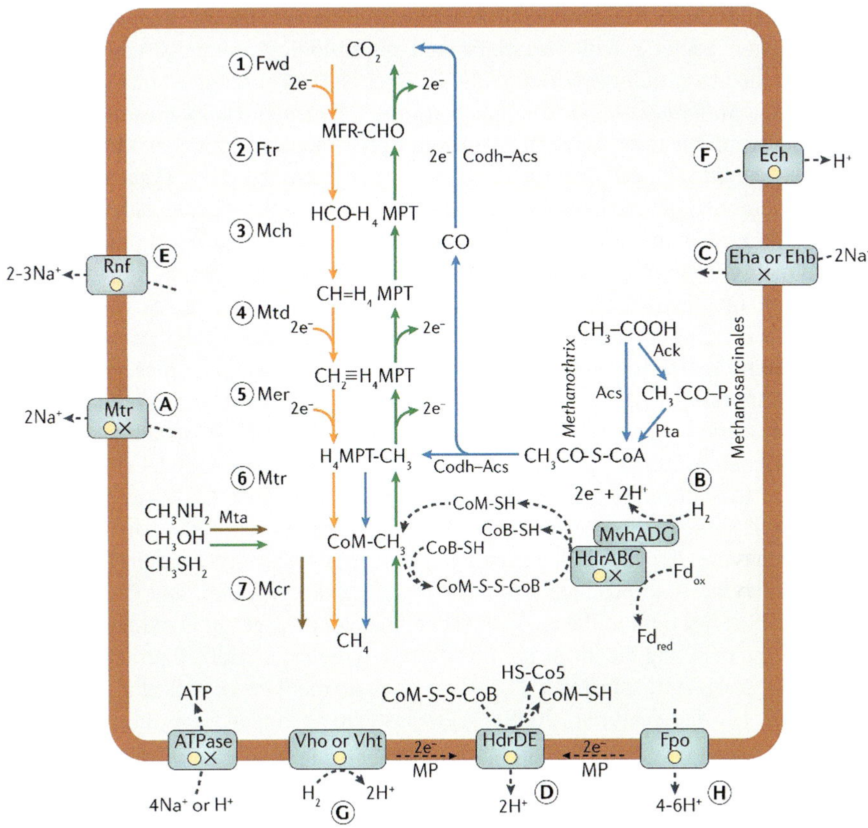

FIGURE 5.7 The main metabolic pathways of methanogenesis (adapted from Evans et al., 2019).

methylene-H_4MPT by the reductant F420, and further reduced to methyl-H_4MPT by dehydration of the formyl group to methyl group, and then to methylene-H_4MPT. In both reduction steps, coenzyme F420 acted as an electron carrier. Then methyl is transferred to coenzyme M (CoM, carrier of methyl) to form methyl-CoM complex, finally, methyl-CoM is reduced to methane by methyl coenzyme M reductase (Mcr) (Evans et al., 2019; Liu and Whitman 2008). In complex systems, hydrogenotrophic methanogenesis coexists closely with acetic acid production, contributing 30–40% of biomethane production.

5.5.2.2 Acetoclastic Methanogenesis

Acetoclastic methanogens consist of two main genera: *Methanothrix* and *Methanosarcina*. These acetoclastic methanogens produce methane by breaking down acetate through a **single** enzyme (Acs (EC 6.2.1.1)) or a combination of enzymes (Pta (EC 2.3.1.8) and Acs (EC 6.2.1.1)). Acetate is both an important nutrient for the growth of these microorganisms and a key intermediate in their eventual production of methane through the fermentation pathway. Mixotrophic *Methanosarcina* are methanogenic bacteria with metabolic and physiological diversity which have a high tolerance to extreme environments. Jiang et al. (2014) reported that under high VFA stress, *Methanosarcina* dominated by the *mixotrophic Methanosarcina* were also able to inhibit the growth of SAOB and promote methanogenesis from acetic acid cleavage through extensive acid catabolism, suggesting that *Methanosarcina* are highly tolerant to pH changes. Some methanogens such as *Methanosarcina flavescens*, *Methanosarcina barkeri*, and *Methanosarcina mazei* can utilize both acetoclastic and methylotrophic pathways (Kern et al., 2016). For the metabolic pathway of acetoclastic methanogenesis, first acetic acid must be converted to acetyl-coenzyme A (acetyl-CoA), which requires energy. The formation of acetyl-CoA occurs through two different reactions. *Methanosarcina* spp. utilizes acetate kinase and phosphotransacetylase to convert CoA-SH to acetyl-CoA and phosphate, and the conversion of acetic acid to acetyl-CoA requires an energy-rich phosphate bond in the ATP of *Methanosarcina* spp. The *Methanosarcina* spp. uses acetyl-CoA synthase to catalyze the conversion of acetic acid to acetyl-coenzyme A via two molecules of ATP. Acetyl-coenzyme A is then catalyzed by carbon monoxide dehydrogenase-acetyl-coenzyme A. The CO dehydrogenase complex catalyzes the cleavage of acetyl-coenzyme A to produce a methyl group, carboxyl group, and coenzyme A, which temporarily bind to the enzyme. Then, the carboxyl group is oxidized to form CO_2, and the resulting electrons are transferred to the iron oxidized protein. Finally, the methyl group is transferred to H_4SPT to generate methane (Evans et al., 2019; He et al., 2022; Liu and Whitman, 2008). Acetoclastic methanogenesis is one of the most factor-sensitive processes in complex fermentation, accounting for approximately 60–70% of methane production.

5.5.2.3 Methylotrophic Methanogenesis

There are many methylotrophic methanogenic bacteria that carry out their methanogenic activities by utilizing methyl-group containing compounds (e.g., methanol, methylamine, methoxylated benzoate), and they carry out this process in two main ways (Amin et al., 2021; Evans et al., 2019; He et al., 2022; Liu and Whitman 2008). The first way is a strictly methylotrophic methanogenic process, in which the methyl group in a methyl compound ($-CH_3$) is oxidized to carbon dioxide (CO_2), releasing electrons and protons in the process, which are then supplied to the other part of the methyl group for reduction and eventual conversion to methane (CH_4). This reaction is called disproportionation of methyl groups and occurs in most methylotrophic methanogens, including *Methanosarcina*, *Methanomethylovorans*, and *Methanococcoides*. Methyltransferases and cochineal-like proteins transfer methyl from these substrates to CoM to form methyl-CoM, which is subsequently reduced to methane by Mcr. The second way is the H_2-dependent methylotrophic process, and *Methanomicrococcus blatticola* and *Methanosphaera* are typical H_2-dependent methylotrophic methanogens. The variations in the process of methylotrophic methanogenesis stem from whether or not cytochromes are present in the methylotrophic methanogens. Those lacking cytochromes rely solely on hydrogen (H_2) utilization, whereas methanogens possessing cytochromes have a more versatile function, converting H_2 and also being capable of oxidizing methyl groups into carbon dioxide (CO_2) using a membrane-bound electron transport chain to facilitate the methanogenesis process (He et al., 2022).

Acetoclastic methanogenesis (contributing 70%) and hydrogenotrophic methanogenesis (contributing 30%) are the main pathways for biomethane production. The contribution of acetoclastic methanogenesis and hydrogenotrophic methanogenesis to methane production depends on the type of intermediate organic acids and their response to stressful conditions, such as temperature changes, substrate loading shocks, shortened residence time, and accumulation of excess VFAs or ammonia (Demirel and Scherer 2008; Zhang et al., 2023). For example, Wang et al. (2023) reported that methanation preferred the methylotrophic pathway over the hydrogenotrophic pathway when the temperature was increased from 65 °C to 75 °C and that the methanogenic bacterial community in fermentation digesters operated at 37 °C had higher diversity compared to 55 °C. When the temperature is reduced to 25 °C, the microbial community may switch from acetoclastic methanogenesis to hydrogenotrophic methanogenesis (Li et al., 2019; Nguyen and Khanal 2018).

5.6 CONCLUSIONS

Highly efficient methanogenesis of organic wastes is a promising method to achieve global sustainable development goals. Unlike in the methanogenesis of organic wastewater, in which microbial cells come into direct contact with the dissolved substrate, in the methanogenesis of organic wastes, a "solid-liquid-solid" mode is formed among microbial cells, water molecules, and solid-state substrate (organics and inorganics). It is inappropriate to directly apply the methanogenesis theory developed for organic wastewater systems to improve the methanogenesis of organic wastes. In addition, based on the conservation law of elements, the Buswell equation can be used to calculate the theoretical methane production of organic wastes. In the methanogenic fermentation of organic wastes, in addition to considering the basic biotic factors, abiotic factors such as substrate bioavailability and interactions between organic particles should also be taken seriously in the future. Furthermore, acetoclastic methanogenesis and hydrogenotrophic methanogenesis are the main metabolic pathways for biomethane production in the methanogenic fermentation of organic wastes, and how to selectively regulate the two main metabolic pathways is a major challenge for future research.

REFERENCES

Agrawal, A.V., Chaudhari, P.K., Ghosh, P. 2023. Effect of mixing ratio on biomethane potential of anaerobic co-digestion of fruit and vegetable waste and food waste. Biomass Convers. Biorefinery, 10.

Ahn, J., Park, J.-K., Higuchi, S., Lee, N. 2014. Effects of Pre-aeration on the Anaerobic Digestion of Sewage Sludge. *Environ. Eng. Res.*, 19(1), 59–66.

Amin, F.R., Khalid, H., El-Mashad, H.M., Chen, C., Liu, G.Q., Zhang, R.H. 2021. Functions of bacteria and archaea participating in the bioconversion of organic waste for methane production. *Sci. Total Environ.*, 763, 21.

An, X.A., Xu, Y., Dai, X.H. 2024. Biohythane production from two-stage anaerobic digestion of food waste: A review. *J. Environ. Sci.*, 139, 334–349.

Angelidaki, I., Sanders, W. 2004. Assessment of the anaerobic biodegradability of macropollutants. *Rev. Environ. Sci. Bio.*, 3(2), 117–129.

Appels, L., Baeyens, J., Degrève, J., Dewil, R. 2008. Principles and potential of the anaerobic digestion of waste-activated sludge. *Prog. Energy Combust. Sci.*, 34(6), 755–781.

Ariunbaatar, J., Panico, A., Esposito, G., Pirozzi, F., Lens, P.N.L. 2014. Pretreatment methods to enhance anaerobic digestion of organic solid waste. *Appl. Energy*, 123, 143–156.

Ariunbaatar, J., Panico, A., Yeh, D.H., Pirozzi, F., Lens, P.N.L., Esposito, G. 2015. Enhanced mesophilic anaerobic digestion of food waste by thermal pretreatment: Substrate versus digestate heating. *Waste Manage.*, 46, 176–181.

Bougrier, C., Albasi, C., Delgenès, J.P., Carrère, H. 2006. Effect of ultrasonic, thermal and ozone pre-treatments on waste activated sludge solubilisation and anaerobic biodegradability. *Chem. Eng. Process.*, 45(8), 711–718.

Campuzano, R., González-Martínez, S. 2016. Characteristics of the organic fraction of municipal solid waste and methane production: A review. *Waste Manage.*, 54, 3–12.

Canziani, R., Spinosa, L. 2019. Sludge from wastewater treatment plants. in: *Industrial and Municipal Sludge*, pp. 3–30.

Chatterjee, B., Mazumder, D. 2019. Role of stage-separation in the ubiquitous development of Anaerobic Digestion of Organic Fraction of Municipal Solid Waste: A critical review. *Renew. Sust. Energ. Rev.*, 104, 439–469.

Chen, Y., Cheng, J.J., Creamer, K.S. 2008. Inhibition of anaerobic digestion process: A review. *Bioresour. Technol.*, 99(10), 4044–4064.

Chew, K.R., Leong, H.Y., Khoo, K.S., Vo, D.V.N., Anjum, H., Chang, C.K., Show, P.L. 2021. Effects of anaerobic digestion of food waste on biogas production and environmental impacts: a review. *Environ. Chem. Lett.*, 19(4), 2921–2939.

Dai, X.H. 2022. *Treatment, disposal and resource utilization of sewage sludge*. China Architecture & Building Press. Beijing. (In chinese)

Dai, X.H., Hu, C.L., Zhang, D., Dai, L.L., Duan, N.N. 2017a. Impact of a high ammonia-ammonium-pH system on methane-producing archaea and sulfate-reducing bacteria in mesophilic anaerobic digestion. *Bioresour. Technol.*, 245, 598–605.

Dai, X.H., Xu, Y., Dong, B. 2017b. Effect of the micron-sized silica particles (MSSP) on biogas conversion of sewage sludge. *Water Res.*, 115, 220–228.

Dai, X.H., Xu, Y., Lu, Y.Q., Dong, B. 2017c. Recognition of the key chemical constituents of sewage sludge for biogas production. *RSC Adv.*, 7(4), 2033–2037.

De Vrieze, J., Devooght, A., Walraedt, D., Boon, N. 2016. Enrichment of *Methanosaetaceae* on carbon felt and biochar during anaerobic digestion of a potassium-rich molasses stream. *Appl. Microbiol. Biotechnol.*, 100(11), 5177–5187.

Demirel, B., Scherer, P. 2008. The roles of acetotrophic and hydrogenotrophic methanogens during anaerobic conversion of biomass to methane: a review. *Rev. Environ. Sci. Bio.*, 7(2), 173–190.

El-Mashad, H.M., Zhang, R. 2010. Biogas production from co-digestion of dairy manure and food waste. *Bioresour. Technol.*, 101(11), 4021–4028.

Erden, G., Filibeli, A. 2010. Improving anaerobic biodegradability of biological sludges by Fenton pre-treatment: Effects on single stage and two-stage anaerobic digestion. *Desalination*, 251(1–3), 58–63.

Evans, P.N., Boyd, J.A., Leu, A.O., Woodcroft, B., Parks, D.H., Hugenholtz, P., Tyson, G.W. 2019. An evolving view of methane metabolism in the Archaea. *Nat. Rev. Microbiol.*, 17(4), 219–232.

Feng, Y.H., Zhang, Y.B., Quan, X., Chen, S. 2014. Enhanced anaerobic digestion of waste activated sludge digestion by the addition of zero valent iron. *Water Res.*, 52, 242–250.

Ferrer, I., Ponsá, S., Vázquez, F., Font, X. 2008. Increasing biogas production by thermal (70°C) sludge pre-treatment prior to thermophilic anaerobic digestion. *Biochem. Eng. J.*, 42(2), 186–192.

Giwa, A.S., Xu, H., Chang, F.M., Zhang, X.Y., Ali, N., Yuan, J., Wang, K.J. 2019. Pyrolysis coupled anaerobic digestion process for food waste and recalcitrant residues: Fundamentals, challenges, and considerations. *Energy Sci. Eng.*, 7(6), 2250–2264.

Gonzalez, A., Hendriks, A., van Lier, J.B., de Kreuk, M. 2018. Pre-treatments to enhance the biodegradability of waste activated sludge: Elucidating the rate limiting step. *Biotechnol. Adv.*, 36(5), 1434–1469.

Haider, M.R., Zeshan, Yousaf, S., Malik, R.N., Visvanathan, C. 2015. Effect of mixing ratio of food waste and rice husk co-digestion and substrate to inoculum ratio on biogas production. *Bioresour. Technol.*, 190, 451–457.

Hassanpourmoghadam, L., Goharrizi, B.A., Torabian, A., Bouteh, E., Rittmann, B.E. 2023. Effect of Fe3O4 nanoparticles on anaerobic digestion of municipal wastewater sludge. *Biomass Bioenerg.*, 169, 8.

He, K.C., Li, W., Tang, L.X., Lv, S.H., Xing, D.F. 2022. Suppressing Methane Production to Boost High-Purity Hydrogen Production in Microbial Electrolysis Cells. *Environ. Sci. Technol.*, 56(17), 11931–11951.

Horiuchi, J.I., Shimizu, T., Tada, K., Kanno, T., Kobayashi, M. 2002. Selective production of organic acids in anaerobic acid reactor by pH control. *Bioresour. Technol.*, 82(3), 209–213.

Izumi, K., Okishio, Y.K., Nagao, N., Niwa, C., Yamamoto, S., Toda, T. 2010. Effects of particle size on anaerobic digestion of food waste. *Int. Biodeterior. Biodegrad.*, 64(7), 601–608.

Jiang, M.M., Qiao, W., Wang, Y.C., Zou, T., Lin, M., Dong, R.J. 2022. Balancing acidogenesis and methanogenesis metabolism in thermophilic anaerobic digestion of food waste under a high loading rate. *Sci. Total Environ.*, 824, 8.

Jiang, Y., McAdam, E., Zhang, Y., Heaven, S., Banks, C., Longhurst, P. 2019. Ammonia inhibition and toxicity in anaerobic digestion: A critical review. *J. Water Process. Eng.*, 32, 9.

Jin, C.X., Sun, S.Q., Yang, D.H., Sheng, W.J., Ma, Y.D., He, W.Z., Li, G.M. 2021. Anaerobic digestion: An alternative resource treatment option for food waste in China. *Sci. Total Environ.*, 779, 23.

Kern, T., Fischer, M.A., Deppenmeier, U., Schmitz, R.A., Rother, M. 2016. Methanosarcina flavescens sp. nov., a methanogenic archaeon isolated from a full-scale anaerobic digester. *Int. J. Syst. Evol. Microbiol.*, 66(3), 1533–1538.

Kim, M.S., Kim, D.H., Yun, Y.M. 2017. Effect of operation temperature on anaerobic digestion of food waste: Performance and microbial analysis. *Fuel*, 209, 598–605.

Le, T.M., Vo, P.T., Do, T.A., Tran, L.T., Truong, H.T., Le, T.T.X., Chen, Y.H., Chang, C.C., Chang, C.Y., Tran, Q.T., Thanh, T., Do, M.V. 2019. Effect of assisted ultrasonication and ozone pretreatments on sludge characteristics and yield of biogas production. *Processes*, 7(10), 12.

Li, L., Xu, Y., Dai, X.H., Dai, L.L. 2021a. Principles and advancements in improving anaerobic digestion of organic waste via direct interspecies electron transfer. *Renew. Sust. Energ. Rev.*, 148, 15.

Li, Y., Chen, Y., Wu, J. 2019. Enhancement of methane production in anaerobic digestion process: A review. *Appl. Energy*, 240, 120–137.

Li, Y.Y., Wang, L.E., Liu, G., Cheng, S.K. 2021b. Rural household food waste characteristics and driving factors in China. *Resour. Conserv. Recycl.*, 164, 13.

Liu, F.H., Conrad, R. 2011. Chemolithotrophic acetogenic H_2/CO_2 utilization in Italian rice field soil. *Isme J.*, 5(9), 1526–1539.

Liu, H.Y., Xu, Y., Geng, H., Chen, Y.D., Dai, X.H. 2022a. Contributions of MOF-808 to methane production from anaerobic digestion of waste activated sludge. *Water Res.*, 220, 10.

Liu, H.Y., Xu, Y., Li, L., Dai, X.H., Dai, L.L. 2021. A review on application of single and composite conductive additives for anaerobic digestion: Advances, challenges and prospects. *Resour. Conserv. Recycl.*, 174, 17.

Liu, S.J., Moon, C.D., Zheng, N., Huws, S.R., Zhao, S.G., Wang, J.Q. 2022b. Opportunities and challenges of using metagenomic data to bring uncultured microbes into cultivation. *Microbiome*, 10(1), 14.

for hydrogen and 0.3 m/s for methane. Moreover, because of more stable combustion, biohythane provides the opportunity to increase the lean burn limit. Moreover, because biohythane burns faster and is less prone to catch fire, it expands the narrow burn cap. Additionally, biohythane doesn't release dangerous compounds like sulfur and benzene. In contrast to diesel cars, the hydrazine tests show that fuel combustion produces very little pollution in exhaust gas. On burning, biohythane emits 2.8 kg CO_2/kg of fuel, whereas diesel emits 3.4 kg CO_2/kg (Bolzonella et al., 2018). Because of these qualities, biohythane has become very appealing to the automobile industry, and several automakers, including Toyota, have already created hythane-powered vehicles (Genovese and Ortenzi, 2016). The most intriguing feature of biohythane is that it can be used with engines that are currently on the market, meaning that vehicle designs don't need to be drastically altered—just minor engine operating and combustion control adjustments are needed. Furthermore, biohythane can be produced at the refueling station directly or obtained via the gas supply network. The reason the use of biohythane is superior to other fuels goes to its ability to generate less particulate matter and sulfur, less greenhouse gas emissions and other toxic amalgams like benzene, highly reactive olefins and higher molecular weight hydrocarbons, etc.

In addition to being utilized as fuel for vehicles, biohythane can also yield useful by-products such as liquid fuels (Rawoof et al., 2020a). Methane is typically utilized to produce methanol, which is then used to activate the methane C–H bond by methanotrophs to produce formaldehyde and methyl butyl ethers (Rawoof et al., 2020a). Methane to methanol conversion, however, is less effective and expensive. The presence of hydrogen in biohythane allows for the production of methanol from it through adsorption and immobilization processes, which can improve the production rate by about 200% when compared to methane as a raw material. Since hydrogen plays a crucial function as an electron donor for the enzymatic reaction involved in the biological methanol pathway, the significance of biohythane can be explained by its presence. Furthermore, hydrogen helps to minimize the inhibitory effects caused by the synthesis of ammonia, carbon dioxide, and hydrogen sulfide. The production of methanol can also be achieved by Fischer-Tropsch (FT) processes, which include feeding hydrogen and carbon monoxide into an FT reactor to produce a variety of straight-chain alcohols and alkanes, including methanol. FT liquids are also distilled to separate the alkanes and olefins, which are then further distilled to hydrocarbons in the diesel and naphtha range (Hu et al., 2012). The polyhydroxyalkanoates produce biohythane through a more innovative approach than these two methods because of the utilization of *Rhodospirillum rubrum*, subsequently consumes CO generated by the steam-reformation in biohythane during the AD process. The biologically mediated water gas shift reaction is thought to consume the remaining 80% of the CO and the bacterial cell biomass is thought to absorb the remaining 20% producing 40% PHA (Bolzonella et al., 2018). Reduced energy consumption for engine ignition and improved fuel and heat efficiency are only a few of the established advantages of biohythane in the automotive industry (Shanmugam et al., 2020). Therefore, the advantages of biohythane such as good calorific value, carbon neutral, environment-friendly, cost-effectiveness, and sustainable fuel make it a promising renewable energy source in the market.

6.3 PROCESS OF BIOHYTHANE PRODUCTION

6.3.1 Hydrogen Production Pathway by Dark Fermentation

The dark fermentation process results in the generation of hydrogen gas (H_2) (Meena et al., 2020). Dark fermentation (DF) primarily utilized glucose as its substrate due to its simplicity as a sugar. The fundamental process comprises the transformation of sugars derived by hydrolysis into acetic acid through the activity of specific microbes, ultimately resulting in the generation of hydrogen. Therefore, the hydrolytic bacteria catalyze the hydrolysis of the complex sugar into simple sugar, which is subsequently metabolized by dark fermentation bacteria via the glycolytic pathway, resulting in the formation of pyruvate. In contrast, facultative and obligate anaerobes utilized distinct mechanisms for hydrogen production.

Pyruvate ferredoxin oxidoreductase, an enzyme in obligate anaerobes, converts pyruvate to acetyl CoA while also reducing ferredoxin (Fd). This is subsequently oxidized by FeFe hydrogenase to create H_2 (Ghosh and Kar, 2022).

$$\text{Pyruvate} + \text{CoA} + 2\,\text{Fd}(\text{ox}) = \text{Acety} - \text{CoA} + 2\,\text{Fd}(\text{red}) + \text{CO}_2 \tag{6.1}$$

$$2\text{H}^+ + 2\,\text{Fd}(\text{red}) = \text{H}_2 + \text{Fd}(\text{ox}) \tag{6.2}$$

In facultative anaerobes, the enzyme formate lyase facilitates the conversion of pyruvate with CoA producing formate and acetyl-CoA. The formate generated in this process is enzymatically cleaved by formate hydrogen lyase, resulting in the production of CO_2 and H_2 (Ghosh and Kar, 2022). The process is carried out utilizing formate hydrogenlyase (FHL) (Hans and Kumar, 2019).

$$\text{Pyruvate} + \text{CoA} = \text{Acetyl} - \text{CoA} + \text{Formate} \tag{6.3}$$

$$\text{HCOOH} = \text{CO}_2 + \text{H}_2 \tag{6.4}$$

In the process of pyruvate oxidation, it has been observed that when acetate is produced as the final product, a stoichiometric ratio of 4 moles of hydrogen can be obtained from 1 mole of glucose oxidized. On the other hand, when considering butyrate, the transformation of one glucose mole results in the production of just 2 moles of hydrogen.

$$\text{C}_6\text{H}_{12}\text{O}_6 + 2\text{H}_2\text{O} = 2\text{CH}_3\text{COOH} + 2\text{CO}_2 + 4\text{H}_2 \tag{6.5}$$

$$\text{C}_6\text{H}_{12}\text{O}_6 + 2\text{H}_2\text{O} = \text{C}_3\text{H}_7\text{COOH} + 2\text{CO}_2 + 2\text{H}_2 \tag{6.6}$$

According to Constant and Hallenbeck (2019), hydrogenase plays a crucial role in the process by which facultative/obligate anaerobes produce hydrogen. *E. coli*, a facultative anaerobe, can synthesize and assemble four different types of nickel-iron hydrogenases: "hydrogenase-1 (Hyd-1), hydrogenase-2 (Hyd-2), hydrogenase-3 (Hyd-3), and hydrogenase-4 (Hyd-4)". The FeFe hydrogenase enzyme found in obligate anaerobes facilitates the generation of hydrogen by promoting electron transport from ferredoxin to protons. Additionally, the potential of anaerobic digestion to decompose many types of feedstocks may contribute to cost reduction in the whole process. Jiraprasertwong et al. (2019) found that introducing cassava residue into cassava wastewater resulted in a hydrogen (H_2) generation rate of 15 mL H_2/gCOD eliminated in a two-stage upflow anaerobic sludge blanket (UASB) system, despite the absence of sufficient nutrients. The outcomes of the investigation indicated that cassava residue has a significant impact on various aspects, including its high starch content. This high starch content was found to contribute to an increase in volatile fatty acid (VFA) generation, as well as affecting H_2 production and chemical oxygen demand (COD) removal during the acidogenesis process. The significance of nutrients in dairy effluent treated with DF is stated by Gadhe et al. (2013). For example, a COD/N ratio of 100.5 and a COD/P ratio of 120 are suitable for dairy wastewater, with an H_2 production rate of 29.91 mmol H_2/g. However, high concentrations of N and P both hinder the synthesis of H_2 and accelerate the growth of organic solvents.

The process of methanogenesis involves a series of biochemical reactions that result in the creation of methane. The waste medium from dark fermentation includes considerable amounts of short-chain fatty acids (SCFAs) including acetate, butyrate, and propionate. These SCFAs have the potential for use as substrates for methanogenic bacteria. Certain hydrogenotrophic methanogenic bacteria may produce methane by using dissolved CO_2 and H_2 in the waste medium. Methanogenesis is the process of producing methane by using methyl-coenzyme M reductase (MCR) as the major enzyme. Anaerobic oxidation is then performed after the process. In this case, the substrates are methyl thioether, methyl-coenzyme M and thiol coenzyme B. As a consequence, these specific substrates undergo reversible digestion, yielding methane and an equal quantity of heterosulfide. The exact catalytic process remains unclear.

Methanogens have three principal C1 unit carriers: methofuran (MFR), tetrahydromethopterin (H_4MPT) and HS-CoM, as well as electron carriers such as N-7 mercaptoheptanoyl-O-phospho-L-threonine (H-S-HTP) and coenzyme F420. The intermediate species play a part in the process of converting CO_2 to CH_4. The intermediates in the reduction process that are coenzyme-bound include CHO-MFR, CHO-H_4MPT, CHO-H_4MPT, CH_2-H_4MPT, CH_3-H_4MPT$^+$, CH_3-S-CoM, methyl coenzyme M (MPT-methylene tetrahydromethanopterin), and MFR-methyl tetrahydrofuran (Ghosh and Kar, 2022). The catalytic mechanism remains unknown.

Anaerobic digestion is divided into two types: single-stage AD and two-stage AD, which are differentiated depending on the resulting output products. Methane predominates in single-stage AD, with some trace products. On the other hand, methane and hydrogen are produced simultaneously during the two-stage AD. The two AD process methods are the main ways that biohythane is produced. First, AD can accomplish waste treatment through two distinct approaches, namely, the wet method and the high solids method. To provide an illustration, the wet AD approach is deemed appropriate for treating TS (total solids) with a concentration below 15%; when the solid concentration exceeds 15% but not 40%, the high solid approach is suggested. The utilization of wet AD has gained popularity in the treatment of waste materials that possess a high moisture content, mostly due to its ability to stimulate microbial activity. However, it is anticipated that the emergence of high solids AD may eventually render wet AD obsolete. The popularity of high solids AD can be attributed to several factors, including increased organic load, reduced installation and running costs, reduced water consumption during digestion, and lowered digester maintenance needs.

6.3.2 Single- and Two-Stage Anaerobic Digestion Processes

The simultaneous occurrence of hydrolysis, acidogenesis, acetogenesis, and methanogenesis within a single reactor is the single-stage AD process. Consequently, the main outcome of the single-stage procedure is the generation of H_2, CO_2, H_2S, and approximately 60–65% of CH_4. Despite the efficacy and natural occurrence of this process, its widespread utilization is hindered by a single downside. One drawback associated with acetogenesis and acidogenesis is the generation of VFAs, which contribute to the acidification of the reactor and the subsequent reduction in pH. This acidic environment poses a threat to the survival of bacteria within the bioreactor. This phenomenon results in a decrease in the methane production output. The hydrogen producers find the implementation of this VFA to be unwarranted and unacceptable. The process of digestion is halted specifically during the stages of acetogenesis and acidogenesis. The two-stage AD method includes two separate manufacturing steps (Hans and Kumar, 2019). The initial stage focuses on the generation of biohydrogen through the activities of microorganisms that produce hydrogen. This stage encompasses hydrolysis, acidogenesis, and acetogenesis, all of which occur in an acidic pH condition. The occurrence of two-stage biomethane generation, subsequent to one-stage biohydrogen production, was facilitated by the pH conditions in the range of 7.0–8.0 and the presence of microbial methanogens involved in methanogenesis (Aashabharathi et al., 2024). Using a two-phase AD process may reduce VFA buildup and increase methanogen bacterial growth, hence facilitating the generation of biomethane during the second step. Furthermore, it has been observed that the process of two-stage AD yields a higher energy output in comparison to the one-stage AD method.

In their investigation, Massanet-Nicolau et al. (2013) conducted a comparative analysis between one-stage AD and two-stage AD using pellets of wheat feed and inoculum derived from a sewage treatment plant (STP) effluent. The researchers employed real-time gas production data to evaluate and contrast the performance of these two AD systems. A significant enhancement in methane production, specifically an increase of approximately 359 L/kg-VS, was recorded while employing a two-stage AD system compared to the single-stage AD system, which yielded approximately 261 L/kg-VS. This improvement was noticed within a 20-day timeframe. The duration necessary to complete a single cycle of one-stage AD is around 28–30 days. In their study, Schievano et al. (2014) conducted an analysis of the production of bio-CH_4 and bio-H_2 from organic biomass utilizing a two-stage AD method. The researcher arrived at the conclusion that there is potential to obtain an increase ranging from around 8% to 43% in recovered energy. The two-step AD method demonstrates a reliable and dependable operation, mostly attributed to its effective self-pH regulating capability. The aforementioned procedure effectively mitigates breakouts resulting from the excessive buildup of VFAs (Hans and Kumar, 2019).

6.3.3 Integration of Biohydrogen and Biomethane Processes for Biohythane Production

Biohydrogen can be produced from biomass through an effective process called dark fermentation. The AD system for methane production comprises four distinct mechanisms: hydrolysis, acidogenesis, acetogenesis, and methanogenesis. The first stage in the biohythane synthesis process is hydrolysis. At the conclusion of this procedure, effluent that is rich in VFAs is collected. In the reactor environment, the presence of a high hydrogen partial pressure leads to the inhibition of activity among hydrogen producing microorganisms (HPM). Only methanogens possess the capacity to thrive in this particular environment. The utilization of a medium that is somewhat rich in the second step of the AD employs a TSAF process to provide VFAs for methane production via methanogenesis (Mozhiarasi et al., 2023). The subsequent stage in the biohythane manufacturing process involves the transformation of lactic acid, alcohols, and VFAs into acetic acid through the process of acetogenesis. This acetic acid is further decomposed into methane and CO_2. The integration of multiple stages in AD has several advantages, including enhanced energy recovery, reduced fermentation time, and greater operational stability, compared to single-step AD *Clostridium*, *Enterobacter*, and other acidogenic bacteria used in biohythane production. Conversely, methanogenic bacteria such as *Methanoculleus* and *Methanosarcina* have been utilized for bio-CH_4 production. The production of biohythane encompasses the blending of biohydrogen and biomethane (Fig. 6.1). In the initial phase, slightly acidic conditions are ideal for hydrogen production, usually with a shorter retention time of 1–3 days and a pH range of 5.5–6.5. Subsequently, in the second stage of AD, acetoclastic and hydrogenotrophic methanogens use the spent media or high-VFA H_2 effluent to produce methane at neutral pH having a retention time of

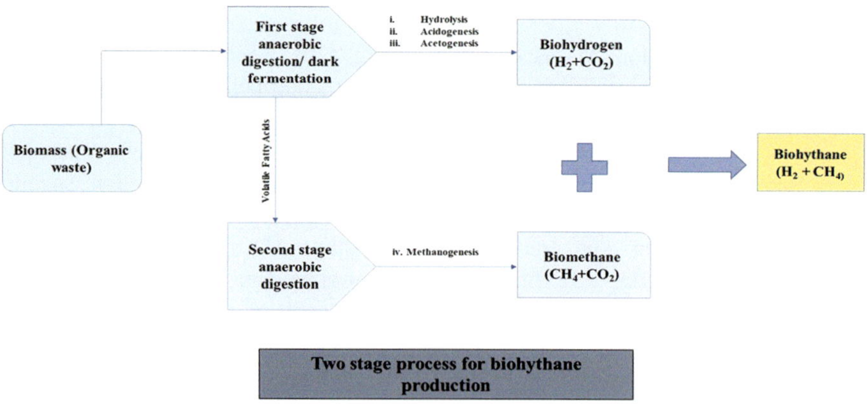

FIGURE 6.1 Two-stage process for biohythane production.

10–15 days. Using two-stage AD, the overall biomass fermentation duration is reduced to around 13–18 days.

Acidogenic bacteria and methanogenic bacteria accumulate in two different digestion tanks having optimal conditions for development in the two stages AD process. The various processes involved in biohythane production are summarized in Table 6.1. Different bacterial strains which are capable of producing H_2 through the fermentation of different carbohydrates are involved in the first-stage reactor. This includes strict anaerobic *Clostridia* sp. to produce higher H_2 yields, such as *C. welchii*, *C. pasteurianum*, *C. beijerinckii*, and *C. butyricum* (Kamalaskar et al., 2010). Several facultative anaerobic species, such as *Enterobacteriaceae*, are also potential H_2 producers, exhibiting resistance to low concentrations of dissolved oxygen. Dark fermentation for H_2 production at thermophilic conditions is observed to have better kinetics and stoichiometry in comparison to mesophilic conditions. Higher temperatures lead to a more thermodynamically favorable metabolism, which is also less affected by the partial pressure of H_2 in the liquid media. *Thermoanaerobacterium* sp., *Clostridium* sp., and *Thermoanaerobacter* sp. are the main microbial species involved in thermophilic dark fermentation (Harnvoravongchai et al., 2020). For example, a moderate thermophilic species of bacteria (60°C), *Thermoanaerobacterium thermosaccharolyticum* ferments acetate and butyrate to create H_2 from carbohydrates (Li et al., 2019). *Caldicellulosiruptor* and *Thermotoga* sp. produce H_2 by dark fermentation in an immensely thermophilic environment. These microbes are additionally anticipated to produce a variety of hydrolytic enzymes that break down cellulose, xylan, and cellobiose. *Thermotoga* sp. is one of the few species that employs elemental sulfur as a source of electrons for generating H_2 (Mozhiarasi et al., 2023). A wide range of archaeal species, including acetoclastic and hydrogenotrophic methylotrophs, produce methane in anaerobic digesters within the second stage of the AD reactor. *Methanococcus* sp., *Methanoculleus* sp., and *Methanospirillum* sp. act as key species in the methanogenesis stage as they use acetic acid, CO_2, and H_2 to produce methane (Mozhiarasi et al., 2023). *Methanothermococcus* sp. and *Methanothermus* sp. use formate, H_2, and CO_2 as energy sources for producing CH_4 in an extremely thermophilic environment (85°C) (Cheng et al., 2010; Mozhiarasi et al., 2023).

6.4 FEEDSTOCK TYPE FOR BIOHYTHANE PRODUCTION

6.4.1 Agriculture Based (Agro Residue)

For the production of biohythane, substrates rich in carbohydrates are most effective, followed by substrates rich in proteins; substrates high in fats are not so ideal (Table 6.2) (Mozhiarasi et al., 2023). Cellulosic biomass breakdown may provide gaseous biofuels. But the main problem regarding employing such types of biomass is that its three-dimensional structure seems rather inflexible, which makes its degradation problematic (Ghosh and Kar, 2022). Due to these issues, research has been conducted on employing this substrate in batch mode. Cassava, sugar beet, maize, sorghum, wheat, cornstalk, sugarcane, peanut shell, rice grain, and hazelnut are examples of agricultural waste that may be used to produce biohydrogen and biogas.

TABLE 6.1
Details of Various Process Involved in Biohythane Production

Process	Equations Associated	Microorganisms Involved	Key Enzymes	References
Dark Fermentation	$C_6H_{12}O_6 + 2H_2O = 2CH_3COOH + 2CO_2 + 4H_2$	*Clostridium*, *Escherichia coli*, *Enterobacter*, *Bacillus*	Glucose-6-phosphate dehydrogenase (zwf), membrane bound hydrogenases (hupSL), carbamoyltransferase (hypF)	Kim et al., 2022; Mozhiarasi et al., 2023
Photofermentation	$C_6H_{12}O_6 + 6H_2O = 6CO_2 + 12H_2$	Phototrophic bacteria	(HoxEFUYH) type of [NiFe] hydrogenase -	Kim et al., 2022; Mozhiarasi et al., 2023
Biophotolysis	$12H_2O = 6O_2 + 12H_2$ $CO + H_2O = CO_2 + H_2$	Photosynthetic bacteria, green algae	(HoxEFUYH) type of [NiFe] hydrogenase), [FeFe]-hydrogenases (hydA)	Abdur Rawoof et al., 2021; Kim et al., 2022
Microbial electrolysis cell	$C_6H_{12}O_6 + 2H_2O = 2CH_3COOH + 2CO_2 + 4H_2$ Cathode: $8H^+ + 8e^- = 4H_2$ Anode: $CH_3COOH + 2H_2O = 2CO_2 + 8H^+ + 8e^-$	*Geobacter*		Ghosh and Kar, 2022
Methanogenesis	$4H_2 + HCO_3^- + H^+ = CH_4 + 3H_2O$ $CH_3COO^- + H_2O = HCO_3^- + CH_4$ $CH_3COO^- + 4H_2O = 4H_2 + HCO_3^- + H^+$	Syntrophic acetate oxidizing bacteria, hydrogenotrophic or acetoclastic methanogens, *Methanosarcina*	Methyl-coenzyme M reductase (MCR), acetate kinase (ACK), phosphate acetyltransferase (PTA)	Abdur Rawoof et al., 2021; Kim et al., 2022

TABLE 6.2
Biohythane Yield from Different Feedstocks

Type of Feedstocks	Temperature (°C)	pH	Operational Parameters	Hydrogen Yield (L-H_2/kg VS)	Methane Yield (L-CH_4/kg VS)	References
Food waste	37 in both the stages	6 and 8	HRT – 3,21 days	212.2	412.6	Zhao et al., 2021
POME	55 and 35	5.5 and 7.5		210.0	315	Mamimin et al., 2017
Wheat straw	70 and 37	6.9 and 7.5		270.0	179	Willquist et al., 2012
Biowaste from the treatment plant	55 in both the stages		HRT – 32 days	51	780	Cavinato et al., 2011
Cassava WW (cassava residue)	55 in both the stages	5.5	OLR – 10.3 kg m^3/d	15	259	(Chavadej et al., 2019
Artificial food waste	37 in both the stages	6.0 and 7.2	–	30.3	190.3	Luo et al., 2021
Mixture of sewage sludge, vinasse, and pig manure	55 in both the stages	5.0 and 5.5	–	27	59	Sillero et al., 2022
Biowaste	55 and 35	5.5 and 8.0	–	41.0	102.0	Cavinato et al., 2012
Food waste	35 in both the stages	–	HRT – 25 days	310	210	Han and Shin, 2004
Cornstalk	37 in both the stages	7.7 and 8.0	–	25.02	95.38	Li et al., 2020

High solid content and poor digestibility both harm bioreactor performance and biogas generation (Meena et al., 2020). These forms of wastes are used in steam boilers. Since agriculture is the largest source of human output worldwide, agro leftovers have the greatest potential to meet this need (Mozhiarasi et al., 2023).

In a study by Li et al. (2020), it was reported that cornstalk in a first-stage fermentation produced 25.02 L/kg TS of H_2 and second-stage fermentation produced 95.38 L/kg TS of methane. Sorghum crops rich in fermentable sugars enabled rapid breakdown and biogas generation. Sorghum may also be found in all seasons and preserved for a long-time using techniques like ensiling (Meena et al., 2020). Peanut shell wastes with high cellulose and sCOD content may have superior microbial growth properties. Furthermore, the peanut shell's buffering capacity affects bioreactor production quality (Meena et al., 2020). As a result, co-digestion is the best alternative for treating agricultural leftovers for energy recovery. For maximal biohydrogen production, use a 5:1 co-digestion ratio of agro-waste, such as rice straw, and pig manure (Mozhiarasi et al., 2023).

6.4.2 Household Food Waste

Food waste may be an excellent fuel for biohythane synthesis. The massive quantity of food waste comes from households including food processing (38%) and from the whole supply chain (20%) (Bolzonella et al., 2018). However, kitchen trash produced less hydrogen than maize starch because of the lack of degradable carbohydrates and inhibitors such as ammonia, oil (fat), and grease. Wang et al. (2014) discovered that when lipid concentrations are above 25% w/w, methane production is reduced for a variety of reasons, including the formation of long-chain fatty acids and, as a result, increased hydraulic retention time (HRT) due to the more resistant nature of these molecules. The management of pH is critical when dealing with this sort of waste (Aashabharathi et al., 2024). The optimal pH has been determined to be about 5.5, which is ideal for maximal hydrogenase enzyme activity. One advanced pH control technique is to recirculate digested effluent from the methanogenic phase, which is rich in buffer agents, to regulate the pH of dark fermentation. Recirculation enables the utilization of digestate's remaining buffer capacity (ammonium, bicarbonate) to deliver nutrients while diluting the feedstock (Bolzonella et al., 2018). Temperature and the kind of reactor to be utilized are other important criteria to consider. Continuous stirred tank reactor (CSTR) is most often employed for dark fermentation, whereas for the methanogenic phase anaerobic fluidized bed reactor (AFBR) is considered (Deheri and Acharya, 2022; Ghosh and Kar, 2022). For instance, for the production of biohythane using food waste, an integrated CSTR and an AFBR were used, providing 115.2 and 334.7 L/kg VS yield, respectively. During the experiment, the effluent from the AFBR was recirculated to the CSTR to assess its impact, which was discovered to help maintain the pH of the first stage reactor used to produce hydrogen.

Fruit and vegetable waste have also been shown to be useful in biohythane recovery. Tomato and potato processing leftovers account for the majority of vegetable processing waste worldwide. Similarly, apple and grape processing contributes to the overall fruit processing waste. The skins and seeds of extracted fruits and vegetables are known as pomace in the industry (Meena et al., 2020). In a study by Achmon et al. (2019), the biohythane yield was examined utilizing grape and tomato pomace wastes. The principal microbial species involved were *Methnosarcina*, *Methanocculleus*, and *Halanaerobiacea*, with H_2 and CH_4 production rates of 201 and 132 mL/g TS, respectively. In another study, Bolzonella et al. (2018) used thermophilic conditions to create biohythane from co-digested waste activated sludge (WAS) and organic waste, yielding 24 and 570 mL/g of H_2 and CH_4, respectively.

6.4.3 Industrial Solid Waste

Industrial effluents and wastewaters are growing increasingly complicated as sophisticated techniques are used in numerous industrial activities. These wastewaters endanger the health of the whole ecosystem, including water, air, and soil. Anaerobic methods for wastewater treatment are gaining popularity due to their environmental friendliness, cost-effectiveness, and capacity to recover clean bioenergy (Dhull et al., 2023). The creation of biohythane from these industrial wastewaters might meet the objective of generating clean energy and managing toxicity by reducing COD and biological oxygen demand (BOD). The potential feedstocks for anaerobic digestion include waste generated and disposed of by industries such as the dairy industry, the paper and pulp industry, the black liquor industry, the packaging industry, the shredded or spoiled vegetable industry, and the animal or poultry manure industry (Mozhiarasi et al., 2023). The black, semisolid form of palm oil mill effluent (POME) indicates a low pH at higher temperatures. According to O-Thong et al. (2008), POME contains high levels of suspended solids (8.5–12 g/L), total nitrogen (0.83–0.92 g/L), total phosphorus (0.097–0.125 g/L), and total solids (35–42 g/L). The POME with VFA (mostly butyrate and acetate) after first-stage dark fermentation is often referred to as acidified POME. The influence of VFA present in acidified POME on biohythane synthesis is noteworthy. A low VFA loading (0.9–1.8 g/L) reportedly shows a 15–20% greater methane yield than a high VFA loading (3.6–4.7 g/L). The presence of large concentrations (8 g/L) of butyric acids has a detrimental influence on the methane production process individually; also, an interaction of lactic, acetic, and butyric acids with propionic acid at concentrations greater than 0.5 g/L results in an inhibitory condition. Under thermophilic conditions, biohythane was synthesized on a lab scale using wastewater from palm oil mills. The study's conclusions showed that 1.93 L/L reactor/day of biohythane, consisting of 52% CH_4, 37% CO_2, and 11% H_2, could be produced. The resultant H_2/CH_4 ratio, which ranged from 0.13 to 0.18, showed that it was suitable for use as motor fuel.

6.4.4 Brewery Processing Wastes

Because of its high organic content, some liquid wastewater created during winemaking prewashing offers a lot of promise for biohythane production. Industrial wastewater containing dissolved organic compounds, such as sugars and lipids, might theoretically be anaerobically digested for the production of biogas and other biofuels like bioethanol (Fu et al., 2017). The beer, whiskey, and tequila industries create three forms of brewery processing waste: brewer's waste grain, pot ale, and vinasse. These wastes include 9–25% cellulose and 7–28% lignin, respectively. These qualities may increase the danger of environmental contamination if disposed untreated (Estevam et al., 2018). The brewery business is one of India's 17 most polluting industries, with an annual emission of 30–40 billion gallons. It has a high COD (60–120 g/L) and BOD (45–60 g/L) and has a distinctive dark brown hue. Melanoidins ($C_{17-18}H_{26-27}O_{10}N$), a color-causing pigment, are responsible for the brown hue (Rawoof et al., 2021). Two-stage fermentation of distillery-spent wash effluent mixed with sewage wastewater can be utilized efficient biohythane synthesis (Estevam et al., 2018). . The greatest amount of biohythane produced under mesophilic conditions (35°C) at an optimal leachate concentration (30% at 60 g/L substrate concentration) was 42 mmol CH_4/L and 67 mmol H_2/L biohydrogen.

6.4.5 Dairy Processing Wastes

In dairy processing, particularly cheese making, significant quantities of liquid by-products are generated, notably cheese whey. Cheese whey and dairy waste permeate (which primarily consists of lactose) are acidic liquids containing mineral salts, phosphates, nitrogen, proteins, oils, and grease. The waste from cheese production also contains proteins, lipids, trace minerals, salts such as NaCl and KCl, and acids like citric and uric acids. Waste with a high organic content, characterized by elevated levels of BOD (up to 50 g/L) and COD (reaching up to 80 g/L), along with an acidic pH and high biodegradability, may pose contamination risks and render it unsuitable for land disposal or aquatic environments. Animal manure from the butcher, dairy, and swine sectors is higher in ash, moisture, and organic matter. Nguyen et al. (2022) produced biohythane by co-digesting swine dung and pineapple trash using gel-entrapped anaerobic microbial consortia. At a 6-hour HRT, the resulting H_2 and CH_4 production rates were 1240 and 812 mL/L reactor/d, respectively.

6.4.6 Municipal Solid Wastes

Municipal solid waste (MSW) contains carbohydrates, lipids, and proteins, making it a possible renewable resource for energy recovery. Domestic garbage is gathered annually in millions of tons, with the majority of it being dumped openly in wastelands or on lake coastlines with no treatment or safety standards in place (Mozhiarasi et al., 2023). In two-stage biohythane production using organic fraction of municipal solid waste (OFMSW) for acidogenic hydrogen generation, followed by CH_4 synthesis, can increase the energy recovery efficiency of the methanogenic reactor by employing the acidogenic reactor's waste materials as a substrate in second-stage reactor for methane production (Prashanth Kumar et al., 2019). The anaerobic co-digestion of WAS and organic fertilizer (OFMSW) in single- and two-stage processes was examined by Bolzonella et al. (2020). While the two-step AD yields 24 L/kg VS of H_2 in the first stage and 570 L/kg VS of CH_4 in the second, the single-stage process produced 490 L/kg VS of feed.

6.5 FACTORS AFFECTING BIOHYTHANE PRODUCTION

The production of biohythane is significantly affected by different parameters such as feedstock composition, pH, microbial consortium or inoculum used, hydrogen concentration, production of VFAs and ammonia, carbon/nitrogen (C/N) ratio, alkalinity, temperature, HRT, reactor configuration, etc.

6.5.1 pH

pH plays a crucial role in maintaining reactor stability in both dark fermentation and methanogenesis processes. It significantly impacts the activity of fermentative bacteria (acid-producing) and acetoclastic methanogens (methane-producing), as well as the oxidation-reduction potential of anaerobic digestion reactions. The optimal pH range for achieving maximum H_2 production is typically considered to be between 5.5 and 6.5 (Bolzonella et al., 2020; Cai et al., 2010). pH control is necessary to maintain the reactor stability as the organic acids produced in the first stage tend to decrease the pH of the AD process. A pH drop below 4.5 affects the activity of the hydrogenase enzyme and has the potential to shift H_2 production to solvent (butanol, ethanol, and acetone) production (Kim et al., 2022). Therefore, appropriate buffering and nutrient dispersal are vital in the fermentation process to ensure vigorous microbial growth and to offset pH fluctuations triggered by the production of organic acids (O-Thong et al., 2008). pH also embraces substantial importance in the AD process, with an optimal range typically falling between 6.5 and 8.0 (Carotenuto et al., 2016). In dark fermentation reactions, pH drop is often observed due to the production of VFAs, particularly under high organic loading rates in the AD reactor. The decreasing pH can significantly impact the production of methane, underscoring the necessity for pH maintenance to achieve effective biohythane yield (Preethi et al., 2019).

6.5.2 Temperature

Temperature is a crucial parameter that exerts a significant influence on microbial development within biochemical processes, playing a key role in maintaining homeostasis and modulating metabolic pathways. Enzymatic reactions, which are primarily characterized by reaction temperature and pH differences, mediate cellular metabolism. Enzyme productivity is maximum at a specific temperature range, and significant temperature variations can cause metabolic and life-sustaining enzymes to become inactive or denatured. Therefore, temperature is important for the generation of H_2 and CH_4 and the microorganisms involved can function in a range of temperatures (Hung et al., 2011). During the initial stage, there are three temperature ranges in which the dark fermentation reactor operates: mesophilic (35°C–45°C), thermophilic (55°C–60°C), and extreme thermophilic (70°C–80°C) conditions. Most fermentation reactors employ mesophilic bacteria, such as *Clostridium* sp. and *Enterobacter* sp., and operate within mesophilic conditions (Mozhiarasi et al., 2023). In contrast to mesophilic reactors, thermophilic H_2-generating reactors are considered to yield the highest amounts of H_2. In comparison to mesophilic conditions, where approximately 1.7 mol of H_2/mol of glucose is produced, thermophilic conditions yield a thermodynamically advantageous H_2 output of around ~2.1 mol of H_2/mol of glucose. Additionally, thermophilic H_2-generating microorganisms exhibit low solubility of carbon dioxide and hydrogen and enhanced substrate hydrolysis and are minimally impacted by H_2 partial pressure. Therefore, temperature has a significant impact on the metabolic product, the properties of microbial cells, and the rate of reaction (Kim et al., 2022). Similar to the first stage, temperature plays a significant role in affecting the stability of the anaerobic digestion process in the second stage too. Thermophilic processes produce more methane than mesophilic processes (An et al., 2024). Thermophilic conditions not only decrease the solubility of carbon dioxide and methane but also enhance the thermodynamic efficiency of methane production and eliminate pathogens present in the anaerobic reactor effluent. However, slight temperature fluctuations can hinder methanogen performance, leading to the accumulation of VFAs and a decline in methane production (Sukphun et al., 2023). While all anaerobic microorganisms experience a notable reduction in methane yield as temperatures decrease, they can recover if suitable temperature stabilization is maintained, which is crucial for efficient biohythane synthesis.

6.5.3 Alkalinity

The alkalinity describes the medium's ability to act as a buffer. The ions such as Ca^{2+}, Mg^{2+}, citrates, phosphates, and carbonates control the buffer solution's strength. The buffering of the medium involves regulating organic acid accumulation. In order to increase the buffering capacity of a dark fermentation reactor, it was filled with the effluent from the second-stage anaerobic reactor having 1000–2000 mg/L alkalinity range as pH drop reduces the output of biohydrogen in a CSTR system (Yeshanew et al., 2016). In a study by Venkata Mohan (2009), the biohydrogen production from dairy effluent was examined in an anaerobic sequencing batch reactor having an alkalinity range of 125–1600 mg/L and an OLR of 3.5 kg COD/m^3/day, and it was found to have steady hydrogen production. Therefore, the concentration of alkalinity has a major impact on hydrogen generation. Therefore, depending on the type of feedstock and the reactor settings, the range of alkalinity needs to be regulated.

6.5.4 Volatile Fatty Acid Production

The intermediate metabolic products known as VFA can be produced by AD during the phases of acetogenesis and acidogenesis. After acidogenesis, switching to a short reaction time might inhibit methanogen activity and promote the buildup of volatile organic compounds (VOCs), i.e., butyric and acetic acids (Algapani et al., 2018). VFAs play a crucial role in affecting the biohydrogen and biomethane yield (Atasoy et al., 2018). The highest possible concentration of acetic acid could be converted to CH_4 (Khan et al., 2016). VFA buildup in digesters results in issues with the environment and operation (Kim et al., 2022). Mesophilic temperature conditions may indicate the creation of acetate and propionate, whereas thermophilic temperature conditions may indicate the formation of butyrate. Moreover, the buildup of VFA is significantly impacted by mesophilic temperature. Irrespective of the composition of the feedstock, longer HRTs lead to the buildup of VFA while shorter HRTs produce less VFA (Atasoy et al., 2018). Khan et al. (2016) found the accumulation of acetic, propionic, butyric, and lactic acids in the case of sugarcane molasses from the sugar manufacturing process. VFAs possess reduced danger of storage and transportation when compared to biogas. As a result, VFA and intermediate products serve as inexpensive raw materials for a variety of industrial uses, including the production of biopolymers, biodiesel, biogas, and many more (Atasoy et al., 2018).

6.5.5 Hydraulic Retention Time

Hydraulic retention time (HRT) is the average amount of time the feedstock needs to stay in the reactor and go through it before being released. It is dependent on the temperature and feedstock content and can be expressed as

$$HRT = \frac{V}{Q} \tag{6.7}$$

where V is the reactor volume and Q is the flow rate of the substrate.

Optimization of HRT is crucial for the reactor's continuous operation in the production of H_2/CH_4 (Dhull et al., 2023). In contrast to methanogenesis, low HRT is advantageous for the H_2 fermentation mechanism; nevertheless, too low HRT may cause active cells to wash out of the reactor. Additionally, microbes, elevated levels of inhibitory intermediary forms, an imbalanced food-to-microorganism ratio, etc. could result from high HRT (García-Depraect et al., 2022). As a result of metabolite inhibition and a decreased reactor pH, there will be less H_2 produced. HRT may also enhance microbial consortium growth. Changes in HRT in the first stage reactor will force slow-growing microorganisms such as methanogens and H_2-consuming bacteria to leave the system, increasing the amount of H_2-producing bacteria. Lowering HRT would thereby increase H_2-producing metabolism, decrease methanogens, and improve H_2 production. To achieve optimum performance and maximized profits, appropriate HRT must be maintained.

6.5.6 Ammonia Production

Ammonia production during the AD process has a significant effect on the yield of hydrogen and methane produced. Protein-rich processing wastes, such as animal effluents, during hydrolysis release ammonia (as free ammonia (NH_3) and ammonium ion (NH_4^+), which accumulate in the reactor and have an inhibitory impact causing process instability that ultimately results in system failure (Polizzi et al., 2018). On the other hand, the amount of ammonia may promote inhibition in methanogenesis as opposed to acidogenesis. Moreover, high OLR causes ammonia inhibition. Methanogens are better able to withstand elevated ammonia levels at thermophilic temperatures than at mesophilic ones. Elevated ammonia levels caused alterations in the transcriptional profiles of bacteria and archaea (Yenigün and Demirel, 2013). In short, the microbial community is poisonous to highly unionized ammonia. By piercing the cell membrane, it destroys the population. It is difficult for these bacteria to survive in an environment with such high ammonia concentrations (Meena et al., 2020).

6.5.7 Feedstock Composition

The feedstock's composition and type are among the important factors that influence the biohythane yield (Dhull et al., 2024). The most appropriate substrates for the creation of hydrogen are those high in carbohydrates, followed by those rich in protein, while the least favored feedstocks are those rich in fat. Generally, the presence of glucose or sucrose is required for H_2 generation during the initial stage of biohythane formation or dark fermentation (Mozhiarasi et al., 2023). Starch and cellulose are two primary substances found in organic solid wastes; glucose is their monomeric unit. Wastes high in carbohydrates yield more hydrogen than other types of feedstocks. For example, it has been observed that food wastes yield 205 H_2/kg VS of H_2, while cellulosic wastes, such as water hyacinth, yield 51.7 L H_2/kg VS of H_2 (Cheng et al., 2010). As animal dung is a complex feedstock, pretreatment is preferred for the production of H_2 before the dark fermentation process. For instance, heat treatment has proven to be more successful for the treatment of swine manure as compared to acid or alkali pretreatment.

6.5.8 Carbon/Nitrogen Ratio

The amount of carbon and nitrogen in a substrate is a major factor in determining reactor stability. The reactor's microbial community primarily uses carbon as an energy source, despite the fact that nitrogen is required for the growth of microbes. Based on the addition of co-substrates, existing studies show that a balanced C/N ratio improves energy production (Velusamy et al., 2020). In the second stage of anaerobic digestion, a C/N ratio in the range between 20 and 30 is thought to be optimal for both steady methane synthesis and stable H_2 production (Mamimin et al., 2017). Furthermore, the amount of phosphate present in raw material is crucial for dark fermentation because it serves as the framework for nucleic acid and ATP and maintains buffer conditions throughout the process. According to Elreedy et al. (2017), the yield of hydrogen increased 1.45 times while the ratio of nitrogen to phosphorus decreased from 8.5 to 4.6.

6.5.9 Microbial Consortium/Inoculum

Management of effective microbial consortia to produce the right H_2/CH_4 ratio is a crucial step in a two-stage fermentation process. In the first stage of the dark fermentation reaction, it is essential to remove H_2-consuming bacteria and develop an improved inoculant, consisting of hydrogen-producing bacteria. Numerous specific techniques are used for this enrichment, such as ultraviolet treatment, ultrasonication, thermal and chemical treatments, etc. (O-Thong et al., 2008). pH plays a decisive role in maintaining the diverse microbial consortia necessary to achieve the desired H_2/CH_4 ratio. During the selection process, H_2-producing bacteria, which are spore-forming, generate endospores (Elreedy et al., 2017). On the other hand, non-spore-forming H_2-consuming bacteria and methanogenic archaea are primarily removed from the system. Applying selective treatments prompts the germination of endospores under favorable conditions, leading to the prevalence of H_2-generating bacteria. Moreover, these bacteria are readily found in various mixed cultures, including sewage sludge, anaerobic sludge, compost, and others. For dark fermentative H_2 production, these cultures can be employed directly as an inoculum. Also, compared to pure cultures, mixed cultures are reported to be more adaptable and degrade waste better (Velusamy et al., 2020).

6.5.10 Partial Pressure of H_2

Another influencing parameter for biohythane production is the partial pressure of H_2 and it has been observed that a high partial pressure deactivates the enzyme hydrogenase. The partial pressure of H_2 in a batch thermophilic fermentation reactor is lowered by periodically sparging nitrogen, which may enhance H_2 production (Velusamy et al., 2020).

6.5.11 Trace Elements

Different kinds of metal ions, such as Ni^{2+}, Co^{2+}, Fe^{2+}, Mg^{2+}, and so forth, are micronutrients that are crucial to microbial metabolism which produces biohydrogen and biomethane. Trace levels of these micronutrients may play important roles as cofactors, transport process facilitators, and structural skeletons of enzymes involved in the biochemistry of dark fermentation-based H_2 generation, such as Fe–Fe hydrogenase and Ni–Fe hydrogenase (Romero-Güiza et al., 2016). According to Lee et al. (2001), adding 200 mg/L of Fe^{2+} ion increased the amount of H_2 produced from 131 to 196 mL H_2/g of sucrose. Further, nickel is a crucial metal that aids in the activities of enzymes such as hydrogenase, methyl coenzyme M reductase, and monoxide dehydrogenase which promote the synthesis of methane. Supplementing with magnesium ions also serves as a cofactor for numerous glycolysis-related enzymes. Therefore, the synthesis of biomethane and biohydrogen depends critically on the presence of micronutrients.

6.5.12 Reactor Configuration

The reactor type and configuration comprise the important position in terms of biohythane yield that allows various anaerobic microorganisms to survive and grow and perform their metabolic functions (Braguglia et al., 2018). While batch mode may be more economically feasible and easier to operate, continuous mode is more beneficial due to ease of monitoring important parameters, design simplicity and steady state continuous production. For biohythane production, two reactors with heads that are either the same type or different types are needed to produce H_2 and CH_4. Nualsri et al. (2016) treated sugarcane syrup to produce methanogenesis in a UASB reactor with a working volume of 24 L and acidogenesis in a CSTR digester operating at a volume of 1 L. Also, during the treatment of POME, anaerobic SBR produces more H_2, while a UASB reactor produces more CH_4 (Seengenyoung et al., 2019). Similarly,

Chavadej et al. (2019) discovered that when treating cassava wastewater with cassava residue, both UASB reactors are expected to boost biohythane production. All things considered, the first reactor is designed for quickly growing acidogenic bacteria, while the second reactor is designed for slowly growing methanogenic bacteria, which prepares the ground for the synthesis of biohythane. When it comes to the creation of biogas on an individual basis, particularly in DF, CSTR contributes to increase H_2 production in the treatment of FPW in comparison to reactors like AFBRs, packed bed reactors, membrane bioreactors, and UASB reactors. While UASB reactors drive the growth in CH_4 production during the treatment of FPW, CSTR is thought to be responsible for the full mixing and suspended form of acidogenic microorganisms in the reactor digestate (Alexandropoulou et al., 2018).

6.6 ENHANCEMENT TECHNIQUE

Some of the proposed techniques for increasing the biohythane yield are discussed further.

6.6.1 Pretreatment

The amount of carbohydrates in the fermentable feedstock is very significant in the output of biohydrogen and biomethane. Organic substrates such as food waste, agricultural waste, algae, etc. are rich sources of carbohydrates. These complex sugars are difficult to ferment in order to produce bioenergy. Therefore, pretreatment is necessary for these substrates in order to improve the fermentation process and the production of biohythane (Rawoof et al., 2020b). A few of the pretreatment methods that are employed to hydrolyze the organic waste include mechanical, thermal, physicochemical, biological, and combination methods. Each of these methods has benefits and drawbacks, and thus the type and composition of the substrate should be taken into consideration when choosing the optimal pretreatment procedure. While pretreatment methods like microwave, ultrasound, hydrodynamic cavitation, pulsed electric field, pyrolysis, and autoclave require less time and are very effective at disrupting cells, they also use a lot of energy and have the potential to cause structural changes and the production of free radicals. A perfect pretreatment method should be able to effectively hydrolyze the substrate and prevent inhibitors from building up throughout the acetogenic and methanogenic stages (Bolzonella et al., 2018). Applying many pretreatment processes might sometimes yield greater results than a single pretreatment process. Pretreatment ensures that the sugar compounds are available to the microorganisms and that they can be used by the enzymes for catalysis (Meena et al., 2020).

6.6.2 Co-digestion

The process of co-digesting food waste with other biowaste enhances the harmful chemical degradation. Co-digestion of substrates, such as compost, fruit and vegetable waste, and cow dung, enhances methane and hydrogen recovery (Rawoof et al., 2020b). In a study, cheese whey mono-digestion produced more biomethane, while cheese whey when co-digested with calf slurry produced more biohydrogen (Aathika et al., 2018). Furthermore, co-digestion can be used to maintain the pH of the fermentation system. Therefore, the digestion time is shortened when substrates like sewage sludge, agricultural residues, animal waste, and phytomass are used in combination with the primary substrate (Khoshnevisan et al., 2021).

6.6.3 Management of Microbial Consortia

The most crucial element in a two-stage process's effective generation of biohythane is maintaining the necessary microbial consortia (Khoshnevisan et al., 2021). To produce biohythane, two groups of microorganisms collaborate during the acetogenic and methanogenic stages. The native microorganisms in the fermentation system are unable to ferment substrates for prolonged period of time and are unable to produce biohydrogen and biomethane simultaneously (Lin and Lu, 2021). Pretreatment is necessary because the native microbial species cannot hydrolyze every type of substrate. Additionally, native microorganisms produce hydrogen at a slower rate than pure cultures (Mamimin et al., 2019). Genetically engineered microbes help to simultaneously hydrolyze and ferment substrates, increasing the production of biohythane. Direct bioprocessing, which involves saccharification and fermentation occurring simultaneously, has drawn increased attention recently due to the potential cost savings associated with using microbial strains that can produce hydrogen and cellulolytic enzymes concurrently. Bioaugmentation is another way to improve the hydrolysis and fermentation of substrates. This can be done by introducing a strain with a sufficient array of enzymes into the consortium or by supplying nutrients outside (Lin and Lu, 2021). Bioaugmentation has several benefits, including (i) the absence of substrate pretreatment requirements; (ii) the ability to decrease the bioreactor's organic loading rate; and (iii) the prevention of VFA accumulation during anaerobic fermentation (Shanmugam et al., 2020). It is critical to identify putative strains and comprehend how they behave in relation to the generation of biohythane. Nicotinamide adenine dinucleotide phosphate and nicotinamide adenine dinucleotide hydride are examples of reducing equivalents that consume the by-products produced during fermentation, increasing the amount of hydrogen produced (Rawoof et al., 2020a). Lactic acid is a significant by-product that uses the electron required for the hydrogenase enzyme and obstructs the neutralization process whereas propionic and butyric acids inhibit the methanogenesis process (Rawoof et al., 2020b). Therefore, the microbial consortia enrichment and the potential microbes are needed to be studied in detail to get more insights into the metabolic pathways and enhance the biohythane yield.

6.6.4 Process Integration

One of the most important aspects of the low-cost production of biohythane is being capable of producing biohydrogen and biomethane independently from a renewable organic feedstock. The two steps in the process of making biohythane are hydrogenesis and methanogenesis, which need to be optimized independently for the production of biohydrogen and biomethane. Generally, fermentation requires energy for heating, digestate recirculation, and separation of gases which increase the operation cost. So, combining the two-stage process with a continuous system for biohythane production can reduce the operational costs and help to understand the reaction kinetics (Rawoof et al., 2020a). The appropriate mixing of digestate in the fermentation mode and bioreactor architecture helps preserve the microorganisms' synergism without interfering with their metabolism (Rawoof et al., 2020b). Furthermore, maintaining a short hydraulic retention period and rotating the digestate help to keep the necessary microbial community in the reactor. During acidogenesis, the hydrogen producers produce VFAs, which further serve as a source of methane during methanogenesis (Rawoof et al., 2020b). During continuous operation, however, the use of the acidogenic phase residue in the methanogenic phase is a little more challenging. For this reason, improving biohythane production requires online monitoring of critical parameters and quality control using an integrated system. Hydrogen producers are also harmed by digestate recirculation from methanogenic to acetogenic reactors. Nevertheless, this can be avoided by using strategies like bioaugmentation, membrane filtering, and an increase in the rate of organic loading (Rajendran et al., 2020). In order to generate biohydrogen during the first stage of the two-stage process, dry anaerobic digestion in a leaching bed reactor is better suitable for substrates with total solids above 15%. The leachate can subsequently be used in high-rate reactors such as expanded granule sludge bed reactors and UASB for the second step of methane generation (Rawoof et al., 2021). CSTRs are better suited for the first stage of processing of substrates with total solids of 2–12%. The kind of digestate produced in the first stage determines the reactor layout for the second stage. Furthermore, UASB works better with liquid substrates whereas CSTR works with any substrate and is energy efficient (Corona and Razo-Flores, 2018). UASB is better suited for producing methane and hydrogen on liquid substrates with a total solids content of less than 2% (Rawoof et al., 2020a). During the process of integration, there are two primary elements that need to be taken into account. Initially, the slurry extracted from the first reactor following the synthesis of hydrogen may be fed directly into the second reactor for the production of methane or into the settling tank. Installing a settling tank in the middle of the two stages has the advantage of restoring the first stage's bacteria and changing the pH needed for the best possible methane output (Vo et al., 2019). The second stage's residue might then be released outside or fed back into the hydrogenic or methanogenic reactor. It may be possible to maintain pH and prevent organic overloading by recirculating slurry to the methanogenic reactor. On the other hand, recirculation to the hydrogenic reactor will increase the amount of hydrogenic bacteria and offer a pH buffer; however, it will also introduce undesirable microbes that consume hydrogen into the reactor. A few methods to overcome this problem include raising the hydrogenic reactor's organic loading rate, domesticating hydrogenic microorganisms, keeping the process temperature-phased, and membrane-filtering the methanogenic slurry from the second stage (Rawoof et al., 2020b). Process poses some drawbacks including expensive initial and ongoing maintenance as well as complicated designs. Microbial management between the two phases and the reactor's efficient integration are critical to achieving continuous production of methane and hydrogen. It is possible to enhance the process performance by managing parameters such as pH, retention time, and reflex ratio. The two-stage process is regulated by the reflex ratio; however, the rate of recirculation to the biomethane reactor can be improved to improve the hydrogen/methane ratio (Hans and Kumar, 2019).

6.7 CHALLENGES AND FUTURE PERSPECTIVE

As biohythane is seen and utilized as a potential biofuel for the increasing demands of energy throughout the world, there are still challenges to be faced in the area. Research on biohythane appears to be in its budding stage, with a number of lab-scale experiments demonstrating its viability and suggesting its advantages when applied to various feedstocks and experimental and operational settings (Yang et al., 2021). The gas distribution network stands in the way of widespread adoption of biohythane as a vehicle fuel. The problem arises with the occurrence of hydrogen in biohythane which entails different alterations of the pipelines used for the distribution. For example, it requires to use steel that is less likely to embrittle hydrogen under pressure, which increases its price (Liu et al., 2023). As an expanding research area, both fundamental and applied research is required for biohythane production in order to attain a sustainable engineering environment. More innovative solutions and strategies are needed to work out to address the obstruction of practical applications. Four major directions to work on in this regard include process scale-up, optimum microbial performance, process configuration and parameter optimization, and economic consideration (Shanmugam et al., 2021). Different kinds of pretreatment methods have been studied to achieve sustainable gaseous biohythane production using various feedstocks. But these methods pose their own pros and cons as no single method is optimal to be used, which requires more research to gain more insights (Shanmugam et al., 2021). Different meta-omics approaches have been considered to improve the microbial strains but they still lack comprehensive understanding to achieve better results. Gene sequencing studies have been the most practiced method to study microbial communities in two-stage anaerobic digestion (TSAD) (García-Depraect et al., 2022). Additionally, high-throughput sequencing (HTS) technologies like PacBio, Nanopore, etc. have significantly increased sequencing throughput and accuracy, enabling researchers to quickly produce massive datasets (Saini et al., 2022). For instance, Siqueira et al. (2022) used next-generation sequencing (NGS) metagenomic evaluation to determine which types of microbes were present in samples retrieved from blanket reactors that treat mixed vinasse and secondary effluent in anaerobic sludge. On the other hand, metagenomic analysis was employed by Greses et al. (2023) to examine the effects of small pH variations on the anaerobic fermentative ethanol generation process from food waste. In another multi-omic investigation, the impact of temperature variations on AD over time was inferred and applied to full-scale anaerobic digesters. The findings elucidate that microbial communities undergo metabolic shifts, accumulate short peptides, and diminish methane production (Sukphun et al., 2023). These technologies offer a means to bridge the knowledge divide between laboratory-scale thermophilic anaerobic digestion (TSAD) and the complex microbial interactions observed in large-scale systems (Basak et al., 2022). Genetic engineering studies have enhanced our understanding of metabolic electron flux for biohydrogen production and TSAD metabolic pathways, enabling the regulation of desirable product formation. In metabolic research focused on biohydrogen, enzymes like hypF and hupSL that hinder hydrogen production are typically suppressed, while hydA, zwf, and mcrA are overexpressed (Sukphun et al., 2023). Recently, genome-editing techniques consisting of CRISPR/Cas9 and CRISPR/AsCas12a have been affordable (Li et al. (2022) facilitating gene insertion, deletion, and manipulation of gene expression. This improvement helps in enhancing TSAD efficiency. Therefore, meta-omics approaches like metagenomics, meta-proteomics, meta-transcriptomics, and metabolomics and genetic engineering tools can be used to achieve a better understanding of the functional dynamics of microbial communities by providing details about their quantifying metabolites and proteins, metabolic interactions, and pathways (Kim et al., 2022). Further, this also helps to improve the efficiency of the microbial community for biohydrogen and biomethane production and offers an in-depth understanding of genes and their expression in TSAD. Another cutting-edge strategy for enhancing large-scale anaerobic digesters is bioaugmentation, which involves adding microorganisms that enhance already existing microbial populations or get beyond these limitations. In a study by Wongfaed et al. (2023), KKU-MC1, a microbial consortium, has been created which comprises Proteobacteria, Bacteroidetes, and Firmicutes responsible for degrading cellulose and hemicellulose. KKU-MC1 bioaugmentation has improved the methane yield up to 42% of lignocellulose feedstocks such as cassava bagasse, Napier grass, etc. According to Liu et al. (2023), bioaugmentation may enhance successive biohydrogen and methane production in TSAD. Finally, real-time monitoring, performance prediction, and process optimization have all benefited from the recent upsurge in popularity of artificial intelligence or machine learning (ML). Different studies have used AI-based algorithms like genetic simulated annealing algorithm (GSA), artificial neural network (ANN), and tree-based pipeline optimization tool (TPOT) and predicted H_2 and CH_4 and other gas compositions (Yang et al., 2021). There are benefits and drawbacks to these models, and further study is necessary for evaluating algorithms and identifying the best combinations of algorithms.

6.8 CONCLUSION

The potential of biohythane as a renewable fuel has been discussed with a focus on its efficient production and enhancement. In this regard, two-stage AD has been considered as a feasible option for the co-production of hydrogen and methane in the form of biohythane. As biohythane has a low inflammability range of methane, less ignition time and temperature, and no NO_x emissions, it proves to be a better alternative than CNG. However, the research on biohythane is still considered to be in the early stages with a pool of lab-scale studies which further needs attention to overcome the challenges faced for its implementation at a pilot scale. This can be done with more focused studies and necessary standardization in the operation of biohythane production and process designing. To address the challenges of practical application, much deeper investigations in terms of reactor design and microbial consortia development should be considered. Also, as a rising field in the renewable energy sector, biohythane production needs to be developed at both the basic and applied research levels. Additionally, the cost-effectiveness is also a crucial aspect and should be addressed when implemented at larger scale.

REFERENCES

Aashabharathi, M., Kumar, S.D., Shobana, S., Karthigadevi, G., Srinidhiy, C.A., Subbaiya, R., Karmegam, N., Kim, W. and Govarthanan, M., 2024. Biohythane production techniques and recent advances for green environment-A comprehensive review. *Process Saf. Environ. Prot.* 184, 400–410. https://doi.org/10.1016/j.psep.2024.01.099

Aathika A. R. S., Kubendran, D., Yuvarani, M., Thiruselvi, D., Amudha, T., Karthik, P., Sivanesan, S., 2018. Enhanced biohydrogen production from leather fleshing waste co-digested with tannery treatment plant sludge using anaerobic hydrogenic batch reactor. Energy Sources, Part A: Recovery, Energy Sources A: Recovery Util. Environ. Eff.40, 586–593. https://doi.org/10.1080/15567036.2018.1435754

Achmon, Y., Claypool, J.T., Pace, S., Simmons, B.A., Singer, S.W., Simmons, C.W., 2019. Assessment of biogas production and microbial ecology in a high solid anaerobic digestion of major California food processing residues. *Bioresour. Technol. Rep.* 5, 1–11. https://doi.org/10.1016/j.biteb.2018.11.007

Alexandropoulou, M., Antonopoulou, G., Trably, E., Carrere, H., Lyberatos, G., 2018. Continuous biohydrogen production from a food industry waste: Influence of operational parameters and microbial community analysis. *J. Clean. Prod.* 174, 1054–1063. https://doi.org/10.1016/j.jclepro.2017.11.078

Algapani, D.E., Qiao, W., di Pumpo, F., Bianchi, D., Wandera, S.M., Adani, F., Dong, R., 2018. Long-term bio-H2 and bio-CH4 production from food waste in a continuous two-stage system: Energy efficiency and conversion pathways. *Bioresour. Technol.* 248, 204–213. https://doi.org/10.1016/J.BIORTECH.2017.05.164

An, X., Xu, Y., Dai, X., 2024. Biohythane production from two-stage anaerobic digestion of food waste: A review. *J. Environ. Sci.* (China). https://doi.org/10.1016/j.jes.2023.04.031

Atasoy, M., Owusu-Agyeman, I., Plaza, E., Cetecioglu, Z., 2018. Bio-based volatile fatty acid production and recovery from waste streams: Current status and future challenges. *Bioresour. Technol.* 268, 773–786. https://doi.org/10.1016/j.biortech.2018.07.042

Basak, B., Patil, S.M., Kumar, R., Ahn, Y., Ha, G.S., Park, Y.K., Ali Khan, M., Jin Chung, W., Woong Chang, S., Jeon, B.H., 2022. Syntrophic bacteria- and Methanosarcina-rich acclimatized microbiota with better carbohydrate metabolism enhances biomethanation of fractionated lignocellulosic biocomponents. *Bioresour. Technol.* 360, 127602. https://doi.org/10.1016/j.biortech.2022.127602

Bolzonella, D., Battista, F., Cavinato, C., Gottardo, M., Micolucci, F., Lyberatos, G., Pavan, P., 2018. Recent developments in biohythane production from household food wastes: A review. *Bioresour. Technol.* https://doi.org/10.1016/j.biortech.2018.02.092

Bolzonella, D., Micolucci, F., Battista, F., Cavinato, C., Gottardo, M., Piovesan, S., Pavan, P., 2020. Producing Biohythane from Urban Organic Wastes. *Waste Biomass Valorization* 11, 2367–2374. https://doi.org/10.1007/s12649-018-00569-7

Braguglia, C.M., Gallipoli, A., Gianico, A., Pagliaccia, P., 2018. Anaerobic bioconversion of food waste into energy: A critical review. *Bioresour. Technol.* 248, 37–56. https://doi.org/10.1016/j.biortech.2017.06.145

Cai, G., Jin, B., Saint, C., Monis, P., 2010. Metabolic flux analysis of hydrogen production network by Clostridium butyricum W5: Effect of pH and glucose concentrations. *Int. J. Hydrog. Energy* 35, 6681–6690. https://doi.org/10.1016/j.ijhydene.2010.04.097

Carotenuto, C., Guarino, G., Morrone, B., Minale, M., 2016. Temperature and pH effect on methane production from buffalo manure anaerobic digestion. *Int. J. Heat Technol.*, 34(2), S425–S429.

Cavinato, C., Bolzonella, D., Fatone, F., Giuliano, A., Pavan, P., 2011. Two-phase thermophilic anaerobic digestion process for biohythane production treating biowaste: Preliminary results. *Water Sci. Technol.* 64, 715–721. https://doi.org/10.2166/wst.2011.698

Cavinato, C., Giuliano, A., Bolzonella, D., Pavan, P., Cecchi, F., 2012. Bio-hythane production from food waste by dark fermentation coupled with anaerobic digestion process: A long-term pilot scale experience. *Int. J. Hydrog. Energy* 37, 11549–11555. https://doi.org/10.1016/j.ijhydene.2012.03.065

Chavadej, S., Wangmor, T., Maitriwong, K., Chaichirawiwat, P., Rangsunvigit, P., Intanoo, P., 2019. Separate production of hydrogen and methane from cassava wastewater with added cassava residue under a thermophilic temperature in relation to digestibility. *J Biotechnol* 291, 61–71. https://doi.org/10.1016/j.jbiotec.2018.11.015

Cheng, J., Xie, B., Zhou, J., Song, W., Cen, K., 2010. Cogeneration of H2 and CH4 from water hyacinth by two-step anaerobic fermentation. *Int. J. Hydrog. Energy* 35, 3029–3035. https://doi.org/10.1016/j.ijhydene.2009.07.012

Constant, P., Hallenbeck, P.C., 2019. *Hydrogenase. Biomass, Biofuels, Biochemicals: Biohydrogen*, 2nd Edition, 49–78. https://doi.org/10.1016/B978-0-444-64203-5.00003-4

Corona, V.M. and Razo-Flores, E., 2018. Continuous hydrogen and methane production from Agave tequilana bagasse hydrolysate by sequential process to maximize energy recovery efficiency. *Bioresour. Technol.* 249, 334–341.

Deheri, C., Acharya, S.K., 2022. Purified biohythane (biohydrogen+biomethane) production from food waste using CaO2+CaCO3 and NaOH as additives. *Int. J. Hydrog. Energy* 47, 2862–2873. https://doi.org/10.1016/j.ijhydene.2021.10.232

Dhull, P., Lohchab, R.K., Kumar, S., Kumari, M., Shaloo, Bhankhar, A.K., 2023. Anaerobic digestion: Advance techniques for enhanced biomethane/biogas production as a source of renewable energy. *BioEnergy Research* 2023 1–22. https://doi.org/10.1007/S12155-023-10621-7

Elreedy, A., Fujii, M., Tawfik, A., 2017. Factors affecting on hythane bio-generation via anaerobic digestion of mono-ethylene glycol contaminated wastewater: Inoculum-to-substrate ratio, nitrogen-to-phosphorus ratio and pH. *Bioresour. Technol.* 223, 10–19. https://doi.org/10.1016/J.BIORTECH.2016.10.026

Estevam, A., Arantes, M.K., Andrigheto, C., Fiorini, A., da Silva, E.A., Alves, H.J., 2018. Production of biohydrogen from brewery wastewater using Klebsiella pneumoniae isolated from the environment. *Int. J. Hydrog. Energy* 43, 4276–4283. https://doi.org/10.1016/j.ijhydene.2018.01.052

Fu, S.F., Xu, X.H., Dai, M., Yuan, X.Z., Guo, R.B., 2017. Hydrogen and methane production from vinasse using two-stage anaerobic digestion. *Process Saf. Environ. Prot.* 107, 81–86. https://doi.org/10.1016/J.PSEP.2017.01.024

Gadhe, A., Sonawane, S.S., Varma, M.N., 2013. Optimization of conditions for hydrogen production from complex dairy wastewater by anaerobic sludge using desirability function approach. *Int. J. Hydrog. Energy* 38, 6607–6617. https://doi.org/10.1016/j.ijhydene.2013.03.078

García-Depraect, O., Martínez-Mendoza, L.J., Diaz, I., Muñoz, R., 2022. Two-stage anaerobic digestion of food waste: Enhanced bioenergy production rate by steering lactate-type fermentation during hydrolysis-acidogenesis. *Bioresour. Technol.* 358, 127358. https://doi.org/10.1016/j.biortech.2022.127358

Genovese, A., Ortenzi, F., 2016. Enriched Methane for city public transport buses. *Green Energy and Technology* 195–213. https://doi.org/10.1007/978-3-319-22192-2_11/COVER

Ghosh, S., Kar, D., 2022. Biohythane: a Potential Biofuel of the Future. *Appl Biochem Biotechnol*. https://doi.org/10.1007/s12010-022-04291-y

Dhull, P., Kumar, S., Yadav, N., Lohchab, R. K. 2024. A comprehensive review on anaerobic digestion with focus on potential feedstocks, limitations associated and recent advances for biogas production. *Environmental Science and Pollution Research*, 1–36. http://dx.doi.org/10.1007/s11356-024-33736-6.

Greses, S., De Bernardini, N., Treu, L., Campanaro, S., González-Fernández, C., 2023. Genome-centric metagenomics revealed the effect of pH on the microbiome involved in short-chain fatty acids and ethanol production. *Bioresour. Technol*. 377, 128920.

Han, S.K., Shin, H.S., 2004. Performance of an Innovative Two-Stage Process Converting Food Waste to Hydrogen and Methane. *J. Air Waste Manage. Assoc*. 54, 242–249. https://doi.org/10.1080/10473289.2004.10470895

Hans, M., Kumar, S., 2019. Biohythane production in two-stage anaerobic digestion system. *Int. J. Hydrog. Energy* 44, 17363–17380. https://doi.org/10.1016/j.ijhydene.2018.10.022

Harnvoravongchai, P., Singwisut, R., Ounjai, P., Aroonnual, A., Kosiyachinda, P., Janvilisri, T., Chankhamhaengdecha, S., 2020. Isolation and characterization of thermophilic cellulose and hemicellulose degrading bacterium, Thermoanaerobacterium sp. R63 from tropical dry deciduous forest soil. *PLoS One* 15, e0236518. https://doi.org/10.1371/journal.pone.0236518

Hu, J., Yu, F., Lu, Y., 2012. Application of Fischer–Tropsch synthesis in biomass to liquid conversion. *Catalysts* 2, 303–326. https://doi.org/10.3390/CATAL2020303

Hung, C.H., Chang, Y.T., Chang, Y.J., 2011. Roles of microorganisms other than Clostridium and Enterobacter in anaerobic fermentative biohydrogen production systems – A review. *Bioresour. Technol*. 102, 8437–8444. https://doi.org/10.1016/j.biortech.2011.02.084

Jiraprasertwong, A., Maitriwong, K., Chavadej, S., 2019. Production of biogas from cassava wastewater using a three-stage upflow anaerobic sludge blanket (UASB) reactor. *Renew Energy* 130, 191–205. https://doi.org/10.1016/j.renene.2018.06.034

Kamalaskar, L.B., Dhakephalkar, P.K., Meher, K.K., Ranade, D.R., 2010. High biohydrogen yielding Clostridium sp. DMHC-10 isolated from sludge of distillery waste treatment plant. *Int. J. Hydrog. Energy* 35, 10639–10644. https://doi.org/10.1016/j.ijhydene.2010.05.020

Khan, M.A., Ngo, H.H., Guo, W.S., Liu, Y., Nghiem, L.D., Hai, F.I., Deng, L.J., Wang, J., Wu, Y., 2016. Optimization of process parameters for production of volatile fatty acid, biohydrogen and methane from anaerobic digestion. *Bioresour. Technol*. 219, 738–748. https://doi.org/10.1016/j.biortech.2016.08.073

Khoshnevisan, B., Duan, N., Tsapekos, P., Awasthi, M.K., Liu, Z., Mohammadi, A., Angelidaki, I., Tsang, D.C., Zhang, Z., Pan, J., Ma, L., 2021. A critical review on livestock manure biorefinery technologies: Sustainability, challenges, and future perspectives. *Renew. Sustain. Energy Rev*. 135, 110033.

Kim, H.H., Saha, S., Hwang, J.H., Hosen, M.A., Ahn, Y.T., Park, Y.K., Khan, M.A., Jeon, B.H., 2022. Integrative biohydrogen- and biomethane-producing bioprocesses for comprehensive production of biohythane. *Bioresour. Technol*. 365. https://doi.org/10.1016/j.biortech.2022.128145

Lee, Y.J., Miyahara, T., Noike, T., 2001. Effect of iron concentration on hydrogen fermentation. *Bioresour. Technol*. 80, 227–231. https://doi.org/10.1016/S0960-8524(01)00067-0

Li, J., He, J., Si, B., Liu, Z., Zhang, C., Wang, Y., Xing, X., 2020. A pilot study of biohythane production from cornstalk via two-stage anaerobic fermentation. *Int. J. Hydrog. Energy* 45, 31719–31731. https://doi.org/10.1016/j.ijhydene.2020.08.253

Li, J., Zhang, L., Xu, Q., Zhang, W., Li, Z., Chen, L., Dong, X., 2022. CRISPR-Cas9 toolkit for genome editing in an Autotrophic CO 2 -fixing methanogenic archaeon. *Microbiol Spectr* 10. https://doi.org/10.1128/SPECTRUM.01165-22

Li, Y., Zhao, Q., Feng, X., Wang, W., Zhang, R., Yan, A., 2019. A variational image segmentation method exploring both intensity means and texture patterns. *Signal Process Image Commun.* 76, 214–230. https://doi.org/10.1016/J.IMAGE.2019.05.002

Lin, C.Y., Lu, C., 2021. Development perspectives of promising lignocellulose feedstocks for production of advanced generation biofuels: A review. *Renew. Sustain. Energy Rev.* 136, 110445.

Liu, Xin, Zhu, X., Yellezuome, D., Liu, R., Liu, X., Sun, C., Abd-Alla, M.H., Rasmey, A.H.M., 2023. Effects of adding *Thermoanaerobacterium thermosaccharolyticum* in the hydrogen production stage of a two-stage anaerobic digestion system on hydrogen-methane production and microbial communities. *Fuel* 342, 127831. https://doi.org/10.1016/j.fuel.2023.127831

Luo, L., Kaur, G., Zhao, J., Zhou, J., Xu, S., Varjani, S., Wong, J.W.C., 2021. Optimization of water replacement during leachate recirculation for two-phase food waste anaerobic digestion system with off-gas diversion. *Bioresour. Technol.* 335, 125234. https://doi.org/10.1016/j.biortech.2021.125234

Mamimin, C., Prasertsan, P., Kongjan, P., O-Thong, S., 2017. Effects of volatile fatty acids in biohydrogen effluent on biohythane production from palm oil mill effluent under thermophilic condition. *Electron. J. Biotechnol.* 29, 78–85. https://doi.org/10.1016/J.EJBT.2017.07.006

Mamimin, C., Probst, M., Gómez-Brandón, M., Podmirseg, S.M., Insam, H., Reungsang, A., O-Thong, S., 2019. Trace metals supplementation enhanced microbiota and biohythane production by two-stage thermophilic fermentation. *Int. J. Hydrog. Energy* 44, 3325–3338. https://doi.org/10.1016/j.ijhydene.2018.09.065

Massanet-Nicolau, J., Dinsdale, R., Guwy, A., Shipley, G., 2013. Use of real time gas production data for more accurate comparison of continuous single-stage and two-stage fermentation. *Bioresour. Technol.* 129, 561–567. https://doi.org/10.1016/j.biortech.2012.11.102

Meena, R.A.A., Banu, J.R., Kannah, R.Y., Yogalakshmi, K.N., Kumar, G., 2020. Biohythane production from food processing wastes–challenges and perspectives. *Bioresour. Technol.* 298, 122449. https://doi.org/10.1016/j.biortech.2019.122449

Mozhiarasi, V., Natarajan, T.S. and Dhamodharan, K., 2023. A high-value biohythane production: feedstocks, reactor configurations, pathways, challenges, technoeconomics and applications. *Environ. Res.* 219, 115094. https://doi.org/10.1016/j.envres.2022.115094

Nguyen, T.T., Ta, D.T., Lin, C.Y., Chu, C.Y., Ta, T.M.N., 2022. Biohythane production from swine manure and pineapple waste in a single-stage two-chamber digester using gel-entrapped anaerobic microorganisms. *Int. J. Hydrog. Energy* 47, 25245–25255. https://doi.org/10.1016/j.ijhydene.2022.05.259

Nualsri, C., Kongjan, P., Reungsang, A., 2016. Direct integration of CSTR-UASB reactors for two-stage hydrogen and methane production from sugarcane syrup. *Int. J. Hydrog. Energy* 41, 17884–17895. https://doi.org/10.1016/j.ijhydene.2016.07.135

Thong, S., Prasertsan, P., Intrasungkha, N., Dhamwichukorn, S., Birkeland, N. Kå., 2008. Optimization of simultaneous thermophilic fermentative hydrogen production and COD reduction from palm oil mill effluent by Thermoanaerobacterium-rich sludge. *Int. J. Hydrog. Energy* 33, 1221–1231. https://doi.org/10.1016/j.ijhydene.2007.12.017

Polizzi, C., Alatriste-Mondragón, F., Munz, G., 2018. The role of organic load and ammonia inhibition in anaerobic digestion of tannery fleshing. *Water Resour. Ind.* 19, 25–34. https://doi.org/10.1016/j.wri.2017.12.001

Prashanth Kumar, C., Rena, Meenakshi, A., Khapre, A.S., Kumar, S., Anshul, A., Singh, L., Kim, S.H., Lee, B.D., Kumar, R., 2019. Bio-Hythane production from organic fraction of municipal solid waste in single and two stage anaerobic digestion processes. *Bioresour. Technol.* 294. https://doi.org/10.1016/j.biortech.2019.122220

Preethi, Usman, T.M.M., Rajesh Banu, J., Gunasekaran, M., Kumar, G., 2019. Biohydrogen production from industrial wastewater: An overview. *Bioresour. Technol. Rep* 7, 100287. https://doi.org/10.1016/j.biteb.2019.100287

Rajendran, K., Mahapatra, D., Venkatraman, A.V., Muthuswamy, S. and Pugazhendhi, A., 2020. Advancing anaerobic digestion through two-stage processes: Current developments and future trends. *Renew. Sustain. Energy Rev.* 123, 109746.

Rawoof, S.A.A., Kumar, P.S., Vo, D.V.N., Devaraj, K., Mani, Y., Devaraj, T., Subramanian, S., 2020a. Production of optically pure lactic acid by microbial fermentation: a review. *Environ. Chem. Lett.* 19, 539–556. https://doi.org/10.1007/S10311-020-01083-W

Rawoof, S.A.A., Kumar, P.S., Vo, D.V.N., Devaraj, T., Subramanian, S., 2021. Biohythane as a high potential fuel from anaerobic digestion of organic waste: A review. *Renew. Sustain. Energy Rev.*152. https://doi.org/10.1016/j.rser.2021.111700

Rawoof, S.A.A., Kumar, P.S., Vo, D.V.N., Subramanian, S., 2020b. Sequential production of hydrogen and methane by anaerobic digestion of organic wastes: a review. *Environ. Chem. Lett.* 19, 1043–1063. https://doi.org/10.1007/S10311-020-01122-6

Romero-Güiza, M.S., Vila, J., Mata-Alvarez, J., Chimenos, J.M., Astals, S., 2016. The role of additives on anaerobic digestion: A review. *Renew. Sustain. Energy Rev.* 58, 1486–1499. https://doi.org/10.1016/J.RSER.2015.12.094

Saini, N., Kumar, S., Deepak, B., Mona, S., 2022. High-Throughput Sequencing Technologies in Metagenomics: Advanced Approaches for Algal Research. In: Kumar, V., Thakur, I.S. (eds) *Omics Insights in Environmental Bioremediation* 545–569. https://doi.org/10.1007/978-981-19-4320-1_23

Schievano, A., Tenca, A., Lonati, S., Manzini, E., Adani, F., 2014. Can two-stage instead of one-stage anaerobic digestion really increase energy recovery from biomass? *Appl. Energy* 124, 335–342. https://doi.org/10.1016/j.apenergy.2014.03.024

Seengenyoung, J., Mamimin, C., Prasertsan, P., O-Thong, S., 2019. Pilot-scale of biohythane production from palm oil mill effluent by two-stage thermophilic anaerobic fermentation. *Int. J. Hydrog. Energy* 44, 3347–3355. https://doi.org/10.1016/j.ijhydene.2018.08.021

Shanmugam, S., Ngo, H.H., Wu, Y.R., 2020. Advanced CRISPR/Cas-based genome editing tools for microbial biofuels production: A review. *Renew. Energy* 149, 1107–1119. https://doi.org/10.1016/j.renene.2019.10.107

Shanmugam, S., Sekar, M., Sivaramakrishnan, R., Raj, T., Ong, E.S., Rabbani, A.H., Rene, E.R., Mathimani, T., Brindhadevi, K., Pugazhendhi, A., 2021. Pretreatment of second and third generation feedstock for enhanced biohythane production: Challenges, recent trends and perspectives. *Int. J. Hydrog. Energy* 46, 11252–11268. https://doi.org/10.1016/j.ijhydene.2020.12.083

Sillero, L., Solera, R., Perez, M., 2022. Anaerobic co-digestion of sewage sludge, wine vinasse and poultry manure for bio-hydrogen production. *Int. J. Hydrog. Energy* 47, 3667–3678. https://doi.org/10.1016/j.ijhydene.2021.11.032

Siqueira, J.C. de, Assemany, P., Siniscalchi, L.A.B., 2022. Microbial dynamics and methanogenic potential of co-digestion of sugarcane vinasse and dairy secondary effluent in an upflow anaerobic sludge blanket reactor. *Bioresour. Technol.* 361, 127654. https://doi.org/10.1016/j.biortech.2022.127654

Sukphun, P., Wongarmat, W., Imai, T., Sittijunda, S., Chaiprapat, S., Reungsang, A., 2023. Two-stage biohydrogen and methane production from sugarcane-based sugar and ethanol industrial wastes: A comprehensive review. *Bioresour. Technol.* https://doi.org/10.1016/j.biortech.2023.129519

Ta, D.T., Lin, C.Y., Ta, T.M.N., Chu, C.Y., 2020. Biohythane production via single-stage anaerobic fermentation using entrapped hydrogenic and methanogenic bacteria. *Bioresour. Technol.* 300. https://doi.org/10.1016/j.biortech.2019.122702

Velusamy, M., Speier, C.J., Michealammal, B.R.P., Shrivastava, R., Rajan, B., Weichgrebe, D., Venkatachalam, S.S., 2020. Bio-reserves inventory—improving substrate management for anaerobic waste treatment in a fast-growing Indian urban city, Chennai. Environ. *Sci. Pollut. Res.* 27, 29749–29765. https://doi.org/10.1007/S11356-019-07321-1/FIGURES/5

Venkata Mohan, S., 2009. Harnessing of biohydrogen from wastewater treatment using mixed fermentative consortia: Process evaluation towards optimization. *Int. J. Hydrog. Energy* 34, 7460–7474. https://doi.org/10.1016/j.ijhydene.2009.05.062

Vo, T.P., Lay, C.H., Lin, C.Y., 2019. Effects of hydraulic retention time on biohythane production via single-stage anaerobic fermentation in a two-compartment bioreactor. *Bioresour. Technol.* 292, 121869. https://doi.org/10.1016/j.biortech.2019.121869

Wang, L., Shen, F., Yuan, H., Zou, D., Liu, Y., Zhu, B., Li, X., 2014. Anaerobic co-digestion of kitchen waste and fruit/vegetable waste: Lab-scale and pilot-scale studies. *Waste Management* 34, 2627–2633. https://doi.org/10.1016/j.wasman.2014.08.005

Willquist, K., Nkemka, V.N., Svensson, H., Pawar, S., Ljunggren, M., Karlsson, H., Murto, M., Hulteberg, C., Van Niel, E.W.J., Liden, G., 2012. Design of a novel biohythane process with high H2 and CH4 production rates. *Int. J. Hydrog. Energy* 37, 17749–17762. https://doi.org/10.1016/j.ijhydene.2012.08.092

Wongfaed, N., O-Thong, S., Sittijunda, S., Reungsang, A., 2023. Taxonomic and enzymatic basis of the cellulolytic microbial consortium KKU-MC1 and its application in enhancing biomethane production. *Sci. Rep.* 13, 2968.

Yang, G., Li, Y., Zhen, F., Xu, Y., Liu, J., Li, N., Sun, Y., Luo, L., Wang, M., Zhang, L., 2021. Biochemical methane potential prediction for mixed feedstocks of straw and manure in anaerobic co-digestion. *Bioresour. Technol.* 326, 124745. https://doi.org/10.1016/j.biortech.2021.124745

Yenigün, O., Demirel, B., 2013. Ammonia inhibition in anaerobic digestion: A review. *Process Biochem.* 48, 901–911. https://doi.org/10.1016/j.procbio.2013.04.012

Yeshanew, M.M., Frunzo, L., Pirozzi, F., Lens, P.N.L., Esposito, G., 2016. Production of biohythane from food waste via an integrated system of continuously stirred tank and anaerobic fixed bed reactors. *Bioresour. Technol.* 220, 312–322. https://doi.org/10.1016/j.biortech.2016.08.078

Zhao, Q., Arhin, S.G., Yang, Z., Liu, H., Li, Z., Anwar, N., Papadakis, V.G., Liu, G., Wang, W., 2021. pH regulation of the first phase could enhance the energy recovery from two-phase anaerobic digestion of food waste. *Water Environ. Res.* 93, 1370–1380. https://doi.org/10.1002/WER.1527

7 Jet Fuel Production
Taking Flight with Microbial Innovation

Rajee Olaganathan

7.1 INTRODUCTION

Air transport plays a vital role in promoting social and business connections globally. In the year 2016, the International Air Transportation Association (IATA) estimated that air carriers transported more than 3.8 billion passengers and 54.9 million tons of cargo; this is approximately 35% of global trade (International Air Transportation Association [IATA], 2017). As per Statista (2023), global passenger numbers peaked at 4.54 billion in 2019. Unfortunately, the industry faced a setback in 2020 due to the COVID-19 pandemic. The passenger numbers fell drastically, and airlines worldwide experienced one of their highest revenue losses (Olaganathan, 2021). In 2021, the aviation industry faced severe obstacles from the pandemic's lasting impacts, with scheduled passenger numbers barely surpassing 2.2 billion – an approximate 50% drop compared to their levels before 2019 (Statista, 2023). Yet the cargo industry continued to prosper during the pandemic, with a value reaching over USD 175 billion during the pandemic's height, showing resilience during crises (Airport Council International [ACI], 2023).

The year 2021 saw real progress within the aviation industry as it recovered from the pandemic. According to the Airport Council International's 2023 annual report, passenger traffic fully rebounded to pre-pandemic levels by 2022 with global passenger numbers surpassing 6.6 billion; an improvement of 72.5% when compared with 2019 pre-pandemic figures; noteworthy is the United States leading in passenger transportation during this time (ACI, 2023). Lastly, the International Air Transport Association (IATA) forecast doubling passenger numbers within two decades by 2037 which in turn would increase fuel consumption as aviation passenger numbers double (IATA, 2017).

The United States continues its position as the global leader in jet fuel consumption, accounting for 25% or more annually (U.S. Energy Information Administration [U.S. EIA], 2015). Transportation sector jet fuel consumption was 11.6% or approximately 5.1 billion barrels according to 2017 U.S. EIA figures (U.S. Energy Information Administration [U.S. EIA], 2022). China was among the fastest-growing consumers of jet fuel between 2008 and 2013, showing an annual consumption increase of 12.0% from their earlier levels. China's domestic passenger market

DOI: 10.1201/9781003585398-7

it more accessible and economical (Rodrigue, 2020). In 2018, 20 US cities provided nonstop flights to at least one Asian destination compared to 13 recorded in 1998. This was something unheard of two decades earlier when travelers would have needed to pass through several larger hub airports on their way. Parallel efforts have also been put forth to introduce supersonic airliners. United Airlines placed orders with Boom Supersonic for 15 aircraft in 2021 (United Airlines, 2022). These new jets, each capable of seating 65–80 passengers and traveling at Mach 17 speeds, are projected to begin flying sometime around 2029 (Rodrigue, 2020). Though aviation's growth may unite nations around the globe, its ecological ramifications continue to escalate rapidly. Jet engines that run on fossil fuels release not only CO_2 but also nitrogen oxides and particulate matter pollutants as they burn (Olaganathan, 2018b).

With increased air traffic worldwide and the expansion of the global fleet, carbon emissions skyrocketed. Since aviation's contributions to climate change became apparent, concerns arose regarding their contribution. As a response, rigorous efforts were made to boost fuel efficiency and decrease emissions; developments such as high-bypass turbofan engines, advancements in aerodynamics, and lightweight materials all contributed to more fuel-efficient aircraft (ICAO, 2022a). Furthermore, stringent environmental regulations and agreements such as the International Civil Aviation Organization's CORSIA program seek to minimize emissions associated with aviation (ICAO, 2022a).

Biofuels mark a historic turning point in aviation history. Recognizing the environmental consequences of fossil fuels, scientists and policymakers identified an urgent need to reduce aviation's environmental impact through sustainable alternatives (Olaganathan, 2018a). Biofuels from renewable resources like plants or microorganisms offer one such promising approach that may rewrite history. In this chapter, we delve deeper into microbial approaches to producing jet fuel production as an avenue toward changing this narrative of air travel.

7.3 IMPACT OF TRADITIONAL JET FUELS ON GREENHOUSE GAS EMISSIONS

Despite the existence of various national and international laws and recommendations addressing the climate crisis, CO_2 levels continued to rise from 387 parts per million (ppm) in September 2008 to 419 ppm by October 2023 (National Aeronautics and Space Administration [NASA], 2023). Pacala and Socolow (2004) estimated global annual CO_2 emissions from fossil fuels at seven gigatons of carbon. Notably, CO_2 emissions in the United States increased by approximately 4% between 1990 and 2016 period (Environmental Protection Agency [EPA], 2018). The surge is attributed to the rapid growth in the economy, population, and the demand for air travel.

Figure 7.1 provides a visual representation of the environmental impacts of countries based on their per capita CO_2 emissions in the year 2018 (Graver et al., 2019). Iceland leads with 3,576.9 kg, driven by its significant reliance on geothermal and

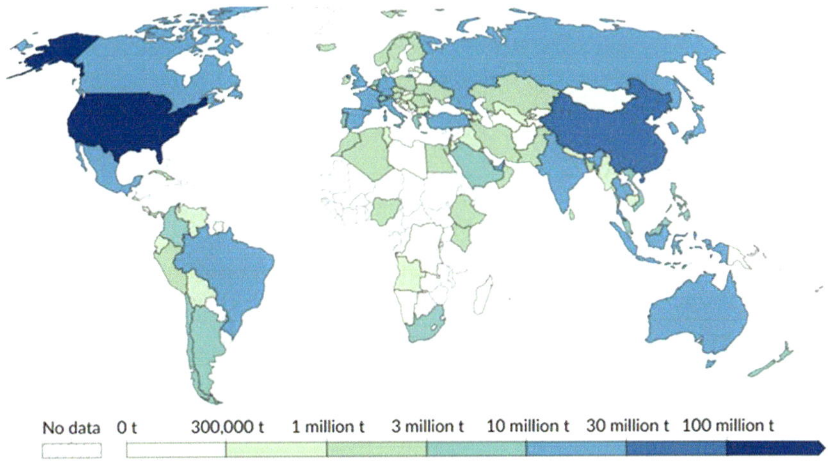

FIGURE 7.1 CO_2 Emissions from Commercial Aviation in 2018 (Adapted from Graver et al., 2019).

hydroelectric power (Graver et al., 2019). Qatar and the United Arab Emirates exhibit higher emissions at 2,472.7 and 2,195.2 kg, respectively, owing to their economic dependence on oil and natural gas industries (Graver et al., 2019). Singapore's urban dynamics result in 1,741.0 kg per capita emissions, while smaller countries like Malta (991.6 kg), Mauritius (608.5 kg), and Ireland (575.9 kg) showcase diverse emissions influenced by their economic structures and populations (Graver et al., 2019). New Zealand's per capita emissions of 814.5 kg are linked to its agricultural practices, and Australia (763.1 kg) is navigating toward sustainability amid resource-intensive endeavors. The United States, with 556.1 kg, represents a complex transition at the national level (Graver et al., 2019). These figures narrate a story about the diverse forces shaping our global environment.

The emissions from the transportation sector, primarily dominated by CO_2 as the most prevalent greenhouse gas (GHG), have surged from 4.6 billion tons in the year 1990 to 8.2 billion tons in 2019, constituting 16% of total emissions. In the year 2020, international aviation, contributing 55.879 kg CO_2 equivalent, accounted for 2% of total GHG emissions (Graver et al., 2019). Despite the limitations imposed by the COVID-19 pandemic in 2020, the demand for air travel within the transportation industry is rapidly rising. The IATA predicts a doubling of air travel demand by the year 2035, with a compound average growth rate (CAGR) of 3.7% (International Air Transport Association [IATA], 2016). In anticipation, by the year 2050, air travel is projected to accommodate 10 billion passengers (Segura et al., 2022).

To address these escalating concerns, IATA member airlines committed to achieving net-zero carbon emissions by the year 2050 during the 77th International Air Transport Association (IATA) Annual General Meeting. This ambitious target necessitates a minimum of 1.8 gigatons of CO_2 reduction by the year 2050, totaling

21.2 gigatons between the current year and 2050 ((International Air Transport Association [IATA], 2021). To achieve this, the aviation industry must collaborate with stakeholders, and governments should provide financial incentives and implement policies to encourage infrastructure providers. Furthermore, the industry needs to embrace risks and experiment with new technologies, while scientists and engineers must play a vital role in developing and designing innovative solutions for widespread adoption by businesses.

7.4 SIGNIFICANCE OF BIOFUELS FOR AVIATION – PATH TOWARD ACHIEVING SUSTAINABILITY

Biofuels play a pivotal role in reshaping the landscape of air travel, especially amid escalating environmental concerns and the imperative for sustainable aviation. Their significance stems from a multitude of reasons, encompassing environmental, economic, social, and strategic considerations. Offering a compelling alternative to conventional jet fuels derived from fossil sources, biofuels have the potential to significantly reduce GHG emissions (Chao et al., 2019). Derived from diverse sources such as plants, algae, and microorganisms, biofuels provide a sustainable alternative, actively sequestering CO_2 during growth through photosynthesis (Jyoti & Wattal, 2019). This unique feature positions biofuels as carbon-neutral, aligning with global efforts to combat climate change and support the aviation industry's drive to decarbonize (Jensen et al., 2023).

Beyond emissions reduction, biofuels contribute to energy security by diversifying sources for aviation fuel. With traditional jet fuels heavily reliant on finite fossil reserves, biofuels offer a more reliable and secure supply, reducing dependence on geopolitically uncertain and volatile fossil energy sources (Cabrera-Jiménez et al., 2022). Furthermore, the utilization of various biofuel feedstocks stimulates local economies and alleviates geopolitical tensions linked to fossil fuel extraction.

Biofuels, however, extend beyond emissions reduction and energy security, embodying a commitment to sustainable aviation. The production process can integrate sustainability principles, emphasizing water conservation, minimal land usage, and avoidance of deforestation. The use of non-food crops and waste as feedstocks aligns with circular economy principles, reinforcing the industry's dedication to responsible practices.

Adopting biofuels allows the aviation sector to demonstrate a holistic commitment to sustainability, addressing environmental, social, and economic factors. This paradigm shift toward sustainable practices is both an ethical imperative and a response to a global community increasingly attuned to ecological concerns. Governments, including the United States of America, the European Union, China, and several other countries, have issued directives mandating an increased share of biofuels in transportation fuels. As such, biofuels play a crucial role in aviation, mitigating GHG emissions, enhancing energy security, and fostering a sustainable future for air travel. The industry is entering a new era marked by the development of microbial technologies for jet fuel production, marking a convergence of environmental responsibility and technological innovation aimed at creating a cleaner and more sustainable future.

7.5 MICROBIAL INSIGHTS ON BIOFUEL PRODUCTION, AND CLASSIFICATION OF BIOFUELS

Microbes possess the potential to transform biomass with its diverse properties and composition into biofuels like biodiesel, ethanol, or biogas. Ko and his research team published a paper in 2020 showing that microbes could efficiently utilize various carbon sources found within biomass for biofuel production (Ko et al., 2020). Biofuel is divided into first, second, or third generations based on factors like ease of use or raw material requirements (Kargbo et al., 2021).

First-generation biofuels, known as conventional biofuels, constitute the predominant and widely utilized category globally, sourced from edible feedstocks such as sugarcane or vegetable oil (Arous et al., 2017; Phukan et al., 2019). The reliance on agricultural resources like starch, sugar, animal fats, and vegetable oil for first-generation biofuel production contributes to unsustainable practices, posing threats to food production and resulting in deforestation, biodiversity loss, and water scarcity in developing countries (Kargbo et al., 2021). The cultivation of crops for biodiesel or bioethanol production, especially outside traditional agricultural settings, also raises concerns about increased carbon emissions and potential threats to food supplies (Kargbo et al., 2021). The ongoing debate questions the actual reduction of GHG and carbon emissions by first-generation biofuels, with some arguing that the production process releases more CO_2 than the feedstocks, resulting in a net energy loss (Lee et al., 2014). Various factors, including agricultural prices, seasonal variations, market speculation, and extreme weather patterns, further impact the growth and production of crops. Despite their widespread use, first-generation biofuels are accompanied by significant challenges, including their impact on food security, unsustainable production practices, elevated production costs, and the ongoing debate regarding their contribution to reducing GHG emissions (Lee et al., 2014). These concerns highlight the necessity for continued exploration and development of more sustainable and efficient biofuel alternatives.

In response to the adverse environmental impacts and elevated costs associated with first-generation biofuels, the adoption of second-generation biofuels has emerged as a more sustainable alternative. Researchers investigating second-generation biofuel production have explored various sources, including agricultural, domestic, industrial, and non-food waste (Bonatto et al., 2020). Second-generation biofuels utilize inedible plant crops, specifically "lingo-cellulosic feedstock," encompassing agricultural and forest residues like wood processing wastes and other dry plant matter (Al-Azzawi & Jassem, 2016; Zhu et al., 2020a; Carmona-Cabello et al., 2021). Vegetative grasses and short rotation forestry crops are identified as viable alternatives in regions with limited lingocellulosic feedstock, contributing to reduced dependency on water supplies and agricultural chemicals, thus supporting environmental preservation (Lee et al., 2014).

However, the cultivation of non-food crops, a characteristic of second-generation biofuels, presents challenges as these crops may become invasive species, competing for nutrients and endangering ecosystems (Elshahed, 2010). Another drawback is the limited compatibility of second-generation biofuels for commercial-scale use. For example, biodiesel derived from jatropha oil, a non-food crop, requires a specialized process called "transesterification" for efficient utilization (Bonatto et al., 2020).

Despite these challenges, second-generation biofuels are gaining attention in the industry. Lignocellulosic biomass, with an annual production of about 2×10^{11} metric tons (Francois et al., 2020), serves as the primary second-generation biomass. It is characterized by a complex structure comprising polysaccharides (cellulose, hemicellulose, and lignin) efficiently used as carbon sources for biofuel production. However, the challenge lies in the assimilation of these sources by microorganisms due to the presence of lignin. Second-generation ethanol is commercially produced using residues from first-generation ethanol production, primarily sugarcane bagasse, straw, and corn straw, yielding approximately 285 million gallons of ethanol annually. For biodiesel production, second-generation biomass includes energy crops (*Jatropha*, *Pongamia*, *Madhuca*, and *Ricinus*), animal fat (bovine tallow, chicken fat, and yellow fat), food waste, slaughterhouse effluents, and waste oils (Pinzi and Dorado, 2012). Incorporating these biomasses into the biodiesel production chain offers the potential to reduce the overall cost of biodiesel manufacturing, given the lower cost of these biomasses compared to vegetable oils. For instance, waste cooking oil is 2.5–3.5 times cheaper than vegetable oils (Balat, 2011).

Considering the diverse agricultural limitations of previous biofuel generations, third-generation biofuels offer a promising solution by exclusively relying on lipids from algae, commonly referred to as algal biofuel. The versatility of algae cultivation in open ponds, closed-loop systems, or photobioreactors makes it a highly feasible alternative (Banerjee et al., 2020). Moreover, algae's product quality and diversity surpass that of any other crop. For instance, algae not only yield up to 20 times more than first and second generations but can also be "genetically manipulated" to produce various types of biofuels from all generations (Olaganathan et al., 2014). Algae exhibit the unique ability to directly convert carbon emissions from industrial plants into usable fuel, contributing to the reduction of air pollution and enhancing the efficiency which has garnered more interest recently (Banerjee et al., 2020).

Fourth-generation biofuels involve hydro refining petroleum or using advanced biochemistry methods like Joule's "solar-to-fuel" system. Unfortunately, such procedures are typically theoretical or limited to laboratory environments (Milledge & Heaven, 2013).

7.6 MICROBIAL CONTRIBUTIONS IN BIOFUEL PRODUCTION PROCESSES – AN ANALYSIS OF BACTERIA, PROTISTS, FUNGI, AND YEASTS

7.6.1 Bacteria

Bacteria are microscopic, prokaryotic, and single-celled organisms that lack a nuclear membrane and contain their genetic information in a DNA segment, possibly with genes in a plasmid, conferring advantages like resistance to adverse conditions (Bakermans et al., 2003). Reproduction occurs through binary fission, resulting in identical clones. Bacteria exhibit versatility in substrate use and some form endospores for defense (Bakermans et al., 2003). Cellulase, a notable bacterial enzyme, is crucial for crystalline cellulose hydrolysis, contributing to biofuel production from various biomasses (Chakraborty et al., 2020). Bacterial enzymes typically exist as multi-enzyme complexes, a feature that enhances their functionality. Some bacteria exhibit stability even in adverse conditions such as acidic environments, extreme temperatures, and high salt concentrations. Additionally, bacteria demonstrate faster growth compared to other microorganisms like fungi. This faster growth rate contributes to the more efficient production of recombinant enzymes (Chakraborty et al., 2020).

Genetic engineering offers the potential to enhance the capabilities of bacteria, leading to higher ethanol yields (Komesu et al., 2020). This application is successfully demonstrated in strains such as *Escherchia coli*, *Klebsiella oxytoca*, *Zymomonas mobilis*, *Clostridium thermocellum*, *Clostridium phytofermentans*, *C. cellulolyticum*, *Thermoanaerobacterium saccharolyticum*, *Caldicellulosiruptor bescii*, and *Bacillus* sp. (Zhu et al., 2020a). These strains exhibit accelerated reproduction and can ferment various sugars and substrates. Notably, the Gram-negative bacterium *Z. mobilis* stands out as a biocatalyst for ethanol production due to its ability to tolerate ethanol concentrations up to 120 g/L in the medium, enabling the production of ethanol with high specificity and productivity. Extremophilic bacteria, with their desirable traits, are potential biofuel producers, particularly in biorefineries where on-site microbial enzymes can be cost-effective (Zhu et al., 2020b). Biogas production involves bacterial and archaeal consortia through stages like hydrolysis and methanogenesis, offering a promising avenue for biofuel production (Angelidaki et al., 2018; Tabatabaei et al., 2020).

Bacteria, like *Serratia* sp., can synthesize fatty acids for biodiesel production, contributing to a sustainable biofuel sector (Kumar et al., 2018). Halotolerant bacteria, such as *Haloarcula* sp., pose challenges but offer benefits, like stable enzymes, for biofuel production in saline environments (Zhu et al., 2020b). However, bacteria may not always synthesize a broad spectrum of sugars from biomass, requiring further biotechnological maturation for full industrial benefits (Yang et al., 2020).

7.6.2 Protists

Protists, single-celled eukaryotic organisms, exhibit limited biodiversity compared to bacteria due to nutritional metabolism constraints (Levandowsky, 2012). Protists, including ciliates, dinoflagellates, algae, diatoms, and amoebae flagellates, may undergo asexual or sexual life cycles, featuring various reproductive methods (Levandowsky, 2012). Protists' unique characteristics, such as flagella, cilia, or pseudopods, make them biotechnologically valuable for survival mechanisms applicable in the biorefinery sector (Vallesi et al., 2020). The marine protist *Thraustochytrium striatum* HB shows potential in lignocellulosic biomass pretreatment, producing biofuels, fatty acids, and carotenoids (Li et al., 2020). Protist enzymes, like α-amylase and superoxide dismutase, offer versatility for various processes (Vallesi et al., 2020).

7.6.3 Microalgae

Microalgae, a subset of protists, stand out for their metabolic plasticity and relevance to biofuel production, exhibiting advantages like adaptation to wastewater and the sea (Kumar et al., 2020). Certain strains, such as *Dunaliella*, *Botryococcus*,

Chlamydomonas, *Chlorella*, and *Arthrospira*, show promise for biodiesel production, offering higher lipid yields than traditional sources (Tasić et al., 2016). Amoebas, another protist group, employ extracellular vesicles for adaptation to environmental stress, with applications in vaccines and liquid biofuel production (Gonçalves et al., 2019). Although the incorporation of protists into biofuel production processes is in its early stages, microalgae, with their significant potential, are emerging as a crucial focus for biotechnological exploration (Anto et al., 2020).

7.6.4 Fungi

Fungi, classified as non-photosynthetic hyphal eukaryotes, form a diverse kingdom with macro- and microorganisms (Hanson, 2008). Fungi play a crucial role in nature, contributing to plant development, complex compound degradation, and organic recycling (Troiano et al., 2020). In biotechnology, fungi, such as *Aspergillus* and *Trichoderma* species, excel in various processes, offering advantages in biomass conversion to high-value products (Troiano et al., 2020).

Filamentous fungi, with their effective colonization and enzymatic secretion capabilities, stand out for breaking down complex structures into absorbable molecules (Weosten, 2019). This ability, particularly relevant in lignocellulosic biomass processes, involves enzymes like amylases, cellulases, xylanases, and more, contributing to efficient pretreatment and hydrolysis (Troiano et al., 2020). Filamentous fungi, including *Aspergillus niger*, *Rhizopus*, and *Trichoderma*, exhibit high enzymatic activity, stability across pH and temperature ranges, and efficiency in biomass conversion (Troiano et al., 2020). Their specialization in degrading the cell wall of lignocellulosic biomasses, facilitated by the excretion of active carbohydrate enzymes (CAZymes), enhances their effectiveness (Huberman et al., 2016).

In pretreatment processes, filamentous fungi, preferred over other microorganisms, maximize enzyme productivity, adapt to complex substrates, and contribute to waste recovery (Troiano et al., 2020). Brown rot, white rot, and soft rot fungi are commonly employed in pretreatment, exhibiting selective lignin degradation capacity (Fang et al., 2020). Biorefineries leverage filamentous fungi in pretreatment systems, providing greater control, specificity, and extraction of high-value compounds like phenolic acids and vanillin from degraded lignin (Arora et al., 2018). Filamentous fungi, through their diverse applications, contribute to reducing biofuel production costs and supporting regional development based on bioeconomics and biotechnology.

7.6.5 Yeast

Yeasts, essential in human civilization since the Neolithic revolution, have evolved into key players in various bioprocesses, extending beyond traditional uses in bread and alcoholic drink production (Tulha et al., 2012). The symbiotic relationship with humans intensified in the 19th century with Louis Pasteur's discovery of yeasts' role in sugar-to-alcohol conversion, followed by Hansen's strain selection (1968) for specific characteristics (Eliodório et al., 2019).

Belonging to the Fungi kingdom, yeasts are single-celled eukaryotic microorganisms with a versatile cell wall composition. They multiply through budding or binary fission, exhibiting both asexual and sexual reproduction (Fraser & Heitman, 2004). Yeasts' success in industrial processes stems from their ease of maintenance, adaptability to various nutritional conditions, and significant tolerance to pH variations and environmental stressors (Walker & White, 2017).

Yeasts, particularly *Saccharomyces cerevisiae*, play a pivotal role in ethanol production, constituting 90–95% of the world's biofuel volume. The United States and Brazil dominate global ethanol production, reflecting its importance as a renewable and less polluting fuel (Sarris & Papanikolaou, 2016). Ethanol production, primarily first-generation, relies on the fermentation of starchy or saccharine substrates, with *S. cerevisiae* being the preferred species due to its efficiency and resistance to osmotic stress (Selim et al., 2018).

While first-generation ethanol production is well-established, efforts to make second-generation (2G) ethanol economically viable involve utilizing residues for fermentation. However, efficient fermentation requires yeast strains capable of converting various carbohydrates found in lignocellulosic hydrolysates. *S. cerevisiae* industrial strains lack the ability to ferment pentoses (xylose and arabinose) and cellobiose, necessitating genetic modifications or the exploration of new yeast species (Walker & Walker, 2018).

Genetically engineered *S. cerevisiae* strains exhibit improved xylose fermentation efficiency, providing a potential solution for 2G ethanol production (Kwak & Jin, 2017). Alternatively, naturally occurring yeasts, such as *Zygosaccharomyces rouxii*, *Ogataea polymorpha*, and *Dekkera bruxellensis*, isolated from environments with lignocellulosic components, exhibit desirable traits for industrial settings (Cadete & Rosa, 2018). Promising species like *Scheffersomyces stipitis* and *Spathaspora arborariae* have demonstrated high ethanol production from xylose, making them potential candidates for optimizing 2G ethanol production (Eliodório et al., 2019).

7.7 MICROBIAL INNOVATIONS FOR SUSTAINABLE AVIATION FUEL PRODUCTION

Microbial innovation plays a central role in the production and advancement of SAF, offering promising avenues to reduce the environmental impact of aircraft (Segura et al., 2022). This innovation is crucial for enabling the utilization of various feedstocks, including agricultural residues, waste oils, algae, lignocellulosic biomass, and other forms of biomass. Through fermentation, enzyme conversion, and metabolic engineering techniques, bacteria, yeast, and algae efficiently convert feedstocks to biofuels (Kumar et al., 2017). Microbes can be precisely tailored and optimized for biofuel production using genetic engineering and synthetic biology techniques. This approach streamlines the production process, enhancing yield and reducing SAF production costs. Biofuels derived from microorganisms boast a lower carbon footprint than fossil fuels by producing lower emissions of GHGs, mitigating climate change, and decreasing overall environmental impacts (Sharma et al., 2020). Microbes also play a critical role in converting organic wastes, such as municipal,

of lignocellulosic biomass as an affordable feedstock is gradually replacing fossil-based energy production (Wang et al., 2021).

The abundant nature of lignocellulosic biomass makes it a compelling source of renewable and sustainable energy, drawing considerable interest in alternative energy strategies (Wang et al., 2021). Comprising biopolymers like cellulose, hemicellulose, and lignin, along with minor elements such as terpenoids, resins, proteins, inorganic minerals, pectin, and chlorophyll, lignocellulosic biomass plays a multifaceted role in biofuel production (Haldar & Purkait, 2021; Wang et al., 2021). Zheng et al. (2014) emphasize the critical role of the significant presence of these three main compounds (cellulose, hemicellulose, and lignin) in evaluating lignocellulosic biomass for energy production. These components, integral to plant cell walls, serve distinct functions, with cellulose imparting strength and hemicellulose linking cellulose and lignin. Lignin, crucial for providing structural support to the cell wall, forms a stable and complex structure that resists breakdown (Yadav et al., 2020). The following sections will provide brief insights into the main components of lignocellulosic biomass.

7.7.2.1 Structure of Cellulose and Its Properties

Cellulose, one of the most abundant organic materials on earth, is found in organisms such as plants, animals, algae, and fungi (Liu et al., 2021). A biodegradable resource and integral component of plant cell walls (Patinvoh et al., 2017), cellulose content varies among different plant species. Wood contains 40–50% cellulose, and bamboo and cotton contain 40–55% and 90%, respectively (Liu et al., 2021). Chemically, cellulose comprises several glucose units in long chains connected by glycosidic bonds. Van der Waals forces and hydrogen bonds contribute to its insolubility in most solvents (Robak & Balcerek, 2018). The properties of cellulose depend on its degree of polymerization and chain length (Irmak, 2019). Operating conditions for breaking down cellulose for biofuel production require a high temperature of 320°C and a pressure of 25 MPa (Robak & Balcerek, 2018).

7.7.2.2 Composition and Characteristics of Hemicellulose

Cellulose fibers are protected by an interwoven matrix of lignin and hemicellulose, forming cross-links to provide strength. Hemicellulose, an intricate polysaccharide, comprises different sugar molecules, including diverse hexoses (mannose, galactose, glucose, rhamnose) and pentoses (l-arabinose, d-xylose), as well as sugar acids known as uronic acids (galacturonic acid, 4-O-methyl-D-glucuronic acid) (Hasanov et al., 2020). Hemicellulose content typically ranges between 20% and 30% of dry mass, depending on wood species (Ehite et al., 2021). Hardwood trees (dicot Angiosperms) mainly consist of xylans, while softwoods (Gymnosperms) primarily utilize glucomannans as primary constituents (Deshavath et al., 2019). Due to its high branching structures and susceptibility to hydrolysis, hemicellulose processes more quickly in acidic environments (Hasanov et al., 2020).

7.7.2.3 Lignin Structure and Its Properties

Lignin, an aromatic heterogeneous polymer, plays a crucial role in binding the lignocellulose matrix together, providing mechanical strength, hydrophobicity, and resistance against hydrolytic attacks on plant cell walls (Zakzeski et al., 2010). Comprising phenolic and nonphenolic compounds, lignin's main constituents include phenylpropanoid units such as p-coumarin (H), sinapyl (S), and coniferyl (G) (Sindhu et al., 2016). Lignin varies depending on its source, with hardwood possessing both G and S units (GS-type lignin), softwood having more G units (G-type lignin), and non-woody biomass containing all three units (Li et al., 2015). Lignin finds applications in direct combustion, heat, and power production, and as a precursor in manufacturing various products (Sindhu et al., 2016). It acts as an essential inhibitor against hydrolytic attacks on plant cell walls. As plants age, lignin production increases exponentially, forming stable structures that thwart hydrolysis (Asgher et al., 2018). Due to its high molecular weight and insoluble properties, lignin is resistant to breakdown (Yamada et al., 2013). Proper pretreatment is necessary for accessibility to enzyme hydrolysis. Reprocessed lignocellulosic biomass, through enzymatic hydrolysis, can be converted to fuels and chemicals, though ongoing efforts are needed to improve economic efficiencies (Bhatia et al., 2021). Engineering alginolytic enzymes are considered as transformative tools due to their potential for low-cost conversion (Bhatia et al., 2021).

7.7.2.4 Current Status of Microbial Biofuel Production

Bioethanol production from lignocellulosic material relies heavily on renewable sources to generate sugars that can then be processed using either catalytic upgrading of sugars to hydrocarbons or biological fermentation, according to Wang and Tao (2016). Two fermentation methods – direct and indirect fermentations – can be employed. In direct fermentation, plant materials are converted directly to sugar for further transformation into alcohol; conversely, indirect fermentation involves the pyrolysis of plant materials followed by conversion of syngas produced through pyrolysis to syngas, which is then finally converted into alcohol using acetogenic bacteria as proposed by Klasson et al. (1992).

The ATJ route involves the conversion of butanol and ethanol into jet fuels using processes such as alcohol dehydration, oligomerization, and hydrogenation. Hydroprocessed esters and fatty acids (HEFA) and hydrotreated renewable jet (HRJ) fuels are produced through the OTJ synthesis, employing various methods such as catalytic thermolysis, hydrothermal liquefaction, and the hydrotreated depolymerized cellulosic jet process (Kallio et al., 2014; Wang & Tao, 2016). GTJ includes the Fischer-Tropsch Synthetic Paraffinic Kerosene (FT-SPK) and gas fermentation pathways. In the FT process, biomass is converted into syngas, which is then transformed into liquid hydrocarbons. Gas fermentation, on the other hand, converts biomass into syngas, ferments it into ethanol or butanol, and finally upgrades it to jet fuel through the ATJ process. The STJ pathway integrates both biological and catalytic conversion methods to transform sugars into hydrocarbons (Karatzos, McMillan, & Saddler, 2014). The American Society for Testing and Materials (ASTM) certifies various conversion pathways for drop-in jet fuels, categorized by feedstock and technology variants, as outlined in Table 7.2 (ICAO, 2023).

Biological fermentation offers significant advantages over catalytic upgrading pathways by eliminating the need for chemical catalysts and high energy/temperature

TABLE 7.2
Conversion Processes Certified by ASTM ICAO Global Framework for Aviation Alternative Fuels (ICAO, 2023)

Conversion Process	Abbreviation	Maximum Blend Ratio by Volume	Commercialization Proposals/Projects
Fischer-Tropsch hydro processed synthesized paraffinic kerosene	FT	50%	Fulcrum Bioenergy, Red Rock Biofuels, SG Preston, Kaidi, Sasol, Shell, Syntroleum
Synthesized paraffinic kerosene from hydro processed esters and fatty acids	HEFA	50%	World Energy, Honeywell UOP, Neste Oil, Dynamic Fuels, EERC
Synthesized iso-paraffins from hydro processed fermented sugars	SIP	10%	Amyris, Total
Synthesized kerosene with aromatics derived by alkylation of light aromatics from non-petroleum sources	FT-SKA	50%	Sasol
Alcohol-to-jet synthetic paraffinic kerosene	ATJ-SPK	50%	Gevo, Cobalt, Honeywell UOP, Lanzatech, Swedish Biofuels, Byogy
Catalytic hydro thermolysis jet fuel	CHJ	50%	Applied Research Associates (ARA)
Synthesized paraffinic kerosene from hydrocarbon-hydro processed esters and fatty acids	HC-HEFA-SPK	10%	IHI Corporation
Co-processing		5%	

reactions. This allows for a streamlined conversion process within one fermentation tank. Because of these advantages, STJ technology stands out as a highly prospective strategy for producing jet biofuel. The STJ process consists of several key steps, such as identifying plant sources, isolating microbes (fungi or bacteria), and designing efficient conversion pathways to transform plant materials into sugars. These sugars are further converted into ethanol by microbes either native to or genetically engineered strains; an essential phase in biofuel production lies in this conversion phase.

Biofuel production requires using various plant materials, including crops, grasses, weeds, and algae that possess different compositions of cellulose, hemicellulose, and other components like lignin (Smith 2006). Subsequently, the selection of microbes as well as optimal incubation conditions and other requirements depends upon the composition makeup of the chosen plant material (Smith 2006). Ideal candidates should exhibit robust tolerance against inhibitors, efficiently use substrates, and possess modified sugar transport pathways that are less subject to regulatory mechanisms, leading to the production of a single fermentation product (Smith 2006).

7.7.2.5 Fermentation Process and Ethanol Production

Numerous research studies aimed to identify potential microorganisms capable of facilitating crucial transformations and to develop technologically and economically viable methods based on this conversion process. Regardless of the plant material utilized, the initial degradation of starch, cellulose, or hemicellulose involves the conversion into pentose and hexoses, followed by fermentation into ethanol. Various fermentation pathways, such as heterolactic acid fermentation by lactic acid bacteria (e.g., Leuconostoc species) and mixed acid fermentation by enteric bacteria, yield ethanol as one of the end products. However, for commercial biofuel production, ethanol must be the primary output.

Notably, the yeast *Saccharomyces cerevisiae* and the Gram-negative bacteria *Zymomonas mobilis* can produce two moles of ethanol for every mole of hexose. Both these microorganisms convert pyruvate to alcohol through pyruvate decarboxylase/alcohol dehydrogenase enzymes. Among them, the industrial-level production of ethanol through the conversion of hexoses using *S. cerevisiae* is widely adopted (Klinke, Thomsen, & Ahring, 2004).

As plant materials comprise both cellulose and hemicellulose (a precursor of xylose, a pentose sugar), microbes efficient in metabolizing these two sugars are essential for industrial alcohol production. Numerous microorganisms, including the fungi *Penicillium capsulatum* and *Talaromyces emersonii* (Filho et al., 1991), the thermophilic actinomycete *Thermomonospor fusca* (Bachmann & McCarthy, 1991), and the hyperthermophile *Caldicellulosiruptor saccharolyticus* (van de Werken et al., 2008), can completely depolymerize hemicellulases to xylose. Typically, xylose is metabolized to pyruvate through the pentose phosphate pathway. Various microbes, such as anaerobic fungi and different groups of mesophilic and thermophilic anaerobic bacteria, can efficiently metabolize xylose (Wisselink et al., 2009).

7.7.2.6 Consolidated Bioprocessing

Consolidated bioprocessing (CBP) represents an economically efficient biofuel production technique. CBP involves using microbes to convert plant material directly to biofuel without employing enzyme treatments (Lynd et al., 2005). Conversion may occur via one microbe alone or multiple ones; one key candidate used in CBP is *Clostridium thermocellum* and another microbe is *C. Bescii* which was modified specifically to produce ethanol (Chung et al. 2014). *Trichoderma reesei* has also been engineered for use as an ethanol production source (Huang et al., 2014). Yamada et al. (2013) proposed another efficient approach by engineering microbes capable of producing ethanol to digest cellulose; Microorganisms like *S. cerevisiae*, *E. coli*, and *Z. mobilis* have all been engineered with cellulolytic enzyme expression capabilities as an efficient means to digest cellulose.

Lynd et al. (2005) utilized thermophilic Gram-positive *Firmicutes* belonging to the orders Clostridiales and Thermoanaerobiales, demonstrating that *C. thermocellum* exhibits the highest cellulose degradation ability among the order Clostridiales. *Clostridium phytofermentans* is a Gram-positive anaerobic bacterium which converts various polysaccharides, oligosaccharides, and monomeric sugars primarily into ethanol (Maki, Leung & Qin 2009). *Clostridium acetobutylicum*, another fascinating species, produces butanol in addition to acetone and ethanol (Durre, 2008). Several Clostridia species in general can degrade cellulose into sugar which can then be turned into ethanol for further fermentation processes. Owing to its potential, however, CBP rarely yields more than 5–6% on a weight/volume scale and commercialization has yet to occur using thermophilic anaerobes such as *Geobacillus thermoglucosidans* (Taylor et al. 2009).

7.7.2.7 Microorganisms That Produce Alkane/Alkene Compounds

According to Fu et al. (2015), most eubacteria intracellularly synthesize a limited quantity of alkanes, making their separation and purification challenging. Conversely, some prokaryotes produce alkanes both inside and outside the cells. Among prokaryotes, anaerobic bacteria, particularly *Clostridium* and *Desulfovibrio* species, predominantly produce alkanes, mainly ranging from C25 to C35 (Bagaeva & Zinurova, 2004). Cyanobacteria, on the other hand, contain alkanes in lipid droplets, packed with hydrophobic energy-dense compounds (Peramuna & Summers, 2014). Han et al. (1968) reported that certain cyanobacteria species can produce hydrocarbons within the appropriate range for jet fuel (C15 to C19). However, the alkane biosynthesis pathway in cyanobacteria remained unclear for an extended period until Schirmer et al. (2010) explained it. In this cyanobacterial pathway, fatty acyl-ACP is transformed into alkane/alkene, especially C17 among C15 to C19.

Beller et al. (2010) studied another pathway in which alkene is produced by *Micrococcus luteus*, a Gram-positive bacterium. They reported that the alkane/alkene pathway results in a long-chain hydrocarbon (C23 to C33) with at least one double bond. This process involves the head-to-head condensation of fatty acids, starting with the conversion of fatty acyl-CoA (or ACP) into β-ketoacyl-CoA through the sequential action of acyl-CoA dehydrogenase, an enoyl-CoA hydratase, and a 3-hydroxy acyl-CoA dehydrogenase, ultimately leading to alkene biosynthesis.

Park et al. (2001) found that *Vibrio furnissii* synthesizes alkanes ranging from C15 to C24. This pathway starts with fatty acids, which are then converted into alkanes, known as the hypothetical vibrio pathway.

7.7.2.8 Genetically Engineering Microbes to Produce Alkanes

Genetic engineering, a field involving the manipulation of an organism's genetic material through gene insertion, deletion, or modification (Khalil, 2020), primarily centers on the intricate manipulation of DNA or RNA. Key methodologies, including gene cloning, CRISPR-Cas9, and recombinant DNA technology, empower scientists to precisely engineer the genetic makeup of organisms (Xu & Li, 2020). This discipline introduces novel genes, refines existing ones, or selectively eliminates specific genes, contributing significantly to shaping the genetic landscape of living organisms, notably in the production of SAF.

While many microorganisms can naturally produce alkanes, the levels are not consistently high. Cyanobacteria, particularly those with photosynthetic capabilities, have caught the attention in the biofuel industry due to their capacity to synthesize alkanes using sunlight and carbon dioxide. Ongoing research employs diverse approaches, involving the modification of growth conditions and genetic engineering of microbes, to enhance alkane production on a commercial scale.

In a study by Wang et al. (2013), the genetic modification of *Synechocystis* sp. PCC 6803 resulted in an 8.3-fold increase in alkane production. This was achieved by deleting the β-ketolyase phaA gene and overexpressing native FAR (fatty acyl-ACP reductase) and ADO (aldehyde-deformylating oxygenase) enzymes. Similarly, Kageyama et al. (2015) doubled alkane production in the nitrogen-fixing cyanobacterium *Anabaena* sp. PCC 7120 through exposure to nitrogen deficiency or salt stress. Peramuna et al. (2015) applied genetic modifications to *Nostoc punctiforme* cells, combined with high-light conditions, resulting in an increase in alkane production from 1.12% to 12.9% of cell dry weight (CDW).

Despite advancements, cyanobacterial alkane production remains relatively modest, prompting exploration into alternative microbial platforms by the biofuel industry. *E. coli*, *Bacillus subtilis*, *A. nidulans*, and the yeast *S. cerevisiae* are preferred for their production capabilities or suitability for industrial use. Schirmer et al. (2010) pioneered the expression of the cyanobacterial pathway in *E. coli* for alkane production. Numerous engineered *E. coli* strains have since been developed to generate advanced biofuels.

The first recombinant yeast *S. cerevisiae* capable of alkane production was reported by Bernard et al. (2012). Buijs et al. (2015) tested the cyanobacterial alkane production pathway in *S. cerevisiae*, achieving production levels of 22 μg/L CDW. Zhou et al. (2016) made several improvements, such as substituting glucose as a substrate, leading to a production of 10.4 g/L of free fatty acids (FFA). *Yarrowia lipolytica*, an oleaginous yeast, is considered a promising strain for industrial biofuel production, with the capability to store up to 36% of CDW as lipids (Beopoulos et al., 2009). Additionally, the black yeast *Aerobasidium pullulans* var. *melanogenum* and other microbial isolates have demonstrated potential in alkane production. However, further research is necessary to comprehensively characterize these alternative microbes and their metabolic pathways for commercial utilization in alkane production, meeting the growing demand for sustainable fuels.

7.7.2.9 Metabolic Engineering Developments as Path to Sustainable Bioeconomy

Metabolic engineering, a process that optimizes or manipulates an organism's metabolic pathways to increase the production of specific compounds, centers around the overall metabolic network of biochemical reactions (Otero-Muras & Carbonell, 2021). This intricate approach often involves modifying the expression of endogenous genes or introducing foreign genes to redirect metabolic flux, with the aim of enhancing the yield of metabolites such as biofuels, pharmaceuticals, or chemicals (Ullah et al., 2021).

As a rapidly advancing field, metabolic engineering is pivotal for constructing robust biofuel-producing microbial hosts, serving as a key component for the future bioeconomy. Widely adopted over the last decade, this approach has played a crucial role in redirecting and overhauling biosynthetic pathways, achieving high titers of target products (Lu et al., 2022). Biotechnologists and metabolic scientists have effectively generated a diverse range of industrially relevant products through the engineering or optimization of metabolic pathways, as highlighted by Moon et al. (2023). This section specifically emphasizes the utilization of metabolically engineered microbes as promising cell factories for the enhanced production of advanced biofuels.

Within the group of alcohol-fermenting microbes, *Zymomonas mobilis* and *S. cerevisiae* stand out for their ability to ferment sucrose and hexose sugars into ethanol. However, their growth is impeded by the accumulation of end products, as noted by Chandel et al. (2010), Martiniano et al. (2013), and Moysés et al. (2016). Similarly, pentose-fermenting platform microbes like *Candida shehatae*, *Pichia stipitis*, and *Pachysolen tannophilus* face challenges in the presence of end products, as reported by Martiniano et al. (2013). Despite the capacity of numerous microbes to ferment lignocellulosic biomass, the complete assimilation of lignocellulosic feedstock for biofuel production is hindered by the absence of ideal microbial strains capable of efficiently converting both hexose and pentose sugars to glucose, as outlined by Sarkar et al. (2012). An optimal host for industrial-scale biofuel production should demonstrate adaptability in utilizing various substrates, yield high product titers, withstand high temperatures and inhibitory compounds, and tolerate increased concentrations of by-products/end products, while also exhibiting enhanced cellulase activity. Additionally, it should display resilience against toxic chemicals and stresses, facilitating the real-time industrialization of advanced biofuels, as highlighted by Avanthi et al. (2017).

In recent years, there has been a noticeable increase in the adoption of metabolic engineering, specifically in the modification of wild-type strains to improve the production of biofuels (refer to Table 7.3). Various starches or lignocellulose-based biomass feedstocks have been explored to produce biofuels by tailoring metabolic pathways in diverse microbial strains, as discussed by Kang and Lee (2015). Metabolic engineering, employing a recombinant DNA approach, has resulted in the

TABLE 7.3
Biofuels Production by Various Genetically Engineered Microbial Platforms (Lu et al., 2022)

Microorganism	Metabolic Engineering Strategy Applied	Advanced Biofuel	Titer	Reference
Clostridium tyrobutyricum	Overexpression of homologous or heterologous heat shock protein	Butanol	12.15 g/L	Fu et al. (2021)
Saccharomyces cerevisiae	Cell surface display of cellulolytic enzymes Display β-galactosidase for lactose utilization	Ethanol	>50 g/L	Cunha et al. (2021)
Pichia pastoris	Overexpressing leucine and valine biosynthesis pathways. Overexpressing keto-acid degradation pathway Down-regulating side-product by CRISPR/Cas9 system	Isopentanol	191 mg/L	Siripong et al. (2020)
S. cerevisiae	Overexpression of three endogenous enzymes, acetolactate synthase, acetohydroxyacid reductoisomerase, and dihydroxy-acid dehydratase Successive deletion of competing pathways genes for 2,3-butanediol synthesis of (BDH1 and BDH2), pantothenate (ECM31), leucine (LEU4 and LEU9), and isoleucine (ILV1)	Isobutanol	2.09 g/L	Wess et al. (2019)
Pichia pastoris	Introduction of a new synthetic pathway Codon-optimization and synthesis of *alsS* and *alsD* genes from *B. subtilis* Integration of three pathway genes into *P. pastoris* genome and expression	2,3-butanediol	45 g/L	Yang and Zhang (2018)

(continued)

TABLE 7.3 (*Continued*)
Biofuels Production by Various Genetically Engineered Microbial Platforms (Lu et al., 2022)

Microorganism	Metabolic Engineering Strategy Applied	Advanced Biofuel	Titer	Reference
Yarrowia lipolytica	Codon optimization and heterologous expression of WS genes encoding wax ester synthases Optimizing promoter for WS overexpression, Elimination of β-oxidation by PEX10 gene deletion Redirecting metabolic flow toward acetyl-CoA	Fatty acid ethyl esters	1.18 g/L	Gao et al. (2018)
Pichia pastoris	Exploitation of yeast's amino acid biosynthetic pathway Redirecting amino acid intermediates toward the 2-keto acid degradation pathway	Isobutanol	284 mg/L	Siripong et al. (2018)
Saccharomyces cerevisiae	Cloning and expression of two wax ester synthase genes Inactivation of negative regulators in phospholipid metabolism Increasing flow toward fatty acyl-CoAs Overexpression of five isobutanol pathway enzymes	Fatty acid short- and branched-chain alkyl esters	230 mg/L	Teo et al. (2015)
Escherichia coli	Integrating fatty acid biosynthesis pathway with branched-chain amino acid pathway Modification of fatty acid pathway	Fatty acid branched-chain esters	273 mg/L	Tao et al. (2015)
Pichia pastoris	Integrating fatty acid biosynthesis pathway with branched-chain amino acid pathway Modification of fatty acid pathway	Fatty acid branched-chain esters	169 mg/L	Tao et al. (2015)
Saccharomyces cerevisiae	Elimination of non-essential fatty acid utilization pathways	Fatty acid ethyl ester	17.2 mg/L	Valle-Rodríguez et al. (2014)

improved production of desired compounds and bioproducts by precisely manipulating biosynthetic routes, regulatory functions, and transport systems within cells, as indicated by Kern et al. (2007). Host strains are manipulated to develop tolerance to elevated levels of end products and inhibiting compounds by knocking down normal genes and enzymes related to the regulation of metabolic pathways, ultimately achieving a high titer of the target product(s).

Microorganisms play a crucial role in biofuel production, but the synthesis of target products by wild-type strains is economically inefficient. Therefore, genetic and metabolic engineering approaches are essential for manipulating and enhancing these strains. Current research activities are focused on diverse metabolic engineering techniques aiming to optimize the economic output of microbial cell factories. There is a growing anticipation regarding the development of more sophisticated metabolic pathways to produce biofuels. The integration of innovative microbial strains with advanced metabolic engineering holds the promise of creating robust cell factories capable of efficiently utilizing lignocellulosic biomass for various bioproducts and biofuels. The exploration of biorefinery approaches is deemed vital, wherein waste from one production stream transforms into a renewable feedstock for another valuable product. The adoption of cutting-edge technologies, such as CRISPR/Cas9 and omic technologies, is advocated for generating unique microbial strains exhibiting superior biofuel production capacity.

Despite the challenges inherent in identifying new strains, microbial metabolic engineering remains crucial for advancing the biofuel-producing industrial sectors and it is crucial to understand important metabolic pathways and the enzymes linked to them. Synthetic biologists and metabolic engineers employ sophisticated approaches to comprehend the limitations of pathways. Technologies like CRISPR/Cas9 play a vital role in accelerating microbial engineering by facilitating efficient genome editing. The ongoing progress in high-throughput technologies, encompassing transcriptomics, metabolomics, and proteomics, is widely leveraged to unravel gene expression patterns and their roles in metabolic pathways for biofuel synthesis. Computational tools of this nature provide mechanistic insights, empowering the screening and engineering of microbial platforms for enhanced biofuel production.

7.8 SUSTAINABLE AVIATION FUELS – COMMERCIALIZATION AND CARBON NEUTRALITY OF SAF

As part of its aim to reach carbon neutrality by 2020, aviation has pledged its use of sustainable biofuels derived from non-competitive organic sources that meet jet fuel standards – this will reduce ecological impact while fueling industry expansion. These biofuels emit less CO_2 over their life cycle than their fossil-based counterparts while positively influencing biodiversity conservation efforts and protecting natural ecosystems from conversion. Biofuel manufacturers, airlines, and non-governmental organizations (NGOs) must collaborate toward commercializing sustainable jet fuels.

Federal Aviation Authority's (FAA) aim is to use one billion gallons of biofuel by 2018 to achieve carbon neutrality for global aviation by 2020 and a 50% reduction of CO_2 emissions before the year 2050. Airlines began using biofuels shortly after

ASTM certification was awarded in 2011 (Olaganathan, 2018a). KLM and Lufthansa conducted successful tests using biofuels; however, biofuel production cost and its supply chain issues still pose challenges to these airlines' use. In 2012, partnerships were formed among airlines and biofuel producers like Lanza Tech and Virgin Atlantic (Olaganathan, 2018a); British Airways collaborated with Solena to convert domestic waste to aviation fuel (British Airways, 2018b). United Airlines has made significant headway since 2009 in sustainable aviation biofuels through its partnerships with AltAir Fuels and Fulcrum BioEnergy (United Airlines, 2022). Despite positive developments in terms of demand and industry expansion, commercialization of SAF remains challenging; SAF production should increase over the coming years; according to the IEA Bioenergy Task 39 Database, there are over 200 production facilities worldwide that are focused on SAF (Figure 7.2) (IEA, 2023a).

Hydrotreatment plants currently in operation are owned by BP (Spain), Eni (Italy), Neste (Netherlands, Finland, Singapore), Pertamina (Indonesia), Preem (Sweden), Repsol (Spain), Total (France), World Energy (USA), and Fulcrum (USA). According to IEA (2023b), manufacturers with functioning FT plants include Fulcrum (USA) and Total (France). Gevo USA owns the ATJ plant, which is currently operational; Haru Oni, a producer of e-fuels via an e-methanol pathway, has already commenced operations. Noteworthy announcements for future SAF production plants include Norsk e-fuel (Kerosene manufactured using air, water, and green electricity – PtL 2023), Nordic Electro Fuels FT-liquids (2024), and GEVO Net-Zero One (2024). It is important to note that not all plants currently produce SAF, and monitoring offtake agreements is one method for estimating current and future SAF capacity, as these agreements provide support to future producers while ensuring a stable supply.

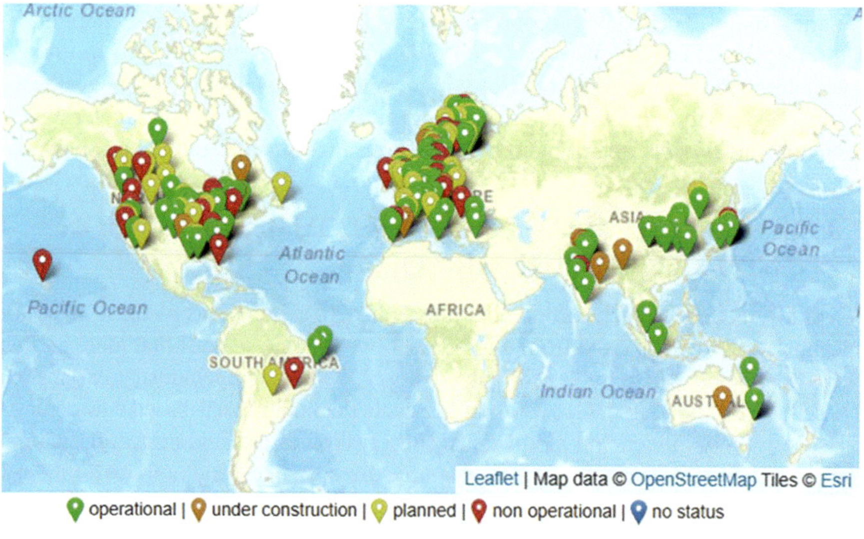

FIGURE 7.2 Biofuel Production Plants Located Worldwide (Screenshot from Bioenergy Task 39, IEA; IEA, 2023a).

In the year 2022, SAF production was estimated to account for 0.1% of current aviation fuel consumption, roughly equivalent to 0.27 million tons (ICAO, 2022b). Initial producers were World Energy of Paramount USA, Neste of Porvoo Finland, Gevo of Silsbee USA, and Total of La Mede, France (ICAO, 2022b). Of all sustainable aviation gasolines worldwide, with HEFA SPK being the leader, 97 offtake agreements totaling approximately 41,800 million liters had already been announced between future producers and mostly airline companies (ICAO, 2022a). The announced offtake volumes in 2022 amounted to 21.715 million liters, as illustrated in Figure 7.3, and volumes increased sharply in the last years and growth is predicted to continue (ICAO, 2023).

7.8.1 Sustainable Aviation Fuel Consumption

By the year 2023, SAF had become integral to aviation, with 37 airports globally having implemented SAF into their offerings, and an additional 15 conducting individual tests using batches of SAF fuels. Figure 7.4 illustrates the geographical distribution of airports that are SAF-enabled, with a predominant presence in North America and Europe (ICAO, 2023). This distribution reflects significant efforts within these regions to incorporate more sustainable practices into aviation operations. The growing presence of SAFs at airports signifies a crucial step in reducing aviation's carbon footprint, demonstrating an industry-wide commitment to sustainability. This trend may potentially encourage global adoption and accelerate the development of additional mechanisms and infrastructure to support the widespread use of SAFs in aviation.

7.8.2 SAF Consumers and Purchasers

Table 7.4 highlights the efforts of several airlines in pursuing ambitious decarbonization goals through SAF, with many making significant strides in this initiative (ICAO, 2023). The table also provides an overview of SAF purchasers and announced

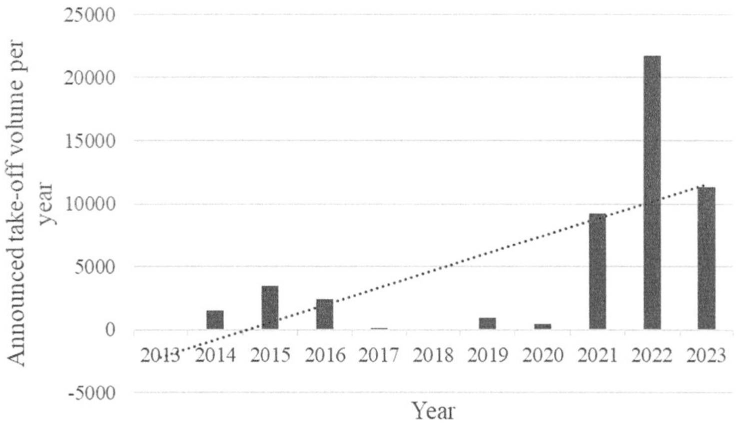

FIGURE 7.3 Announced Take-off Volumes of Biofuel per Year (ICAO, 2023).

FIGURE 7.4 Map of airports' use of SAF (green color represents ongoing deliveries, and the blue color represents batch delivery of SAF) (Screenshot, ICAO, 2023).

TABLE 7.4
Lists of Purchasers of SAF According To Announced Offtake Agreements. Most Quantities Are Sold by Contract Even Before They Are Produced (ICAO, 2023)

SAF Consumer and Purchaser (Top 10)	Total Offtake Volume in Million Liters (Number of Take-off Agreements)
United Airlines	14299.39 (7)
Southwest Airlines	4163.95 (3)
Delta	3862.25 (8)
One World	3785.41 (1)
Lufthansa	3399.4 (4)
Air France - KLM	3094.54 (5)
AirBP	2192.71 (2)
American Airlines	2134.21 (5)
Japan Airlines	1661.95 (2)
Cathy Pacific	1477.07 (2)

commitments. A growing number of these agreements involve selling SAF quantities through contracts even before production takes place. This proactive approach underscores the industry's dedication to transitioning toward environmentally friendly aviation practices, showcasing collaborative arrangements between airlines and SAF manufacturers. Pre-production contracts signify our shared commitment to greener aviation practices, setting a precedent for the broader adoption of SAF as an environmentally friendly aviation fuel (ICAO, 2023).

7.8.3 Case study: United Airlines' Initiative in Promoting Sustainable Aviation Fuel Adoption

United Airlines, one of the world's largest airlines, has always led by example in steering the industry toward a low-carbon, sustainable future. Figure 7.5 depicts their roadmap toward zero emissions. United Airlines achieved an extraordinary milestone by operating its inaugural flight using algae biofuel in 2011 (Olaganathan, 2018a). Following that achievement, United signed a strategic agreement with AltAir Fuels in 2013 for purchasing 15 million gallons of SAF over three years (Olaganathan, 2018b). United Airlines made a $30 million investment, signaling their long-term commitment toward decarbonization operations, further cemented by their pledge of cutting emissions by half by the year 2050 compared to 2005 levels (Olaganathan 2018a). They also launched their Eco-Skies Alliance Program in the spring 2020 and has so far attracted 29 corporate partner companies as part of it to purchase approximately 5.7 million gallons SAF collectively and eliminate 53,000 tons of GHGs by 2023 (United Airlines 2022).

United Airlines reached another significant milestone in 2021 by operating its inaugural commercial flight using 100% drop-in SAF fuel for one of two aircraft

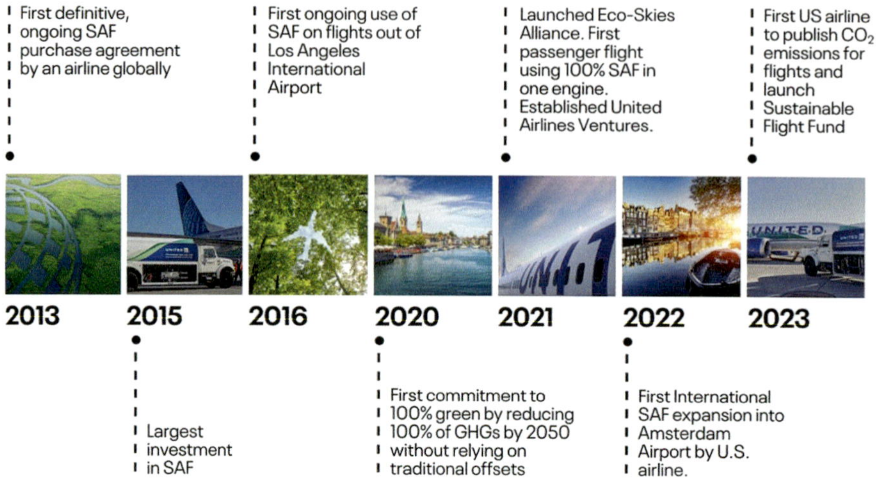

FIGURE 7.5 United Airline's Road to Net Zero (United Airlines Corporate Responsibility Report, 2022).

engines, carrying 115 passengers onboard. Furthermore, it launched United Airlines Ventures as a program emphasizing innovation within their organization during that same year. United Airlines is seeking startups and technologies aligned with its goals of reaching zero emissions by 2050 (United Airlines, 2022). UAV now encompasses SAF producers as well as various sustainability technologies like carbon capture, hydrogen-electric engines, electric regional aircraft, and urban air mobility (United Airlines, 2022). United Airlines and UAV announced in June 2022 their intent to make a strategic investment into Dimensional Energy to purchase at least 300,000,000 gallons of SAF within 20 years (United Airlines, 2022). This investment marks United's fourth move into SAF technology-related innovations that drive positive changes within aviation, highlighting their dedication to driving positive transformation within the aviation industry.

7.9 UNLOCKING ECONOMIC OPPORTUNITIES IN SUSTAINABLE AVIATION FUEL INDUSTRIES

SAF have created significant economic opportunities across multiple industries beyond aviation, producing positive impacts throughout society. Construction and planning have been influenced, with increased SAF demand leading to the construction of new storage and processing facilities. As a result, these sectors have thrived and provided the necessary infrastructure to accommodate SAF needs (ICAO, 2017).

The agricultural sector is experiencing significant expansion due to the demand for raw materials essential for SAF production. This growth creates opportunities for farmers and agricultural businesses as the production of feedstock for SAF manufacturing increases (DBRS Morningstar, 2023). The transportation infrastructure has advanced due to demand from this industry, requiring efficient transport of both raw materials and finished SAF products. Investments in transportation and logistics systems provide opportunities for innovation and expansion within this field (U.S. Department of Energy, 2022).

Conversion and biorefinery facilities have experienced rapid expansion due to the intricate processes associated with SAF production. Skilled personnel and technological improvements are vital, providing businesses specializing in these areas with opportunities (U.S. Department of Energy, 2022). The blending and distribution of SAF into traditional aviation fuels necessitate a robust supply chain system, presenting ample business prospects in fuel distribution, blending, logistics support, etc. (U.S. Department of Energy, 2022).

SAF has become an invaluable source of employment. About 80–90% of jobs generated come from the agriculture, forestry, and transportation sectors alone, further strengthening these sectors and contributing to overall economic development (DBRS Morningstar, 2023). The expansion of the SAF industry provides more benefits than just its direct value chain. The demand for expertise from administrative support services and research/monitoring agencies/entities auditing sustainable practices continues to rise, benefiting the overall economy. Thus the rise of SAF offers numerous economic benefits, such as job creation and influence across various sectors. As the SAF becomes an established industry, these opportunities will significantly increase, with positive ramifications locally and internationally.

7.10 CHALLENGES ASSOCIATED WITH SUSTAINABLE AVIATION FUEL COMMERCIALIZATION

SAF, one component of sustainable aviation, has emerged as an indispensable asset in aviation industry efforts to reach carbon neutrality by the year 2050 and reduce greenhouse emissions by 50% by the same date. Unfortunately, commercializing SAF has its challenges; currently, only 2% of total aviation fuel consumption is SAF, representing a significant hurdle. This low percentage highlights the need for large production plants equipped with cost-efficient production methods to ensure the success of its commercialization (ICAO, 2017).

Many SAFs rely on bio-based materials like animal fats, vegetable oils, and waste biomass as feedstocks; it can be difficult to secure an abundant and diverse supply that does not compete with food production or cause environmental degradation (ICAO, 2017). Innovation will be vital in meeting commercialization challenges for SAFs, specifically improving conversion technologies, refining processes, and exploring alternative feedstock sources (U.S. Department of Energy, 2022).

Lacking an effective and coherent regulatory framework hinders investment, certification, and approval processes, creating an unfavorable environment for investments to flourish. International collaboration is crucial to establish uniform standards, regulations, and best practices (ICAO, 2017). Economic viability can be challenging due to higher SAF production costs compared to fossil fuel alternatives. To enhance economic feasibility, implementing cost reduction strategies, such as incentives, subsidies, and market mechanisms, is essential. Significant improvements are required to effectively integrate SAF fuels into an aviation infrastructure designed for conventional fuels, and current infrastructure gaps impede widespread adoption.

Public perception and awareness of SAF play a crucial role in the commercialization of SAF technology. Education and communication efforts should address concerns related to feedstock competition, environmental impacts, and land-use changes. Successful navigation through these obstacles requires collaboration among aviation stakeholders, including government agencies, research institutions, and private entities.

7.11 CONCLUSION – EXPLORING SUSTAINABLE FLIGHT PATH SOLUTIONS

This chapter covers various facets of aviation's journey toward sustainability, ranging from intricate metabolic engineering techniques to the commercialization of SAF. Genetic and metabolic engineering, crucial for biofuel production, leverages technology to optimize efficiency and cost-effectiveness. Advances in CRISPR/Cas9 technologies and genomics may lead to new strains with enhanced production capacities.

Aviation companies actively seek carbon neutrality, with the biofuel industry emerging as a promising eco-friendly alternative. Pioneers like United Airlines, Singapore Airlines, Air France, British Airways, Japan Airlines, Virgin Atlantic, Air India, Fin Air, Malaysian Airlines, etc. are making significant strides. However, industry stakeholders face obstacles when aiming for large-scale production, such as securing access to diverse feedstock sources and navigating regulatory frameworks. The current industrial infrastructure for SAF production includes hydrotreatment and Fischer-Tropsch facilities, complemented by the emerging presence of electro-fuel plants. Despite progress, challenges in commercialization persist, requiring innovative solutions for scaling and cost-effectiveness within the existing infrastructure.

The economic impacts of SAF consumption extend across various sectors, each playing a crucial role in fostering economic growth and sustainability. Challenges persist, encompassing feedstock availability, economic viability, and adapting regulatory frameworks. Therefore, cooperation among airlines, research institutes, and stakeholders is crucial for addressing these challenges and guiding aviation toward a sustainable future.

This chapter concludes by exploring the delicate balance of microbial science with technological innovation, economic opportunity, and the greening of aerospace sectors. Reflecting on both progress and challenges encountered, the ongoing collaboration, innovation, and commitment to more sustainable aviation practices are essential.

REFERENCES

Airport Council International [ACI], 2023. ACI World confirms top 20 busiest airports worldwide.

Al-Azzawi, A.M., Jassem, E.K., 2016. Synthesis and characterization of several new succinimides linked to phenyl azo benzothiazole or thiazole moieties with expected biological activity. *Iraq. J. Sci.* 57, 534–544.

Angelidaki, I., Treu, L., Tsapekos, P., Luo, G., Campanaro, S., Wenzel, H., Kougias, P. G., 2018. Biogas upgrading and utilization: Current status and perspectives. *Biotechnology Advances*, 36(2), 452–466.

Anto, S., Mukherjee, S. S., Muthappa, R., Mathimani, T., Deviram, G., Kumar, S. S., Verma, T. N., Pugazhendhi, A., 2020. Algae as green energy reserve: Technological outlook on biofuel production. *Chemosphere (Oxford)*, 242, 125079–125079.

Arvidsson, R., Persson, S., Fröling, M., Svanström, M., 2011. Life cycle assessment of hydrotreated vegetable oil from rape, oil palm and Jatropha. *Journal of Cleaner Production*, 19(2), 129–137.

Arora, R., Sharma, N. K., Kumar, S., 2018. Valorization of By-Products Following the Biorefinery Concept: Commercial Aspects of By-Products of Lignocellulosic Biomass. In Arora, R., Sharma, N. K., Kumar, S., (Eds.), *Advances in Sugarcane Biorefinery*, Elsevier Inc., pp. 163–178.

Arous, F., Azabou, S., Triantaphyllidou, I.-E., Aggelis, G., Jaouani, A., Nasri, M., Mechichi, T., 2017. Newly isolated yeasts from Tunisian microhabitats: lipid accumulation and fatty acid composition. *Eng. Life Sci.* 17, 226–236.

Asgher, M., Wahab, A., Bilal, M., Iqbal, H. M. N., 2018. Delignification of lignocellulose biomasses by Alginate–Chitosan immobilized laccase produced from trametes versicolor IBL-04. *Waste and Biomass Valorization*, 9(11), 2071–2079.

Atsumi, S., Hanai, T., Liao, J. C., 2008. Non-fermentative pathways for synthesis of branched-chain higher alcohols as biofuels. *Nature*, 451(7174), 86–89.

Avanthi, A., Kumar, S., Sherpa, K. C., Banerjee, R., 2017. Bioconversion of hemicelluloses of lignocellulosic biomass to ethanol: an attempt to utilize pentose sugars. *Biofuels (London)*, 8(4), 431–444.

Bachmann, S. L., McCarthy, A. J., 1991. Purification and Cooperative Activity of Enzymes Constituting the Xylan-Degrading System of *Thermomonospora fusca*. *Applied and Environmental Microbiology*, 57(8), 2121–2130.

Bagaeva, T. V., Zinurova, E. E., 2004. Comparative characterization of extracellular and intracellular hydrocarbons of *Clostridium pasteurianum*. *Biochemistry. Biokhimiia*, 69(4), 427–428.

Bakermans, C., Tsapin, A. I., Souza-Egipsy, V., Gilichinsky, D. A., Nealson, K. H., 2003. Reproduction and metabolism at − 10°C of bacteria isolated from Siberian permafrost. *Environmental Microbiology*, 5(4), 321–326.

Balat, M., 2011. Potential alternatives to edible oils for biodiesel production—a review of current work. *Energy Convers. Manage.* 52(2), 1479–1492.

Balat, M., Balat, M., 2009. Political, economic and environmental impacts of biomass-based hydrogen. *International Journal of Hydrogen Energy*, 34(9), 3589–3603.

Banerjee, S., Banerjee, S., Ghosh, A. K., Das, D., 2020. Maneuvering the genetic and metabolic pathway for improving biofuel production in algae: Present status and future prospective. *Renewable & Sustainable Energy Reviews*, 133, 110155–.

Bauen, A., 2020. Sustainable Aviation Fuels. *Johnson Matthey Technology Review*, 64(3), 263–278.

Beller, H. R., Goh, E. B., Keasling, J. D., 2010. Genes involved in long-chain alkene biosynthesis in Micrococcus luteus. *Applied and Environmental Microbiology*, 76(4), 1212–1223.

Beopoulos, A., Cescut, J., Haddouche, R., Uribelarrea, J. L., Molina-Jouve, C., Nicaud, J. M., 2009. Yarrowia lipolytica as a model for bio-oil production. *Progress in Lipid Research*, 48(6), 375–387.

Bernard, A., Domergue, F., Pascal, S., Jetter, R., Renne, C., Faure, J. D., Haslam, R. P., Napier, J. A., Lessire, R., Joubès, J., 2012. Reconstitution of plant alkane biosynthesis in yeast demonstrates that Arabidopsis ECERIFERUM1 and ECERIFERUM3 are core components of a very-long-chain alkane synthesis complex. *The Plant Cell*, 24(7), 3106–3118.

Bhatia, S. K., Jagtap, S. S., Bedekar, A. A., Bhatia, R. K., Rajendran, K., Pugazhendhi, A., Rao, C. V., Atabani, A. E., Kumar, G., Yang, Y.-H., 2021. Renewable biohydrogen production from lignocellulosic biomass using fermentation and integration of systems with other energy generation technologies. *The Science of the Total Environment*, 765, 144429–144429.

Bilal, M., & Iqbal, H. M. N., 2020a a. Recent Advancements in the Life Cycle Analysis of Lignocellulosic Biomass. *Current Sustainable/Renewable Energy Reports.*, 7(3), 100–107.

Bilal, M., Iqbal, H. M. N., 2020b b. Ligninolytic Enzymes Mediated Ligninolysis: An Untapped Biocatalytic Potential to Deconstruct Lignocellulosic Molecules in a Sustainable Manner. *Catalysis Letters*, 150(2), 524–543.

Bilal, M., Iqbal, H.M.N., 2021. Ligninolysis potential of ligninolytic enzymes: a green and sustainable approach to bio-transform lignocellulosic biomass into high-value entities. *Alternat. Energy Resour.: Way Sustain. Mod. Soc.* (2021), pp. 151–171.

Bonatto, C., Camargo, A. F., Scapini, T., Stefanski, F. S., Alves S. L., Müller, C., Fongaro, G., Treichel, H., 2020. Biomass to bioenergy research: current and future trends for biofuels. In: Gupta, V. G., Treichel, H., Kuhad, R. C., & Rodriguez-Couto, S. (Eds.), *Recent Developments in Bioenergy Research*. Elsevier, pp. 1–17.

Buijs, N. A., Zhou, Y. J., Siewers, V., Nielsen, J., 2015. Long-chain alkane production by the yeast Saccharomyces cerevisiae. *Biotechnology and Bioengineering*, 112(6), 1275–1279.

Cabrera-Jiménez, R., Mateo-Sanz, J. M., Gavaldà, J., Jiménez, L., Pozo, C., 2022. Comparing biofuels through the lens of sustainability: A data envelopment analysis approach. *Applied Energy*, 307, 118201.

Cadete, R.M., Rosa, C.A., 2018. The yeasts of the genus Spathaspora: potential candidates for second generation biofuel production. *Yeast*, 35(2), 191–199.

Carmona-Cabello, M., García, I.L., Papadaki, A., Tsouko, E., Koutinas, A., Dorado, M.P., 2021. Biodiesel production using microbial lipids derived from food waste discarded by catering services. *Bioresour. Technol.* 323, 124597.

Chandel, A. K., Singh, O. V., Chandrasekhar, G., Rao, L. V., Narasu, M. L., 2010. Key drivers influencing the commercialization of ethanol-based biorefineries. *Journal of Commercial Biotechnology*, 16(3), 239–257.

Chakraborty, D., Sarkar, N., Biswas, I., Jacob, S., 2020. Molecular aspects of prokaryotic and eukaryotic cellulases and their modulation for potential application in biofuel production. In: *Genetic and Metabolic Engineering for Improved Biofuel Production from Lignocellulosic Biomass*. Elsevier, pp. 81–95.

Chao, H., Agusdinata, D. B., DeLaurentis, D. A., 2019. The potential impacts of Emissions Trading Scheme and biofuel options to carbon emissions of U.S. airlines. *Energy Policy*, 134, 110993.

Choi, Y. J., Lee, S. Y., 2013. Microbial production of short-chain alkanes. *Nature*, 502(7472), 571–574.

Choubey, G., Tiwari, M., 2022. *Scramjet combustion: fundamentals and advances.* Butterworth-Heinemann.

Chung, D., Cha, M., Guss, A. M., Westpheling, J., 2014. Direct conversion of plant biomass to ethanol by engineered Caldicellulosiruptor bescii. *Proceedings of the National Academy of Sciences of the United States of America*, 111(24), 8931–8936.

Cunha, J. T., Gomes, D. G., Romaní, A., Inokuma, K., Hasunuma, T., Kondo, A., Domingues, L., 2021. Cell surface engineering of Saccharomyces cerevisiae for simultaneous valorization of corn cob and cheese whey via ethanol production. *Energy Conversion and Management*, 243, 114359.

DBRS Morningstar., 2023. The Green Take-off (Part 2 of 2): Challenges and Opportunities for Sustainable Aviation Fuel.

Deshavath, N. N., Veeranki, V. D., Goud, V. V., 2019. Lignocellulosic feedstocks for the production of bioethanol: availability, structure, and composition. M. Rai, A.P. Ingle (Eds.), *Sustainable bioenergy*, Elsevier, Amsterdam, The Netherlands, 1–19.

d'Espaux, L., Mendez-Perez, D., Li, R., Keasling., 2015. Synthetic biology for microbial production of lipid-based biofuels. *Current Opinion in Chemical Biology*, 29(C), 58–65.

Dürre P., 2008. Fermentative butanol production: bulk chemical and biofuel. *Annals of the New York Academy of Sciences*, 1125, 353–362.

Ehite, E.H., Drumm, E., Abdoulmoumine, N., 2021. The effect of hemicellulose on the interparticle frictional behavior of lignocellulosic biomass particulates. *Particuology*, 55, 16–22.

Elshahed, M. S., 2010. Microbiological aspects of biofuel production: Status and future directions. *Retrieved from Journal of Advanced Research*, 1, 103–111.

Eliodório, K. P., Cunha, G. C. de, Müller, C., Lucaroni, A. C., Giudici, R., Walker, G. M., Alves, J., Basso, T. O. 2019. Advances in yeast alcoholic fermentations for the production of bioethanol, beer and wine. *Adv. Appl. Microbiol.* 109, 61–119.

Environmental Protection Agency (EPA), 2018. Inventory of U.S. Greenhouse Gas Emissions and Sinks:1990-2016.

Fang, W., Zhang, X., Zhang, P., Carol Morera, X., van Lier, J. B., Spanjers, H., 2020. Evaluation of white rot fungi pretreatment of mushroom residues for volatile fatty acid production by anaerobic fermentation: Feedstock applicability and fungal function. *Bioresource Technology*, 297, 122447–122447.

Filho, E. X., Tuohy, M. G., Puls, J., Coughlan, M. P., 1991. The xylan-degrading enzyme systems of Penicillium capsulatum and Talaromyces emersonii. *Biochemical Society Transactions*, 19(1), 25S.

Foo, J. L., Susanto, A. V., Keasling, J. D., Leong, S. S., Chang, M. W., 2017. Whole-cell biocatalytic and de novo production of alkanes from free fatty acids in Saccharomyces cerevisiae. *Biotechnology and Bioengineering*, 114(1), 232–237.

Francois, J.M., Alkim, C., Morin, N., 2020. Engineering microbial pathways for production of bio-based chemicals from lignocellulosic sugars: current status and perspectives. *Biotechnol. Biofuels*, 13 (1), 118.

Fraser, J.A., Heitman, J., 2004. Evolution of fungal sex chromosomes. *Mol. Microbiol.* 51(2), 299–306.

Fu, W. J., Chi, Z., Ma, Z. C., Zhou, H. X., Liu, G. L., Lee, C. F., Chi, Z. M., 2015. Hydrocarbons, the advanced biofuels produced by different organisms, the evidence that alkanes in petroleum can be renewable. *Applied Microbiology and Biotechnology*, 99(18), 7481–7494.

Fu, H., Hu, J., Guo, X., Feng, J., Yang, S. T., Wang, J., 2021. Butanol production from Saccharina japonica hydrolysate by engineered Clostridium tyrobutyricum: The effects of pretreatment method and heat shock protein overexpression. *Bioresource Technology*, 335, 125290.

Gao, Q., Cao, X., Huang, Y. Y., Yang, J. L., Chen, J., Wei, L. J., Hua, Q., 2018. Overproduction of fatty acid ethyl esters by the oleaginous yeast yarrowia lipolytica through metabolic engineering and process optimization. *ACS Synthetic Biology*, 7(5), 1371–1380.

Gonçalves, D. de S., Ferreira, M. da S., & Guimarães, A. J. (2019). Extracellular Vesicles from the Protozoa Acanthamoeba castellanii : Their Role in Pathogenesis, Environmental Adaptation and Potential Applications. *Bioengineering (Basel)*, *6*(1), 13–29.

Graver, B., Zhang, K., Rutherford, D., 2019. CO2 emissions from commercial aviation, 2018. The International Council of Clean Transportation.

Guimarães, P. M. R., Teixeira, J. A., Domingues, L., 2010. Fermentation of lactose to bio-ethanol by yeasts as part of integrated solutions for the valorisation of cheese whey. *Biotechnology Advances*, 28(3), 375–384.

Haldar, D., Purkait, M. K., 2021. A review on the environment-friendly emerging techniques for pretreatment of lignocellulosic biomass: Mechanistic insight and advancements. *Chemosphere (Oxford)*, 264(Pt 2), 128523–128523.

Han, J., McCarthy, E. D., Calvin, M., Benn, H., 1968. Hydrocarbon constituents of the blue-green algae Nostoc muscorum, Anacystis nidulans, Phormidium luridum and Chlorogloea fritschii. *J Chem Soc.*, 40, 2785–2791.

Hanson, J.R., 2008. Fungi and the development of microbiological chemistry. In: Hanson, J.R. (Ed.), *Chemistry of Fungi*, 1st ed. Royal Society of Chemistry, pp. 1–17.

Harger, M., Zheng, L., Moon, A., Ager, C., An, J. H., Choe, C., Lai, Y. L., Mo, B., Zong, D., Smith, M. D., Egbert, R. G., Mills, J. H., Baker, D., Pultz, I. S., Siegel, J. B., 2013. Expanding the product profile of a microbial alkane biosynthetic pathway. *ACS Synthetic Biology*, 2(1), 59–62.

Hari, T.K., Yaakob, Z., Binitha, N. N., 2015. Aviation biofuel from renewable resources: Routes, opportunities, and challenges. *Renewable & Sustainable Energy Reviews*, 42, 1234–1244.

Hasanov, I., Raud, M., Kikas, T., 2020. The role of ionic liquids in the lignin separation from lignocellulosic biomass. *Energies*, 13, 4864.

Hemighaus, G., Boval, T., Bosley, C., Organ, R., Lind, J., Brouette, R., et al., 2006. *Alternative jet fuels, a supplement to Chevron's aviation fuels technical review*. Chevron Corporation.

Hileman, J., Ortiz, D., Bartis, J., Wong, H., Donohoo, P., Weiss, M., Waitz, I., 2009. Near-Term Feasibility of Alternative Jet Fuels. RAND.

Howard, T. P., Middelhaufe, S., Moore, K., Edner, C., Kolak, D. M., Taylor, G. N., Parker, D. A., Lee, R., Smirnoff, N., Aves, S. J., Love, J., 2013. Synthesis of customized petroleum-replica fuel molecules by targeted modification of free fatty acid pools in Escherichia coli. *Proceedings of the National Academy of Sciences of the United States of America*, 110(19), 7636–7641.

Huang, J., Chen, D., Wei, Y., Wang, Q., Li, Z., Chen, Y., Huang, R., 2014. Direct ethanol production from lignocellulosic sugars and sugarcane bagasse by a recombinant Trichoderma reesei strain HJ48. *The Scientific World Journal*, 2014, 798683.

Huberman, L. B., Liu, J., Qin, L., Glass, N. L., 2016. Regulation of the lignocellulolytic response in filamentous fungi. *Fungal Biology Reviews*, 30(3), 101–111.

International Energy Association [IEA], 2023a. Task 39. Bioenergy. Database on facilities for the production of advanced liquid and gaseous biofuels for transport.

International Energy Association [IEA], 2023b. Task 63. Sustainable Aviation Fuels – Status quo and national assessments. A Report from the Advanced Motor Fuels Technology Collaboration Programme.

International Air Transport Association [IATA], 2016. IATA Forecasts Passenger Demand to Double over 20 Years.

International Air Transport Association [IATA], 2017. Carbon offsetting for international aviation.

International Air Transport Association [IATA], 2021. Net-Zero Carbon Emissions by 2050.

International Civil Aviation Organization [ICAO], 2017. Sustainable Aviation Fuels Guide.

International Civil Aviation Organization [ICAO], 2022a. Environmental Report – Innovation for a Greener Transition.

International Civil Aviation Organization [ICAO], 2022b. Report on the feasibility of a long-term aspirational Goal (LTAG) for international civil aviation CO2 Emission reductions.

International Civil Aviation Organization [ICAO], 2023. SAF Offtake Agreements.

Irmak, S., 2019. *Biomass for Bioenergy – Recent Trends and Future Challenges*. IntechOpen, London, 1–11.

Jensen, L. L., Bonnefoy, P. A., Hileman, J. I., Fitzgerald, J. T., 2023. The carbon dioxide challenge facing U.S. aviation and paths to achieve net zero emissions by 2050. *Progress in Aerospace Sciences*, 141, 100921.

Jyoti, S., Wattal, D. D., 2019. Overview of carbon capture technology: Microalgal biorefinery concept and state-of-the-art. *Front. Mar. Sci.*, (6):29.

Kageyama, H., Waditee-Sirisattha, R., Sirisattha, S., Tanaka, Y., Mahakhant, A., Takabe, T., 2015. Improved alkane production in nitrogen-fixing and halotolerant cyanobacteria via abiotic stresses and genetic manipulation of alkane synthetic genes. *Current Microbiology*, 71(1), 115–120.

Kallio, P., Pásztor, A., Akhtar, M. K., Jones, P. R., 2014. Renewable jet fuel. *Current Opinion in Biotechnology*, 26, 50–55.

Karatzos, S., McMillan, J. D., Saddler, J. N., 2014. The Potential and Challenges of Drop-in Biofuels (IEA Bioenergy Task 39. Report T39-T1).

Kang, A., Lee, T. S., 2015. Converting sugars to biofuels: Ethanol and beyond. *Bioengineering (Basel, Switzerland)*, 2(4), 184–203.

Kargbo, H., Harris, J.S., Phan, A.N., 2021. "Drop-in" fuel production from biomass: Critical review on techno-economic feasibility and sustainability. *Renew. Sustain. Energy Rev.*, 135, 110168.

Khalil, A.M., 2020. The genome editing revolution: review. *J. Genet. Eng. Biotechnol.* 18, 68.

Kern, A., Tilley, E., Hunter, I. S., Legisa, M., Glieder, A., 2007. Engineering primary metabolic pathways of industrial micro-organisms. *Journal of Biotechnology*, 129(1), 6–29.

Kohn, R. A., Kim, S.W., 2015. Using the second law of thermodynamics for enrichment and isolation of microorganisms to produce fuel alcohols or hydrocarbons. *Journal of Theoretical Biology*, 382, 356–362.

Kichonge, B., Kivevele, T., 2023. Viability of non-edible oilseed plants and agricultural wastes as feedstock for biofuels production: A techno-economic review from an African perspective. *Biofuels, Bioproducts and Biorefining*, 17(5), 1382–1410.

Kinney, J. R., 2017. The power for flight: NASA's contributions to aircraft propulsion. *Headquarters*, 1–316. https://ntrs.nasa.gov/api/citations/20180003207/downloads/20180003207.pdf

Klasson, K. T., Ackerson, M. D., Clausen, E. C., Gaddy, J. L., 1992. Bioconversion of synthesis gas into liquid or gaseous fuels. *Enzyme. Microb. Technol.*, 14(8), 602–608.

Klinke, H. B., Thomsen, A. B., Ahring, B. K., 2004. Inhibition of ethanol-producing yeast and bacteria by degradation products produced during pre-treatment of biomass. *Applied Microbiology and Biotechnology*, 66(1), 10–26.

Ko, J. K., Lee, J. H., Jung, J. H., Lee, S.-M., 2020. Recent advances and future directions in plant and yeast engineering to improve lignocellulosic biofuel production. *Renewable & Sustainable Energy Reviews*, 134, 110390.

Komesu, A., Oliveria, J., Neto, J.M., Penteada, E.D., Diniz, A.A.R., Martins, L.H.D., 2020. Xylose fermentation to bioethanol production using genetic engineering microorganisms. In: *Genetic and Metabolic Engineering for Improved Biofuel Production from Lignocellulosic Biomass*. Elsevier, pp. 143–154.

Kumar, D., Singh, B., Korstad, J., 2017. Utilization of lignocellulosic biomass by oleaginous yeast and bacteria for production of biodiesel and renewable diesel. *Renewable and Sustainable Energy Reviews*, 73, 654–671.

Kumar, M., Sundaram, S., Gnansounou, E., Larroche, C., Thakur, I. S., 2018. Carbon dioxide capture, storage and production of biofuel and biomaterials by bacteria: A review. *Bioresource Technology*, 247, 1059–1068.

Kumar, M., Sun, Y., Rathour, R., Pandey, A., Thakur, I. S., Tsang, D. C. W., 2020. Algae as potential feedstock for the production of biofuels and value-added products: Opportunities and challenges. *The Science of the Total Environment*, 716, 137116–137116.

Kwak, S., Jin, Y.-S., 2017. Production of fuels and chemicals from xylose by engineered Saccharomyces cerevisiae: a review and perspective. *Microb. Cell Fact.* 16(1), 1–15.

Lane, J., 2015. Biofuels Mandates Around the World 2015. Biofuels Dig.

Lee, A., Tong, D., Jun Cheston, M., & Xuan Yi, Z., Olaganathan, R., 2014. Is biofuel a feasible long-term chief energy source? A global perspective. *International Journal of Innovative Research in Science, Engineering and Technology*, 3(6), 85–100.

Levandowsky, M., 2012. Physiological adaptations of Protists. In: *Cell Physiology Source Book*. Elsevier, pp. 873–890.

Li, X., Li, M., Pu, Y., Ragauskas, A. J., Zheng, Y., 2020. Simultaneous depolymerization and fermentation of lignin into value-added products by the marine protist, Thraustochytrium striatum. *Algal Research (Amsterdam)*, 46, 101773.

Li, C., Zhao, X., Wang, A., Huber, G. W. Zhang, T., 2015. Catalytic transformation of lignin for the production of chemicals and fuels. *Chem. Rev.*, 115, 11559–11624.

Liu, L., Pan, A., Spofford, C., Zhou, N., Alper, H. S., 2015. An evolutionary metabolic engineering approach for enhancing lipogenesis in Yarrowia lipolytica. *Metabolic Engineering*, 29, 36–45.

Liu, K., Du, H.S., Si, C.L., 2021. Recent advances in cellulose and its derivatives for oilfield applications. *Carbohydr. Polym.* 259.

Lu, H., Yadav, V., Zhong, M., Bilal, M., Taherzadeh, M. J., Iqbal, H. M. N., 2022. Bioengineered microbial platforms for biomass-derived biofuel production - A review. *Chemosphere*, 288(Pt 2), 132528.

Lynd, L. R., van Zyl, W. H., McBride, J. E., Laser, M., 2005. Consolidated bioprocessing of cellulosic biomass: An update. *Current Opinion in Biotechnology*, 16(5), 577–583.

Maki, M., Leung, K. T., Qin, W., 2009. The prospects of cellulase-producing bacteria for the bioconversion of lignocellulosic biomass. *International Journal of Biological Sciences*, 5(5), 500–516.

Malik, S., Dhasmana, A., Preetam, S., Mishra, Y. K., Chaudhary, V., Bera, S. P., … Rajput, V. D., 2022. Exploring microbial-based green nanobiotechnology for wastewater remediation: a sustainable strategy. *Nanomaterials*, 12(23), 4187.

Manikandan, S., Subbaiya, R., Biruntha, M., Krishnan, R. Y., Muthusamy, G., Karmegam, N. 2022. Recent development patterns, utilization and prospective of biofuel production: Emerging nanotechnological intervention for environmental sustainability – A review. *Fuel (Guildford)*, 314, 122757.

Martiniano, S. E., Chandel, A. K., Soares, L. C., Pagnocca, F. C., da Silva, S. S., 2013. Evaluation of novel xylose-fermenting yeast strains from Brazilian forests for hemicellulosic ethanol production from sugarcane bagasse. *3 Biotech*, 3(5), 345–352.

Milledge, J.J., Heaven, S., 2013. A review of the harvesting of microalgae for biofuel production. *Rev. Environ. Sci. Biotechnol.* 12, 165–178.

Moon, S. Y., Son, S.H., Baek, S., Lee, J.Y., 2023. Designing microbial cell factories for programmable control of cellular metabolism. *Synthetic Biology*. In Press.

Moysés, D. N., Reis, V. C. B., de Almeida, J. R. M., de Moraes, L. M. P., Torres, F. A. G., 2016. Xylose Fermentation by Saccharomyces cerevisiae: Challenges and Prospects. *International Journal of Molecular Sciences*, 17(3), 207–207.

National Aeronautics and Space Administration [NASA], 2023. Global Climate Change – Vital Signs of the Planet.

Olaganathan, R., 2018a. Algal biofuel: the future of green jet fuel in air transportation. *Journal of Applied Biotechnology & Bioengineering*, 5(2), 133–135.

Olaganathan, R., 2018b. Research of Sustainable Jet Fuel Production Using Microbes. *CPQ Microbiology*, 1(4), 01–12.

Olaganathan, R., 2021. Impact of COVID-19 on airline industry and strategic plan for its recovery with special reference to data analytics technology. *Global Journal of Engineering and Technology Advances*, 7(1), 033–046.

Otero-Muras, I., Carbonell, P., 2021. Automated engineering of synthetic metabolic pathways for efficient biomanufacturing. *Metabolic Engineering*, 63, 61–80.

Olaganathan, R., Ko Qui Shen, F., & Jun Shen, L. (2014). Potential and Technological Advancement of Biofuels. *International Journal of Advanced Scientific and Technical Research*, 4(4), 12–29. https://commons.erau.edu/cgi/viewcontent.cgi?article=1918&context=publication

Pacala, S., Socolow, R., 2004. Stabilization wedges: solving the climate problem for the next 50 years with current technologies. *Science*, 305(5686), 968–972.

Park, M. O., Tanabe, M., Hirata, K., Miyamoto, K., 2001. Isolation and characterization of a bacterium that produces hydrocarbons extracellularly which are equivalent to light oil. *Applied Microbiology and Biotechnology*, 56(3–4), 448–452.

Patinvoh, R. J., Osadolor, O. A., Chandolias, K., Sárvári Horváth, I., Taherzadeh, M. J., 2017. Innovative pretreatment strategies for biogas production. *Bioresource Technology*, 224, 13–24.

Peramuna, A., Morton, R., Summers, M. L., 2015. Enhancing alkane production in cyanobacterial lipid droplets: A model platform for industrially relevant compound production. *Life (Basel, Switzerland)*, 5(2), 1111–1126.

Peramuna, A., Summers, M. L., 2014. Composition and occurrence of lipid droplets in the cyanobacterium Nostoc punctiforme. *Archives of microbiology*, 196(12), 881–890.

Perera F., 2017. Pollution from Fossil-Fuel Combustion is the Leading Environmental Threat to Global Pediatric Health and Equity: Solutions Exist. *International Journal of Environmental Research and Public Health*, 15(1), 16.

Phukan, M.M., Bora, P., Gogoi, K., Konwar, B.K., 2019. Biodiesel from Saccharomyces cerevisiae: Fuel property analysis and comparative economics. *SN Appl. Sci.* 153, 1–10.

Pinzi, S., Dorado, M.P., 2012. Feedstocks for advanced biodiesel production. In Rafael Luque and Juan A. Melero (Eds.), *Advances in Biodiesel Production: Processes and Technologies*. Woodhead Publishing Limited, pp. 69–90.

Probstein, R. F., Hicks, R. E., 2006. *Synthetic Fuels*. Dover Publications.

Robak, K., Balcerek, M., 2018. Review of second-generation bioethanol production from residual biomass. *Food Technol. Biotechnol.*, 56 (2), 174–187

Rodrigue, J.-P., 2020. *The Geography of Transport Systems* (5th ed.). Routledge.

Sarkar, N., Ghosh, S. K., Bannerjee, S., Aikat, K., 2012. Bioethanol production from agricultural wastes: An overview. *Renewable Energy*, 37(1), 19–27.

Sarker, T. R., Nanda, S., Dalai, A. K., Meda, V., 2021. A review of torrefaction technology for upgrading lignocellulosic biomass to solid biofuels. *Bioenergy Research*, 14(2), 645–669.

Sarris, D., Papanikolaou, S., 2016. Biotechnological production of ethanol: biochemistry, processes and technologies. *Eng. Life Sci.* 16 (4), 307–329.

Schirmer, A., Rude, M. A., Li, X., Popova, E., del Cardayre, S. B., 2010. Microbial biosynthesis of alkanes. *Science (New York, N.Y.)*, 329(5991), 559–562.

Schirmer, A. W., Rude, M. A. Brubaker, S. A., 2014. Methods and compositions for producing alkanes and alkenes. US 8846371 B2.

Segura, A., Jiménez, L., Molina, L., 2022. Can microbiology help to make aviation more sustainable? *Microbial Biotechnology*, 16(2), 190–194.

Selim, K., El-Ghwas, D., Easa, S., Abdelwahab Hassan, M., 2018. Bioethanol a Microbial Biofuel Metabolite; New Insights of Yeasts Metabolic Engineering. *Fermentation (Basel)*, 4(1), 16.

Sharma, H.K., Xu, C., Qin, W., 2019. Biological pretreatment of lignocellulosic biomass for biofuels and bioproducts: An overview. *Waste Biomass Valorizat*, 10(2), 235–251.

Sharma, P., Gaur, V. K., Kim, S. H., Pandey, A., 2020. Microbial strategies for bio-transforming food waste into resources. *Bioresource Technology*, 299, 122580.

Sims, R., Taylor, M., Saddler, J., & Mabee, W., 2008. From 1st- to 2nd-Generation biofuel Technologies: An overview of current industry and R&D activities. Retrieved from International Energy Agency.

Sindhu, R., Binod, P., Pandey, A., 2016. Biological pretreatment of lignocellulosic biomass - an overview. *Bioresource Technol*. 199, 76–82.

Siripong, W., Wolf, P., Kusumoputri, T. P., Downes, J. J., Kocharin, K., Tanapongpipat, S., Runguphan, W., 2018. Metabolic engineering of Pichia pastoris for production of isobutanol and isobutyl acetate. *Biotechnology for Biofuels*, 11(1), 1–16.

Smith, W. H. M., 2006. Agriculture as a Producer and Consumer of Energy. In: Outlaw, J. L., Collins, K. J., Duffield, J.A. (Eds.). Trowbridge, United Kingdom: Cromwell Press, Experimental Agriculture, 42(2), 252–252. Cambridge University Press.

Song, X., Yu, H., Zhu, K., 2016. Improving alkane synthesis in Escherichia coli via metabolic engineering. *Applied Microbiology and Biotechnology*, 100(2), 757–767.

Statista. 2023. Number of scheduled passengers boarded by the global airline industry from 2004 to 2022(in millions).

Tabatabaei, M., Aghbashlo, M., Valijanian, E., Kazemi Shariat Panahi, H., Nizami, A.-S., Ghanavati, H., Sulaiman, A., Mirmohamadsadeghi, S., Karimi, K., 2020. A comprehensive review on recent biological innovations to improve biogas production, Part 1: Upstream strategies. *Renewable Energy*, 146, 1204–1220.

Tasić, M. B., Pinto, L. F. R., Klein, B. C., Veljković, V. B., Filho, R. M., 2016. Botryococcus braunii for biodiesel production. *Renewable & Sustainable Energy Reviews*, 64, 260–270.

Tao, H., Guo, D., Zhang, Y., Deng, Z., Liu, T., 2015. Metabolic engineering of microbes for branched-chain biodiesel production with low-temperature property. *Biotechnology for Biofuels*, 8, 92.

Taylor, M. P., Eley, K. L., Martin, S., Tuffin, M. I., Burton, S. G., Cowan, D. A., 2009. Thermophilic ethanologenesis: future prospects for second-generation bioethanol production. *Trends in Biotechnology*, 27(7), 398–405.

Teo, W. S., Ling, H., Yu, A.-Q., Chang, M. W., 2015. Metabolic engineering of Saccharomyces cerevisiae for production of fatty acid short- and branched-chain alkyl esters biodiesel. *Biotechnology for Biofuels*, 8(177), 177–177.

Troiano, D., Orsat, V., Dumont, M.J., 2020. Status of filamentous fungi in integrated biorefineries. *Renew. Sustain. Energy Rev*. 117, 109472.

Tulha, J., Faria-Oliveira, F., Lucas, C., Ferreira, C., 2012. Programmed cell death in Saccharomyces cerevisiae is hampered by the deletion of GUP1 gene. *BMC Microbiology*, 12(1), 80–80.

Ullah, N., Shahzad, K., Wang, M., 2021. The role of metabolic engineering technologies for the production of fatty acids in yeast. *Biology*, 10(7), 632.

United Airlines. 2022. United Airline's Corporate Responsibility Report.

United Airlines. 2023. Our Commitment to Environment.

U.S. Department of Energy, 2022. SAF Grand Challenge Roadmap - Flight Plan for Sustainable Aviation Fuel.

U.S. Energy Information Administration, 2015. International Energy Statistics.
U.S. Energy Information Administration, 2017. Annual Energy Outlook, 2017.
U.S. Energy Information Administration. 2022. Monthly Energy Review. 1–282.
Vallesi, A., Pucciarelli, S., Buonanno, F., Fontana, A., Mangiagalli, M., 2020. Bioactive molecules from protists: Perspectives in biotechnology. *European Journal of Protistology*, 75, 125720–125720.
Valle-Rodríguez, J. O., Shi, S., Siewers, V., Nielsen, J., 2014. Metabolic engineering of Saccharomyces cerevisiae for production of fatty acid ethyl esters, an advanced biofuel, by eliminating non-essential fatty acid utilization pathways. *Applied Energy*, 115, 226–232.
van de Werken, H. J., Verhaart, M. R., VanFossen, A. L., Willquist, K., Lewis, D. L., Nichols, J. D., Goorissen, H. P., Mongodin, E. F., Nelson, K. E., van Niel, E. W., Stams, A. J., Ward, D. E., de Vos, W. M., van der Oost, J., Kelly, R. M., Kengen, S. W., 2008. Hydrogenomics of the extremely thermophilic bacterium Caldicellulosiruptor saccharolyticus. *Applied and Environmental Microbiology*, 74(21), 6720–6729.
Walker, G.M., Walker, R.S.K., 2018. Enhancing yeast alcoholic fermentations. *Adv. Appl. Microbiol.*, 87–129.
Walker, G.M., White, N.A., 2017. Introduction to fungal physiology. In: *Fungi*. John Wiley & Sons, Inc, Hoboken, NJ, USA, pp. 1–35.
Wang, C.-L., Li, Y., Xin, F.-H., Liu, Y.-Y., Chi, Z.-M., 2014. Evaluation of single cell oil from Aureobasidium pullulans var. melanogenum P10 isolated from mangrove ecosystems for biodiesel production. *Process Biochemistry* (1991), 49(5), 725–731.
Wang, W., Liu, X., Lu, X., 2013. Engineering cyanobacteria to improve photosynthetic production of alka(e)nes. *Biotechnology for Biofuels*, 6(1), 69.
Wang, W.-C., Tao, L., 2016. Bio-jet fuel conversion technologies. *Renewable & Sustainable Energy Reviews*, 53(C), 801–822.
Wang, F., Ouyang, D., Zhou, Z., Page, S. J., Liu, D., Zhao, X., 2021. Lignocellulosic biomass as sustainable feedstock and materials for power generation and energy storage. *Journal of Energy Chemistry*, 57, 247–280.
Wei, H., Liu, W., Chen, X., Yang, Q., Li, J., Chen, H., 2019. Renewable jet fuel production for aviation – A review. *Fuel*, 254, p. 115599.
Winchester, N., Malina, R., Staples, M. D., Barrett, S. R. H., 2015. The impact of advanced biofuels on aviation emissions and operations in the U.S. *Energy Economics*, 49(C), 482–491.
Wisselink, H. W., Toirkens, M. J., Wu, Q., Pronk, J. T., van Maris, A. J., 2009. Novel evolutionary engineering approach for accelerated utilization of glucose, xylose, and arabinose mixtures by engineered Saccharomyces cerevisiae strains. *Applied and Environmental Microbiology*, 75(4), 907–914.
Weosten, H.A.B., 2019. Filamentous fungi for the production of enzymes, chemicals and materials. *Curr. Opin. Biotechnol.* 59, 65–70.
Wess, J., Brinek, M., Boles, E., 2019. Improving isobutanol production with the yeast Saccharomyces cerevisiae by successively blocking competing metabolic pathways as well as ethanol and glycerol formation. *Biotechnology for Biofuels*, 12(1), 173–173.
Xu, Y., Li, Z., 2020. CRISPR-Cas systems: Overview, innovations and applications in human disease research and gene therapy. *Computational and Structural Biotechnology Journal*, 18, 2401–2415.
Yang, X., Yuanyuan, W., Zhang, Y., Yang, E., Yuan, Q., Huini, X., Chen, Y., Irbis, C., Yan, J., 2020. A thermo-active laccase isoenzyme from Trametes trogii and its potential for dye decolorization at high temperature. *Frontiers in Microbiology*, 11. https://doi.org/10.3389/fmicb.2020.00241

Yadav, V., Wang, Z., Wei, C., Amo, A., Ahmed, B., Yang, X., Zhang, X., 2020. Phenylpropanoid pathway engineering: An emerging approach towards plant defense. *Pathogens (Basel)*, 9(4), 312.

Yamada, R., Hasunuma, T., Kondo, A., 2013. Endowing non-cellulolytic microorganisms with cellulolytic activity aiming for consolidated bioprocessing. *Biotechnology Advances*, 31(6), 754–763. https://doi.org/10.1016/j.biotechadv.2013.02.007

Yang, Z., Zhang, Z., 2018. Production of (2R, 3R)-2,3-butanediol using engineered Pichia pastoris: strain construction, characterization and fermentation. *Biotechnology for Biofuels*, 11(1), 35–35.

Young, N. N., 2014. The future of biofuels: U.S. (and Global) airlines & aviation alternative fuels. In: EIA conference.

Zakzeski, J., Bruijnincx, P. C., Jongerius, A. L., Weckhuysen, B. M., 2010. The catalytic valorization of lignin for the production of renewable chemicals. *Chem. Rev.*, 110, 3552–3599.

Zheng, Y., Zhao, J., Xu, F., Li, Y., 2014. Pretreatment of lignocellulosic biomass for enhanced biogas production. *Progress in Energy and Combustion Science*, 42, 35–53.

Zhou, Y. J., Buijs, N. A., Zhu, Z., Qin, J., Siewers, V., Nielsen, J., 2016. Production of fatty acid-derived oleochemicals and biofuels by synthetic yeast cell factories. *Nature Communications*, 7, 11709.

Zhu, P., Abdelaziz, O.Y., Hulteberg, C.P., Riisager, A., 2020a a. New synthetic approaches to biofuels from lignocellulosic biomass. *Curr. Opin. Green Sustain. Chem.* 21, 16–21.

Zhu, D., Adebisi, W. A., Ahmad, F., Sethupathy, S., Danso, B., Sun, J., 2020b b. Recent development of extremophilic bacteria and their application in biorefinery. *Front. Bioeng. Biotechnol.* 8, 1–18.

8 Drop-in Fuel Production
Transforming Biomass into Next-Generation Fuels

Eunice O. Babatunde, Yura Kim, and Sujata Mandal

8.1 INTRODUCTION

With increasing global emphasis on environmental protection, climate change, and environmental health comes increased research and investment in the processing of bio-based fuels and materials. It is important that the materials produced are renewable, eco-friendly, have a low carbon footprint, and low environmental impact. To achieve this, research and governmental initiatives have increased in recent times to develop resources that can match current human needs in a planet-friendly way. To sustain and improve modern civilization, energy, power, and transportation are very key to economic growth (Geleynse et al., 2018). These rely heavily on fuels used in energy generation systems as well as industrial machinery and transportation vehicles of different types. Currently, the aviation and shipping industries account for 8% of all anthropogenic CO_2 emissions; expansion in tourism and international trade is expected to boost this percentage even more (Schäppi et al., 2022). Thus, these industrial sectors among others serve to benefit greatly from the use of bio-based fuels (Mascal & Dutta, 2020).

Renewable gasoline or drop-in gasoline or liquid bio-hydrocarbon is a fuel made from biomass sources using a range of biological, thermal, and chemical processes. The fuel satisfies ASTM D4814 specifications and is chemically like petroleum gasoline. Drop-in fuels eliminate the need for engine or fuel system modifications because they are a synthetic and 100% interchangeable replacement for traditional petroleum-derived hydrocarbons (Table 8.1) like gasoline, jet fuel, and diesel (Abdellatief et al., 2024; Mascal & Dutta, 2020). Engines and infrastructure that are already in place can be powered by renewable fuel and thus do not require a redesign of engines and fuel systems. This renewable form of gasoline has gained increasing attention due to its potential environmental advantages and the compatibility with existing production facilities and application infrastructure. Its chemical makeup, compatibility with engines and infrastructure, improved energy security, reduced emissions, and enhanced flexibility are some of its distinguishing features

DOI: 10.1201/9781003585398-8

TABLE 8.1
Comparison of Biofuel Properties

	Biogas	Bioethanol	Biobutanol	Biodiesel	Gasoline
Density (kg/L)	0.00115	0.79	0.81	0.88	0.74
Viscosity at 20°C (mm^2/s)	–	0.5	3.3	5.1–7.5	0.6
Octane number	–	>100	87	–	92
Cetane number	–	8	17	56	15–20
Lower heating value (MJ/kg)	23.3	27	36	37	43.9
Heating value (MJ/L)	0.016–0.028	21.06	29	32.65	32.48
Flash point (°C)	–	13	35	160/120	<21
Fuel equivalence (L)	–	0.65	–	0.91	1

(Zinsmeister et al., 2023). In the US, renewable gasoline is not being used for commercial purposes because the emphasis is on electrifying the light-duty vehicle market (US Department of Energy).

The US Energy Information Administration projects that global aviation fuel utilization will increase from 13 quadrillion British thermal units (quads) in 2018 to 29 quads in 2050. It is anticipated that commercial aviation will continue to develop (U.S. EIA, 2019). Conventionally, kerosene made from crude oil is used to supply aviation fuel. However, using fossil fuels is not sustainable and the emission of greenhouse gases (GHGs) has had a negative influence on the environment. An airplane engine burning one kilogram (kg) of jet fuel yields 3.16 kg of CO_2. 2.5% of CO_2 emissions come from worldwide aviation, which includes both domestic and international travel as well as freight (Ritchie & Roser, 2024). Thus, one of the main factors driving the development of biofuels is the objective of lowering CO_2 emissions in the transportation sector (EA IEA, 2011) and the United States Energy Information Administration anticipates biofuels could contribute significantly to reducing emissions by increasing from 2% of total transport energy today to 27% by 2050 (U.S EIA, 2019). Therefore, it is of great interest to develop efficient and sustainable processes for the synthesis of renewable diesel and renewable aviation fuel also known as bio-jet fuel. The main composition of fossil jet fuel is 20% paraffins, 20% naphthene, 20% aromatics, and 40% iso-paraffins, and the atmospheric pollution of fuel combustion products is also strongly related to their composition (Bernabei et al., 2003). "Jet fuel" is a general name for aviation fuels used in gas-turbine aircraft. Conventionally, jet fuel (or "kerosene") relates to the kerosene distillation fraction of crude oil (~150–275°C). The main components are linear and branched alkanes and cycloalkanes with a typical carbon chain-length distribution of C6–C16 (Benavides et al., 2021; Cookson et al., 1987; Kallio et al., 2014).

In the United States, there are three categories of Jet fuel Jet A, Jet A-1, and Jet B. Jet A is the major commercial jet fuel (Boldt, 1977; Gaughan et al., 2019). The ASTM Standard for Aviation Turbine Fuels D 1655 (Pires et al., 2018) is the fundamental specification for civil jet fuel used in the United States of America (Pires et al., 2018). Jet A is a fuel of the kerosine type with a maximum freeze point of –40°C. Similar to Jet A, Jet A-1 is a kerosine-type fuel with a maximum freeze point of –47°C. Fuel of the wide-cut variety is Jet B which contains a wider range of hydrocarbons, allowing for superior performance in low temperature, but is less commonly used due to its higher flammability. Both domestic and international airlines operate within the United States using Jet A. Some areas of northern Canada employ Jet B because of its increased volatility and lower freeze point, which are advantageous for cold starting and handling. Current Chinese regulations cover five types of jet fuel. The UK's standard civil jet fuel is called DEF STAN 91-91, and the specification for DEF STAN 91-91 is quite close to ASTM D 1655's Jet A-1 specification (Diederichs et al., 2016; Shell Global; Wang & Tao, 2016). Some of the important jet fuel specifications are given in Table 8.2.

TABLE 8.2
Jet Fuel A-1 Specifications

Jet A-1	ASTM D1655-04a	IATA	Def Stan 91–91	ASTM D7566
Composition				
Acidity, total (mg KOH/g)	0.1, max	0.015, max	0.012, max	0.1, max
Aromatics (vol%)	25, max	25, max	25, max	25, max (8, min)
Sulfur, total (wt%)	0.3, max	0.3, max	0.3, max	0.3, max
Volatility				
Distillation temperature:				
10% Recovery (°C)	205, max	205, max	205, max	205, max
20% Recovery (°C)	–	–	–	–
50% Recovery (°C)	–	–	–	(15, min)
90% Recovery (°C)	–	–	–	(40, min)
Final BP (°C)	300, max	300, max	300, max	300, max
Flash point (°C)	38, min	38, min	38, min	38, min
Density @ 15°C (kg/m^3)	775–840	775–840	775–840	775–840
Fluidity				
Freezing point (°C), max	–47	–47	–47	–40 Jet A –47 Jet A-1
Viscosity @ –20°C (cSt)	8, max	8, max	8, max	8, max

(Continued)

TABLE 8.2 (*Continued*)
Jet Fuel A-1 Specifications

Jet A-1	ASTM D1655-04a	IATA	Def Stan 91–91	ASTM D7566
Combustion				
Net heat of comb. (MJ/kg)	42.8, min	42.8, min	42.8, min	42.8, min
Smoke point (mm or min	25, min	25, min	25, min	25, min
Smoke point (mm + min	19 (min)	19 (min)	19 (min)	18 (min)
Naphthalenes (vol%)	3 (max)	3 (max)	3 (max)	3 (max)
Thermal stability				
Tube deposit rating	<3	<3	<3	<3
Conductivity				
Conductivity (pS/m)	50–450	50–450	50–600	–
Lubricity				
BOCLE wear scar diameter (mm)	–	0.85, max	0.85, max	0.85, max

Note: BOCLE stands for "Ball on Cylinder Lubricity Evaluator" and measures the lubricity of jet fuel by simulating friction between a stationary steel ball and a rotating cylinder submerged in the fuel sample.

Biobased jet fuel has become a crucial factor in the aviation sector's strategy to reduce operation and environmental impacts by reducing CO_2 emissions. A completely renewable resource, biomass can be utilized for energy production, manufacturing, building materials, food and feed, chemicals, and other environmentally friendly uses that help reduce GHG emissions. We give an overview, background information on the many kinds of biofuels, details on the processing of biomass, and possible uses of biofuels in the green aviation sector.

8.2 BIOFUEL

In the global scenario, today's liquid biofuels are considered as an unavoidable invention for environmental sustainability reasons. Climate changes, global population growth, fossil fuel depletion, and the global need for increasing food and energy demand are some of the most pressing issues that human beings face in the 21th century (Thanigaivel et al., 2022). Non-renewable crude oil contributes to climate change and increases GHG emissions, and hence the need to find a long-term replacement for crude petroleum is a vital requirement in today's world. Biofuels can be gaseous, liquid, or solid. Examples of biofuels include alcohols, various types of wood, carbon monoxide or biogas, and various biodiesel fuels (Figure 8.1). The high energy density, homogeneous composition, and superior manageability of liquid biofuels are absolute benefits that result in less storage space needed, extended operation without refueling, and easy-to-use technology for handling and storing. These elements are particularly important for applications involving moving machines and automobiles.

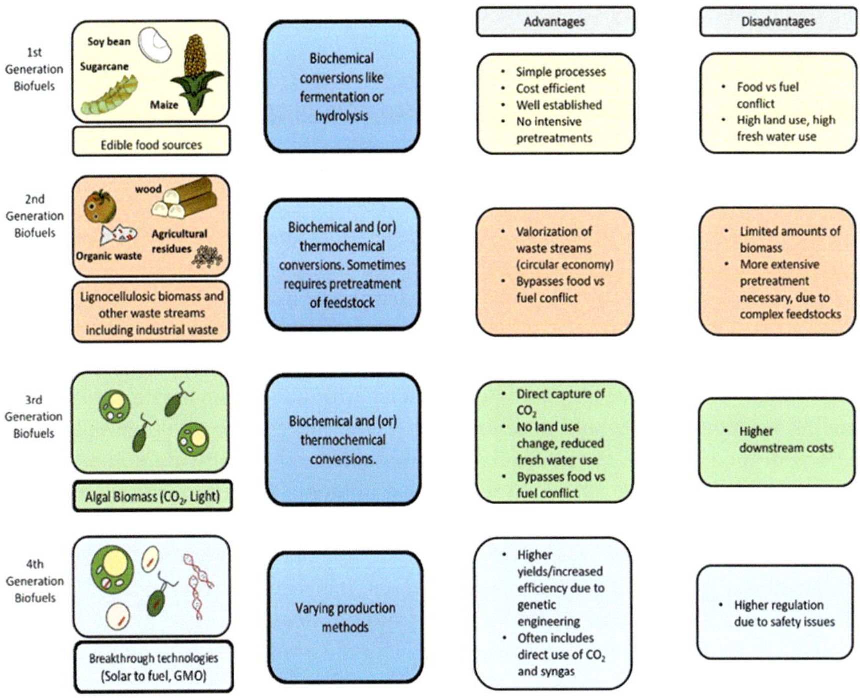

FIGURE 8.1 Biofuels – sources, advantages, disadvantages.

8.2.1 First Generation

Biofuels can be classified as first generation (Fatih Demirbas, 2009), second generation (Naik et al., 2010), third generation (Lee & Lavoie, 2013), and fourth generation (Fairley, 2011). Some major feedstocks for the first-generation biofuels are corn, wheat, barley, sugarcane, palm, and rapeseed which are usually synthesized by transesterification of organic fats and oils. Given that they are also a source of food, it results in competition between fuel and food, which can also raise food commodity prices (Kumar et al., 2020). Rapeseed oil, which makes up almost two thirds of the total input used in the manufacturing of biodiesel in the EU, is one of the raw materials. Animal fats, recycled vegetable oil, soybean oil, and palm oil are further crucial feedstock components (Jokiniemi & Ahokas, 2013). Drawing upon ancient practices such as agriculture and brewing, grain ethanol emerged as the primary fuel for spark-ignition engines in early motor vehicles during the late 19th and early 20th centuries. However, with the widespread availability of petroleum by the 1920s, ethanol was largely relegated to a mere fuel additive. The modern bioethanol industry gained momentum in the 1970s amidst rising gasoline prices and the phasing out of the octane enhancer tetraethyllead. Subsequent events, including the ban on MTBE in California and New York in the early 2000s, coupled with the establishment of the renewable fuel standard by the United States Energy Policy Act of 2005, significantly boosted ethanol demand. Nevertheless, the ethanol market faces limitations due to the blend wall, typically set at 10% ethanol (E10), and the limited adoption of flex-fuel vehicles capable of utilizing higher ethanol blends such as 85% (E85). This hampers its potential to contribute significantly to the movement toward net-zero GHG emissions. Biobutanol has been proposed as an alternative; however, both ethanol and butanol, as "first-generation" biofuels, are costly and entangled in the food-versus-fuel debate as they are primarily derived from cereal grains. In contrast, cellulosic alcohols, derived from cellulose-based sugars, have struggled to gain market traction due to heavy reliance on subsidies to compete commercially with grain-based alcohol (Mascal & Dutta, 2020).

The next viable option for biofuel production involves lipid transesterification with methanol to produce "biodiesel." This biodiesel can be blended with petroleum diesel or used in its pure form in diesel engines under appropriate temperature conditions. While biodiesel could be produced affordably from waste cooking oil, scaling up production would necessitate access to fats that could potentially compete with food sources or require the use of arable land to cultivate non-edible oil crops, thereby reigniting the food versus fuel dilemma. Transesterification's difficulty lies in the relatively high oxygen content it creates, which (a) makes the biodiesel hazy (partially freeze) at relatively high temperatures, (b) reduces its stability, and (c) makes it less energy dense than diesel obtained from petroleum (Manikandan et al., 2022). Since the large production of biofuel demands more arable agricultural lands, which results in diminished lands for the production of food for humans and animals, first-generation biofuels have significant economic, environmental, and political challenges (Naik et al., 2010). Moreover, the production process of first-generation biofuels has also contributed to environmental decline. An alternative approach involving the cultivation of microalgae for lipid production has been explored for decades, yet significant economic obstacles persist, hindering the widespread adoption of this technology (Sembiring et al., 2018). Nonetheless, the biodiesel industry has achieved modest success in the US, although its growth has been hindered by the periodic expiration and reinstatement of federal tax credits initiated in 2005. The challenges associated with ethanol and biodiesel have spurred advocacy for second-generation or "advanced biofuel" production from non-food biomass, despite relatively low blending mandates set by the United States Environmental Protection Agency compared to first-generation fuels. Non-food biomass, comprising cellulose, hemicellulose, and lignin, serves as the raw material for this approach. Advanced biofuel production can be realized through biotechnological, thermolytic, or chemocatalytic processes utilizing raw biomass, isolated lignin, or isolated carbohydrates as feedstocks. Because comparatively little plant oil can be produced per acre of farmland in comparison to other biofuel sources, the primary drawback of plant oil-based biofuels is the high price of the plant oil.

8.2.2 Second Generation

Building upon the challenges associated with first-generation biofuels, the focus has shifted toward second-generation or "advanced biofuels," derived from non-food biomass. Unlike their predecessors, which primarily relied on edible crops, second-generation biofuels utilize non-nutritive carbohydrates such as cellulose, hemicellulose, and lignin (Diederichs et al., 2016). This approach addresses concerns regarding competition with food sources and land use, offering a more sustainable solution to the biofuel dilemma. Advanced biofuel production can be achieved through various processes, including biotechnological, thermolytic, or chemocatalytic methods (Karatzos et al., 2017). These technologies utilize raw biomass or isolated components such as lignin or carbohydrates as feedstocks, paving the way for a more environmentally friendly and economically viable biofuel industry.

More focus is now being placed on second-generation biofuels made from lignocellulosic feedstocks, including non-edible crops and agricultural and forest wastes, as a result of the recently discovered shortcomings of first-generation biofuels made from food crops (perhaps with the exception of sugarcane ethanol) (Ben et al., 2013; R. A. Lee & Lavoie, 2013; Naik et al., 2010). Two very different processing methods can be used to produce biofuels from lignocellulosic feedstocks: biochemical, which uses enzymes and other microorganisms to convert the feedstocks' cellulose and hemicellulose components to sugars before they are fermented to produce ethanol; thermochemical, also called biomass-to-liquids, which uses pyrolysis and gasification to produce a synthesis gas ($CO + H_2$) which yields a variety of long carbon chain biofuels (synthetic diesel, aircraft fuel, or ethanol) using the Fischer–Tropsch (FT) conversion method (Chang, 2018; Kim et al., 2024). However, they also require land for cultivation, thus resulting in competition with food crops (R. E. H. Sims et al., 2010). The lignin component of the biomass is a byproduct of the biochemical enzymatic hydrolysis mechanism and may therefore be used to generate heat and power, which is one important distinction between the biochemical and BTL approaches. The biomass components cellulose and hemicellulose are transformed into synthesis gas during the BTL process, along with the lignin. Despite this variation, both methods have the capacity to convert one dry ton of biomass (20 GJ/t) into approximately 6.5 GJ/t of biofuels, or an overall biomass to the biofuel conversion efficiency of about 35% (Mabee et al., 2006). Currently, there isn't a certain technological or financial benefit to using the thermochemical or biological routes (Sims et al., 2010).

Second-generation fuels are produced more sustainably than some first-generation fuels and have greater land use prospects, including the ability to produce in marginal areas. They are typically based on crop and forest wastes or high yielding, non-food energy crops developed expressly for feedstocks (Sims et al., 2010). The full commercialization of thermochemical and biochemical conversion pathways needs more research, though. Additionally, the process of producing second-generation biofuels is costly and technical, and the commercial production of biofuels is not profitable (Nigam & Singh, 2011). More knowledge of the feedstocks that are currently accessible is needed, as well as information on their geographic distribution and the costs associated with production, transportation, storage, and processing. It is necessary to gain more experience producing diverse energy crop feedstocks in multiple regions in order to comprehend the features of biomass, yield potential, and production costs. To optimize the conversion efficiency of feedstocks into liquid biofuels, the optimal attributes for each feedstock type must be determined. In multi-feedstock plants, the variances in wood, straw, stover, and vegetative grass properties can pose distinct challenges for bioconversion (R. E. H. Sims et al., 2010). Compared to hardwood species, the hemicellulose found in softwoods has a higher mannose/galactose composition and a lower xylose content. Additionally, the fundamental phenyl propane units that make up lignin are only found in softwoods (coumaryl and guaiacyl), whereas extra syringyl units are found in hardwoods and herbaceous plants. Due to the lignin's increased stability in condensed form when exposed to acidic circumstances, its chemical structure makes delignification more difficult, which makes lignocellulosic materials from woody biomass problematic for biochemical conversions (N. Kumar et al., 2020; Naik et al., 2010).

8.2.3 Third Generation

In actuality, advanced biofuels are distinguished by their resemblance to modern jet, diesel, and gasoline fuels. Advanced biofuels are energy dense and compatible with infrastructure. The low energy density of even cellulosic ethanol—which is ethanol's energy content regardless of its source—and incompatibility with current automobile engines, oil pipelines, storage tanks, refineries, etc. are its two drawbacks. For these two reasons, hydrocarbon biofuels—which have the same gas mileage as currently used gasoline and diesel fuels and are fully compatible with the existing oil infrastructure—have been the focus of recent research and development initiatives in the United States (Cabrera-Jiménez et al., 2022; Johnson et al., 2021; Malode et al., 2021; National Research Council, 2011; Sarisky-Reed, 2022).

Thus, an investigation into potential substitutes was started in order to reduce the rivalry between the food crop and the non-food crop. The mission focused on the microbial realm, as certain microorganisms possessed the capacity to generate biofuels and possessed specific beneficial characteristics including accelerated development, uncomplicated nutrient uptake, and simpler bio-machinery related to them. We refer to ethanol made from algae as a third-generation biofuel. Algae are a varied collection of aquatic photoautotrophs (unicellular and multicellular) that use photosynthesis to store solar energy. According to reports, many terrestrial crops have photosynthetic efficiencies of 0.5%, but algae have efficiencies ranging from 3% to 8% (Demirbas & Demirbas, 2012). Seaweed and microalgae are the two types of algae that can be distinguished (Lardon et al., 2009). "Seaweeds," also known as macro-algae, are multicellular organisms with distinct cell structures and functions. They resemble plants in general. They have rudimentary reproductive mechanisms and lack vascular tissue, leaves, stalks, and real roots. They can reach a maximum length of 60 meters and develop relatively quickly in both fresh and saltwater environments. Seaweed does not compete with the food industry for resources and has a shorter life cycle and lower farming costs (Bowles, 2007).

Algae are of particular importance among the diverse range of bacteria because of their abundance in biofuels, their incapacity to thrive in a variety of aquatic environments, and their year-round growth potential (Bowles, 2007; Lardon et al., 2009;

Ziolkowska & Simon, 2014). Microscopic unicellular or multicellular organisms, microalgae are capable of effectively converting solar energy into biomass. Like all biomass, algal biomass is made up of organic matter, which is made up of simpler substances like sugars and amino acids as well as complex polymeric macromolecules including polysaccharides, peptides, lipids, and nucleic acids. Methane, carbon dioxide, fresh biomass, and inorganic wastes are the end products of the anaerobic decomposition of organic matter (Bohutskyi & Bouwer, 2013; Thanigaivel et al., 2022). Algae are currently harvested using chemical, mechanical, biological, and electrical processes. Because algae have relatively small cells, chemical flocculation is frequently used as a pretreatment to make the algae's particles larger before the algae is harvested via flotation. Centrifugation is the fastest and most dependable approach in mechanical processes, and it is frequently employed to recover suspended algae. The negative charge characteristics of the algal cells are exploited in the electrically based method to separate the cells. The movement in an electric field can concentrate these cells (Kumar et al., 1981). Not every species of microalgae is suitable or practical for producing biodiesel. Because of its high lipid content, relatively fast growth rate, and ease of cultivation, *Chlorophyceae* is the most promising family of microalgae. Other major microalgae classifications for the generation of biofuels are *Bacillariophyceae*, *Eustigmatophyte*, *Charophyceae*, *Haptophyceae*, and *Cyanophyceas*. While biodiesel is a renewable fuel, it is not a completely clean-burning substitute for diesel engines. It is designed specifically for use in diesel engines. Through the process of transesterification, it can be made from vegetable oil, animal oil, fats, and leftover cooking oil. Triglycerides, such as fats or oils, combine with alcohol to produce glycerol and esters, a process known as transesterification. Triglyceride and alcohol react during the esterification process when a catalyst—typically potassium or sodium hydroxide—is present. Usually, excess alcohol is utilized to guarantee that all of the fat or oil is converted to its esters (Fatih Demirbas, 2009). The mono-alkyl ester, the primary component of biodiesel, is created when alcohol combines with fat or oil, also producing glycerol in the process. When the glycerol layer and the oil phase containing the esters separate at the conclusion of the reaction time, the transesterification reaction has been effective. Given their drawbacks, including their reliance on sunshine, geographic location, and economic performance, the mentioned fuels cannot entirely replace fossil fuels (Dutta et al., 2014). Genetic engineering of algae for the production of clean fuel is a considerable solution which leads to the development of fourth-generation biofuels (Lü et al., 2011).

When compared to conventional fossil fuels, a variety of advanced biofuels offer considerable reductions in GHG emissions. Cellulosic ethanol can reduce emissions by an astounding 60% when it is made from cellulose, hemicellulose, or lignin. Diesel derived from biomass, such as biodiesel and renewable diesel, is particularly noteworthy since it reduces emissions by at least 50%. Other advanced biofuels that help reduce emissions by at least 50% include drop-in biofuels made from renewable biomass, renewable jet fuels, and butanol. These cutting-edge biofuels provide sustainable substitutes for traditional fuels and are essential in reducing the effects of climate change.

8.2.4 Fourth-Generation Biofuel

Genetic manipulation is used to increase the productivity and quality of microalgae, resulting in the fourth-generation biofuel. Because of their high lipid contents, ease of cultivation, and quick growth rate, microalgae are gaining popularity as a biofuel production alternative. The selection of microalgae strains is critical to the overall success of the production of algal biofuel. The conditions and growth rate of the microalgae strain, their chemical makeup, and their digestibility are some of the most crucial factors that determine how well-suited the strain is for producing biofuel. One of the first microbes, cyanobacteria are freshwater strains with a variety of uses, including wastewater treatment, agricultural biofertilizer, and nutrition sources (Shin et al., 2018). Moreover, because of their straightforward genetic makeup, low nutrient requirements, and wide tolerance to a variety of environmental conditions, cyanobacteria can produce ethanol, biogas, hydrogen, and biodiesel (Figure 8.2) (Agarwal et al., 2022; Parmar et al., 2011; Quintana et al., 2011; Robles-Medina et al., 2009).

The primary component of microalgae utilized to produce biodiesel is lipid. Any type of polar or neutral lipid can be found in microalgae's biochemical composition. Polar lipids bind into the membranes of organelles or the bilayer structure of cell membranes to form structural lipids. Free fatty acids (FFAs), triglycerides, diglycerides, and monoglycerides are examples of neutral lipids, which are storage lipids. One of the most attractive renewable energy sources is biofuel produced from microalgae, which sequesters CO_2 in addition to having lower GHG emissions. Microalgae

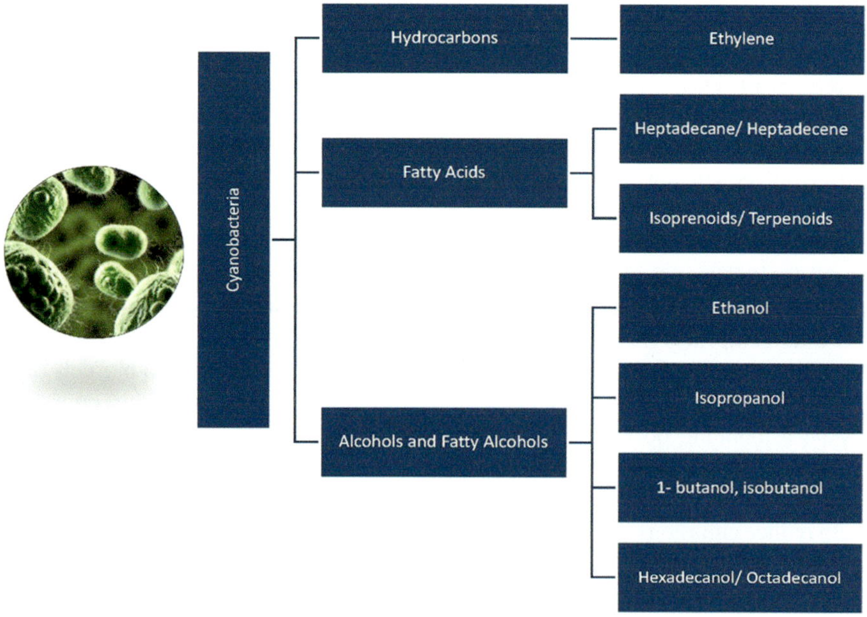

FIGURE 8.2 Fourth-generation biofuels from cyanobacteria.

sequester 10–50 times more CO_2 than many terrestrial plants, and increased CO_2 concentrations lead to larger lipid yields (Bowles, 2007; Shin et al., 2018). Algal sequestration through horticulture is still hampered by the low ambient CO_2 concentration. An effective and economical method of removing organic and inorganic pollutant wastes from wastewater is algal-based wastewater treatment. It is possible to separate the organic and inorganic components found in the effluent from these sources. Organic waste mostly consists of carbon-containing biodegradable materials, but inorganic waste also includes heavy metals, phosphate, and nitrate (Bhadha et al., 2015; Lü et al., 2011; Sekomo et al., 2012).

Biomass yield estimates from algae grown in open pond-raceway systems with freshwater or saltwater inputs came from a biophysical model that was adjusted with data on algal production and included costs from a reliable techno-economic model (Chisti, 2016; Paul et al., 2021). When co-located with ethanol production facilities, coal-fired power plants, and natural gas-fired power plants, the potential for algal biomass across the country is contingent upon various parameters, including the type of algae strain, growth medium, local weather patterns, and projected productivity rates (Lardon et al., 2009; Paul et al., 2021). Under current operating conditions and productivity assumptions, the biomass potential of *Chlorella sorokiniana* in freshwater environments co-located with CO_2 from ethanol production facilities, coal-fired electric generating units (EGUs), and natural gas EGUs is estimated to be 12 million, 19 million, and 15 million dry tons, respectively (Langholtz, 2024). The productivity of these sources is likewise influenced by their distances. Simulators take these productivity levels into account. The amount of pond liner coverage affects the price related to possibly accessible biomass. Although co-location can provide cost savings in many areas of the nation, these savings are not as great as those that can be attained through higher productivity or lower liner costs. Algae have higher minimum pricing per ton than terrestrial feedstocks, but they can yield more fuel per dry ton of biomass than their terrestrial equivalents (Parmar et al., 2011; Shin et al., 2018). Research continues to be focused on finding ways to lower the cost of producing algal feedstock. However, algae can provide other benefits including reduced land requirement for equal yields, flexibility in co-product selection, and flexibility in water and land usage. Algal biofuels have considerable obstacles in their large-scale cultivation and commercialization, even though they are usually thought to be more environmentally friendly than fossil fuels. The possible health and environmental problems that come with mass microalgae production have been generally disregarded as one of the difficulties in commercializing algal biofuels (Bohutskyi & Bouwer, 2013).

8.2.5 Conventional Biofuel

Sugar crops including sugarcane, sugar beet, and sweet sorghum are used to produce sucrose, which is then fermented to produce ethanol. After that, a number of procedures are used to recover and concentrate the ethanol. Hydrolyzing starch into glucose is a necessary extra step in the conversion process of starch crops, and it uses more energy than the sugar-to-ethanol pathway. The value of co-products like fructose and dried distiller's grains with soluble (DDGS) has a significant impact on the overall commercial and environmental effectiveness of starch-based processes (Malode et al., 2021). The cost of producing sugar and starch is highly dependent on feedstock prices, which have been erratic in recent years. By using more efficient amylase enzymes, reducing the cost of ethanol concentration, and making better use of co-products, efficiency might be increased while costs could be reduced (Cabrera-Jiménez et al., 2022). Although conventional biofuels are quite developed, there is still room to improve the technologies' overall sustainability by lessening their negative effects on the environment, the economy, and society. Enhancements in conversion efficiency will benefit traditional biofuels' environmental performance and land-use efficiency in addition to their economic viability.

8.2.6 Conventional Biodiesel

Animal fats used in cooking and raw vegetable oils from sunflower, canola, soybean, or oil palm are used to make biodiesel. Either methanol or ethanol is used to transform these oils and fats into biodiesel. Although vegetable oils can be utilized as untreated raw oils on occasion, it is not advised because of the potential for engine damage and lubricant gelling. The total economics of the biodiesel generation process depends on co-products, primarily glycerin and protein meals. The price of feedstock has an impact on the profitability of producing biodiesel the conventional way (Sims et al., 2008). The purification of the co-product glycerin, increased feedstock flexibility, and more effective catalyst recovery are important areas for improvement in conventional biodiesel. Improved DDGS nutritional value, newer, more effective enzymes, and increased energy efficiency can lower production costs and increase conversion efficiency for conventional ethanol. Optimizing value-added co-product solutions and strengthening the integration of upstream and downstream processes could lead to further cost savings. The production of conventional and/or advanced biofuels in biorefineries would encourage the more economical and environmentally beneficial use of biomass (Chavez-Rodriguez & Nebra, 2010). Table 8.3 lists the benefits and drawbacks of the various biofuel generations.

8.3 Current Production Methods of Biofuels

Various biomass sources can be converted into drop-in gasoline using pyrolysis, gasification, biological sugar upgrading, hydrothermal processing, conventional hydrotreating, and catalytic sugar conversion. Even if there aren't many commercial production techniques at the moment, research and development are always being attempted to increase efficacy and affordability. Drop-in biofuels, which are oxygen-free and functionally identical to petroleum-based transportation fuels, can be produced in a number of ways. These include the following: biochemical processes (such as the biological conversion of biomass to longer chain alcohols and hydrocarbons); oleochemical processes (such as the hydroprocessing of lipid feedstock from algae, oil crops, or tallow); and thermochemical processes (such as the thermal conversion of biomass to fluid intermediates gas or oil that are then hydrolyzed to

TABLE 8.3
Benefits and Drawbacks of the Various Biofuel Generations

Biofuel Generation	Advantage	Disadvantage
First	Greenhouse gas savings, simple and low-cost conversion process	Low yield, uses food crops, uses land meant for food cultivation
Second	Greenhouse gas savings, uses food waste and plant waste as feedstock. Energy crops can be grown on non-arable land	High pretreatment costs, requires advanced technologies for biomass conversion to fuel
Third	Uses algae feedstock, which is easy to cultivate, uses no food crops, versatile (seawater, freshwater)	High energy consumption required to cultivate algae in controlled systems, low lipid content, biomass contamination in open pond system
Fourth	High yield from lipid-containing algae, CO_2 capture ability	High costs of photobioreactor, high initial investment due to initial research efforts

hydrocarbon fuels via catalysis). A brief overview of a fourth category is also provided, which includes combined thermochemical and biological techniques such as synthesis gas fermentation and catalytic carbohydrate reforming. Up to this point, the majority of drop-in biofuels that have been considered for commercial use by industries like aviation have come from oleochemical-based methods. For these methods to work and produce hydrocarbon mixtures that resemble diesel, a basic hydroprocessing step is needed to catalytically remove oxygen from the fatty acid chains in the lipid feedstock. Compared to other new drop-in biofuel production pathways, this technology is well-developed, maturing, and involves comparatively low technological risk and cheap capital expense. Since the majority of lipid feedstocks have 10–11% oxygen by weight, upgrading them to liquid transportation fuels requires less hydrogen. However, the feedstock for lignocellulosic biomass is sometimes expensive and only available in limited quantities. Additionally, competition from other value-added consumers may occasionally limit the feedstock's supply (food and cosmetics sectors). The sustainable production of vegetable oils faces continuous problems due to its resource-intensive and land-intensive nature. Several businesses are running commercial oleochemical feedstock-to-biofuels facilities around the world, despite the likelihood that the "food vs. fuels" argument and related concerns will persist.

Biochemical or thermochemical processing can be used to transform biomass into liquid biofuels. Thermal degradation and chemical reformation are the processes used in thermochemical processing to transform biomass into a variety of products. The complete conversion of the biomass's organic components—while the biochemical process just concentrates on polysaccharides—is the primary benefit of the thermochemical process over the biochemical one (Gomez et al., 2008). The traditional method of "burning" biomass without oxygen to produce charcoal, a product with a higher calorific value, is where the several thermochemical

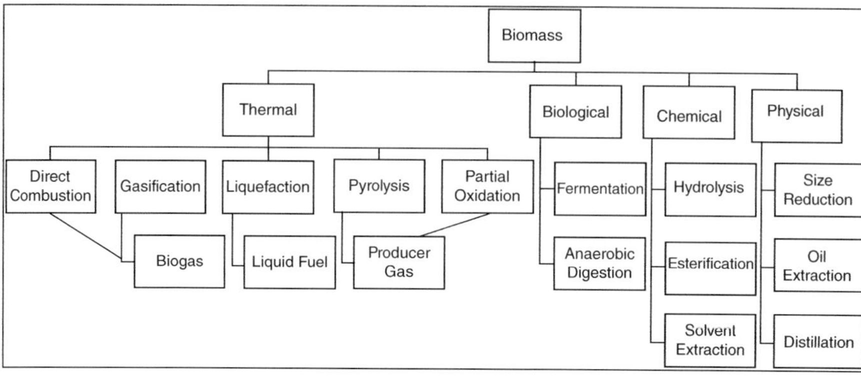

FIGURE 8.3 Thermochemical routes for biomass conversion to fuels.

processes for biofuel production currently under evaluation have their roots. The three primary products of bio-oil, synthesis gas, and char can have their ratios altered by optimizing the circumstances of thermochemical processing. The two primary methods for producing drop-in biofuels are gasification and pyrolysis. Since the early 1980s, fast pyrolysis—basically, heating biomass to 500°C for a brief period of time—has been thoroughly investigated. Using a variety of biomass feedstocks, bio-oil yields of up to 75 weight percent are generally possible (Karatzos et al., 2014; Watson et al., 2024).

Gasification is the other main thermochemical pathway to drop-in biofuels. Synthesis gas, or syngas, is produced by gasifying biomass or bio-oil and is mostly used to power stationary heat and power plants. Syngas is predominantly composed of H_2 and CO. Moreover, syngas can be catalytically condensed to drop-in liquid biofuels by the Fischer–Tropsch process (FT), which was developed in Germany in the 1920s when oil supply was scarce (Karatzos et al., 2014). The Fischer–Tropsch Synthetic Paraffinic Kerosene (FT-SPK) pathway involves a set of catalytic processes of syngas conversion via gasification of biomass feedstocks to a mixture of liquid hydrocarbons with a carbon range of C9–C15 over heterogeneous catalysts (Emmanouilidou et al., 2023). CO and H_2, the building elements of FT liquid hydrocarbons, make up the majority of syngas. 2009 saw the certification of FT synthesis in conjunction with gasification in accordance with ASTM D7566 (Afonso et al., 2023). Figure 8.3 illustrates various methods for turning biomass into energy (Naik et al., 2010).

8.4 THERMOCHEMICAL ROUTES FOR BIOMASS CONVERSION TO FUELS

In exploring thermochemical conversion techniques for biomass in biofuel synthesis, we delve into methods like pyrolysis, gasification, and chemical conversion with heated gases. The chosen synthesis technologies, pyrolysis and Hydroprocessed Esters and Fatty Acids (HEFA) each bring unique strengths to the table. Pyrolysis offers flexibility and energy efficiency in handling diverse biomass. HEFA relies on plant-derived oils, recognized for efficiency in biofuel synthesis, and its compatibility with aviation fuel standards. These processes, conducted at high temperatures, transform biomass into energy sources: gasification turns biomass into gas, pyrolysis decomposes biomass, and chemical conversion produces fuel. Understanding these approaches is crucial for grasping the diversity of strategies in biomass-to-biofuel conversion.

8.4.1 Pyrolysis

Gasification and pyrolysis are two critical thermal conversion technologies used in alternative aviation fuel production. These processes enable the efficient conversion of biomass feedstocks into valuable fuel components. Gasification is the controlled partial combustion of biomass in an environment with limited oxygen or air supply. Biomass is exposed to high temperatures in a gasifier, resulting in the production of syngas or synthesis gas. Pyrolysis is the thermochemical process that uses a partial oxidation process at high temperatures to convert carbonaceous material into gaseous products. The process of thermally breaking down biomass without the presence of air is known as pyrolysis, and it produces methane and bio-oil along with other byproducts (Probstein & Hicks, 2006). The pyrolysis process, also known as gasification, produces biofuels, chemicals, and charcoal while generating CHP with the aid of turbines, motors, and boilers (Fattah et al., 2014). This procedure typically involves three important steps: the removal of moisture, the breakdown of organic structures, and the gradual disintegration of remaining solids (Kumar et al., 2017). It has been demonstrated that the temperature, heating rate, and residence time during the pyrolysis of biomass have a significant impact on the yields and composition of the final products (Demirbas & Arin, 2002). The pyrolysis process can be divided into three subclasses based on operation conditions: slow, fast, and flash pyrolysis. The lignocellulosic cell structure of biomass is thermally broken during biomass gasification or pyrolysis to produce carbon monoxide (CO), hydrogen (H_2), and CO_2 as the primary components of syngas, along with trace amounts of methane (CH_4) and other gases. At 470–530 K, hemicellulose begins to decompose, followed by cellulose at 510–620 K, and lignin at 550–770 K, which is the final component to undergo pyrolysis. The general equation for the process of pyrolysis is as follows:

$$\text{Biomass} \longrightarrow \text{Char} + \text{Tar} + NH_3 + H_2S + CO + CO_2 \quad (8.1)$$

8.4.1.1 Slow Pyrolysis

The well-known process of slow pyrolysis often takes place in conventional charcoal kilns. High charcoal content is linked to slow biomass pyrolysis (Jahirul et al., 2012). The range of 300–700°C is the operating temperature for the slow pyrolysis process (Demirbaş, 2000). The use of catalysts in the presence of Na_2CO_3 may affect how proteins and carbohydrates break down and make pyrolysis easier at low temperatures. Their higher heating values, or HHVs, differ (Jahirul et al., 2012). Consequently, biochar, CH_4, and CO_2 gaseous products are produced when a microalgal stream is slowly pyrolyzed. In 2013, K. Chaiwong conducted research on the slow pyrolysis of

Spirulina sp., resulting in maximum biochar and bio-oil yields of less than 35 and 30 weight percent, respectively, at an optimum temperature of approximately 773–823 K.

8.4.1.2 Fast Pyrolysis

Biomass is thermolyzed at temperatures between 577°C and 977°C in an inert environment during the rapid pyrolysis process. According to Panwar et al. (2012), rapid pyrolysis yields 60–75 weight percent liquid bio-oil, 15–25 weight percent solid char, and 10–20 weight percent non-condensable gases. Rapid pyrolysis is advised since traditional pyrolysis has drawbacks with secondary cracking, condensation, and product polymerization that lower the high heat value and energy needed. A fast-pyrolytic analysis of *Scenedesmus* sp., *Chlorella protothecoides*, and *M. aeruginosa* revealed 18–24 weight percent of bio-oil production with HHV of 29 MJ/kg. Furthermore, recent studies demonstrated that heterotrophic microalgae may be more advantageous for rapid pyrolysis than autotrophic ones (Demirbaş, 2000).

8.4.1.3 Flash Pyrolysis

Different flash pyrolysis techniques have been developed recently to optimize the generation of liquid products for use as chemical feedstocks or fuels. According to Chaiwong et al. (2013), the flash pyrolysis operating temperature ranges from 777°C to 1027°C. With a 70% output efficiency, this technique yields biomass crude oil that is equal to petroleum. A typical bio-oil yield of 75% and gas and char yields of 12–13% are obtained using flash pyrolysis, which necessitates feed particle sizes of no more than 200 mm and higher temperatures of about 800°C–1000°C (Aboelela et al., 2023; Mohan et al., 2006). A summary of the final product yields is shown in Figure 8.5.

Pyrolysis or gasification involves the treatment of pyrolysis products with air or steam to produce syngas, which is a mixture of hydrogen and carbon monoxide. The gasification process is the conversion of solid fuel into gaseous fuel at the temperature range of 800–1300°C (Demirbaş, 2000). The biomass gasification process is used for production of biofuels such as “green” gasoline and electricity. The design of the gasifier depends upon the type of fuel used, air introduction in the fuel column, and the type of combustion bed (Figure 8.6). Gasifier combustion, also known as partial combustion of solid fuel, occurs in a restricted amount of oxygen or air. Producer gas and other impurities including carbon dioxide, nitrogen, sulfur, alkali

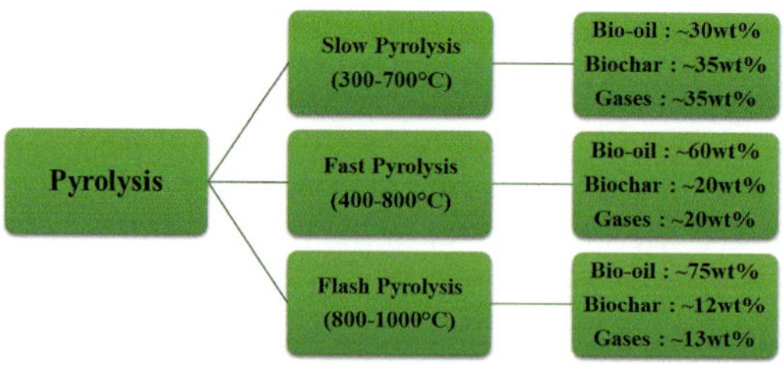

FIGURE 8.4 Yield percentage of various pyrolysis processes.

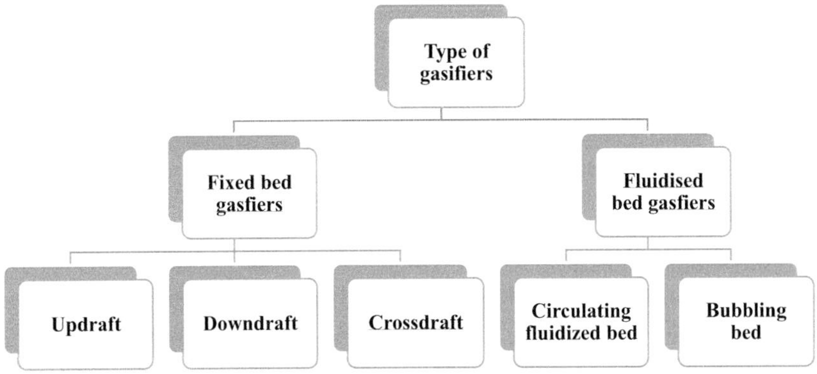

FIGURE 8.5 Classification of gasifiers.

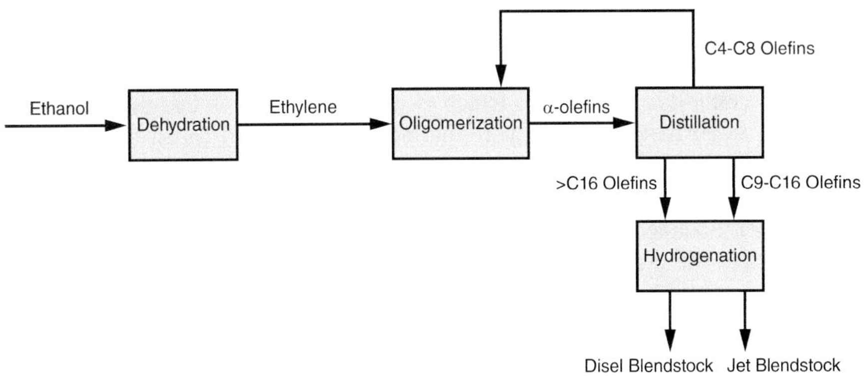

FIGURE 8.6 Conversion pathway of ethanol.

compounds, and tars are the principal results of the gasification process (Undavalli et al., 2023). Biomass pyrolysis stands out as a promising method for bio-oil production, but the resulting bio-oil often exhibits challenges such as low stability, elevated water content, and increased viscosity. To overcome these inherent limitations and enhance the overall efficiency of the bio-oil, catalysts play a pivotal role by effectively removing oxygen and functional groups during the upgrading process.

8.5 CONVERSION PATHWAYS TO BIO-JET FUEL

8.5.1 Alcohol-to-Jet Fuel

The alcohol-to-jet (ATJ) fuel process has high expectations due to its performance and for revalorizing alcohols from fermentation. The conversion of ethanol to long-chain alkanes requires a multi-step carbon-chain growth reaction—dehydration, oligomerization, hydrogenation, and separation (Su & Lin, 2023). In general, ethanol is first converted to medium-chain length aldehydes or alkenes as intermediates, followed by further aldol condensation of aldehydes or polymerization of medium-chain length alkenes to long-chain molecules. For the majority of gasoline-powered cars now on the road, ethanol use is limited to 10–15%, which results in a blend wall that makes it challenging to obtain greater market penetration of ethanol as a blend stock for gasoline. Consequently, converting ethanol to jet fuel blend stock offers a possible route for creating fungible or drop-in fuels for the aviation fuel industry (Xie et al., 2024).

The conversion of bioethanol (Figure 8.6) into liquid fuel for transportation, notably gasoline-range, jet-range, and diesel-range hydrocarbons, has received increasing interest with time(Kim et al., 2018). Oligomerization has been shown to be an efficient method for converting bioethanol molecules into higher-molecular-mass oxygen-containing fuel precursors. The gasoline to diesel range alkanes can be produced following hydrogenation by feeding hydrogen over a 5% Pd/C or Pt/C catalyst (Mohamed et al., 2022).

8.5.2 Oil-to-Jet (OTJ) Fuel

The OTJ conversion pathway comprises three distinct processes: hydrotreated depolymerized cellulosic jet (HDCJ), which is also referred to as fast pyrolysis with upgrading to jet fuel; catalytic hydro-thermolysis (CH), also called hydrothermal liquefaction; and hydro-processed renewable jet (HRJ), also called HEFA. As of right now, only items with a clear ASTM specification and approval for mixing come via the HRJ pipeline. Triglyceride-based feedstocks are used in both HRJ and CH processes, but the way the FFAs are generated is different. While FFAs in the CH method are created by heat hydrolysis, FFAs in the HRJ process are created by propane cleaving of glycerides. The biomass feedstock is pyrolyzed in the HDCJ process to produce the bio-oil (Mohamed et al., 2022; Xie et al., 2024).

Hydro-processed Esters and Fatty Acids-Synthetic Paraffinic Kerosene ("HEFA-SPK") which involves hydro-processing of various oleochemical feedstocks like vegetable oils, animal fats, waste grease, and algal oil is considered a fully developed and commercially available technology for producing bio-jet fuel (Pires et al., 2018; Wang & Tao, 2016). In 2011, it received ASTM certification for blending bio-jet fuel with conventional jet fuel up to a maximum of 50 vol%. High cetane and thermal stability HRJ fuels have lately been employed to make jet fuel for military planes. Lipid feedstocks are hydrotreated, hydrocracked, and hydrogenated in the presence of Pd, Pt, Pt–Re, NiMo, CoMo over γ-Al_2O_3, or activated carbon catalyst during the hydro-processing process.

Hydro-thermolysis catalytic (CH) triglycerides are converted into a mixture of straight chain, branched, and cyclic hydrocarbons through a series of reactions known as cracking, hydrolysis, decarboxylation, isomerization, and cyclization in the novel process known as catalytic hydro-thermolysis (CH), also known as hydrothermal liquefaction. The CH process is carried out with water and a catalyst (or not) at temperatures between 600 and 800 K and 475°C and pressures between 10 and 20 bar. According to reports, a range of triglyceride-based feedstocks, including soybean oil, jatropha oil, camelina oil, and tung oil, can be converted into bio-jet fuels using the CH process (Diederichs et al., 2016; Xie et al., 2024).

TABLE 8.4
CH Process of Bio-jet Fuels

	JP-8	Camelina	Carinata	Soybean	Jatropha	Tung	MIL-DTL-83133H Standard
Aromatics	18.8 vol%	24.2 wt%	16.8 wt%	2.6 wt%	10.8 wt%	61.7 wt%	≤25.0
Paraffins	–	–	–	40.0 vol%	32.8 wt%	16.2 wt%	–
Olefins	0.8	1.3	1.8	–	–	–	≤5.0
Cycloparaffins, vol%	–	–	–	52.0 vol%	39.2 wt%	16.7 wt%	–
Dicycloparaffins, vol%	–	–	–	5.9 vol%	–	–	–
Heat of combustion, MJ/kg	43.3	42.9	43.2	43.4	43.4	42.3	≥42.8
Smoke point	22	22	26	>30	28	–	≥19
Freeze point, °C	–51	–54	–57	<–47	–39	<–66	≤–47
Flash point, °C	51	48	46	>38	45	39	≥38
Distillation	–	–	–	–	195–229	187–252	–
Density, kg/L	0.804	0.818	0.802	>0.775	0.804	0.839	0.775–0.840

8.6 DROP-IN BIOFUELS: THE POTENTIAL AND CHALLENGES

To upgrade crude oil feedstocks of diminishing quality (becoming heavier), the petroleum industry worldwide is predicted to need more hydrogen in the upcoming years. This is especially true in regions like Alberta and Venezuela where heavy oils are sourced. A significant portion of this hydrogen will probably come from natural gas in the near future. In order to deoxygenate biomass, which includes lignin and carbs, and create drop-in hydrocarbon biofuels, there will also be a simultaneous increase in the demand for hydrogen (Karatzos et al., 2014).

Oil refineries employ hydrogen to enhance the quality of low-grade crude oil by eliminating sulfur and other impurities (hydrotreating) and breaking down longer carbon chains into shorter ones, while simultaneously enriching them with hydrogen. This process results in increasing the hydrogen to carbon ratio of low-grade crude oils, which is indicative of their suitability for fuel production. Conversely, biomass-derived syngas, while less energy-dense than natural gas, contains more impurities and has a lower H/C ratio. Therefore, biomass syngas requires hydrogen enrichment and purification to remove impurities like tar, nitrogen, and other heteroatoms that can hinder synthesis catalysts. Hydrogen is often generated from the syngas itself through the "water-gas shift" reaction (Bauen et al., 2020; Ben et al., 2013). However, this reaction consumes carbon from the feedstock, diminishing overall biomass-to-fuel yields. This process is inefficient when contrasted against natural gas hydrogen sourcing. In conventional gasoline production, minimizing feedstock consumption is preferred, especially during periods of high crude oil prices. To produce higher-grade liquid transportation fuels like gasoline, diesel, and jet fuel, the H/C ratio must be raised by additional hydrogen inputs and processing stages. By eliminating carbon in the form of tar and char, non-hydrogen-consuming procedures like thermal or catalytic cracking can also raise the H/C ratio of petroleum feedstocks. Platforms for pyrolysis have a great deal of potential to be integrated with oil refineries, which would lower the startup and ongoing expenses of producing biofuels (Abdellatief et al., 2024; Ben et al., 2013; Bohre et al., 2015). One significant way to cut costs is to purchase hydrogen straight from the oil refinery. However, a major increase in United States refinery hydrogen capacity of 2.9 billion cubic feet per day will be necessary to achieve the 2025 United States Renewable Fuel Standard (RFS) cellulosic advanced biofuel mandate of 3.35 billion gallons using diesel/gasoline blend stock derived from the pyrolysis platform (Melvin, 2023; Statista, 2024). Although existing hydrocracking units within refineries can theoretically co-process petroleum and hydrotreated pyrolysis oils, this practice is not yet commercialized. Challenges include adapting catalyst designs to accommodate two dissimilar feedstocks. Two major hurdles hindering the development of pyrolysis-derived drop-in biofuels are the availability of low-cost sustainable hydrogen and the technological advancements required to adapt hydrotreating catalysts to bio-oil feedstocks (Geleynse et al., 2018; Karatzos et al., 2014).

While drop-in gasoline presents itself as a more environmentally friendly alternative to traditional fossil fuels, it still carries environmental consequences. Its production from biomass, municipal waste, and carbon capture technology can lead to issues such as deforestation, habitat destruction, increased water use and pollution, and competition with food crops. Moreover, the production process from municipal waste may result in the release of harmful pollutants and unwanted byproducts (Lee et al., 2021).

The main advantage of sustainable aviation fuel (SAF) is the significant reduction in carbon emissions compared to conventional fuels. Studies indicate that despite their environmental potential, renewable fuels face technological and economic challenges. Large-scale production is still a barrier, as manufacturing processes are often complex and expensive. In addition, it is imperative to develop robust infrastructure for the distribution and storage of these new fuels, which requires significant investments (Teoh et al., 2022). Another challenge faced by biofuel is the availability of feedstock since feedstocks currently accessible to the market are predominantly farmed materials. Due to the lack of supply of feedstock, feedstock prices are high, and it has an adverse impact on the economy of feedstock conversion process of biofuel. There is a certain requirement to support the building of these feedstock markets by lowering the cost of cellulosic sugars (Scheelhaase et al., 2019) by scaling up the production and investment to increase the availability of feedstock.

The creation of favorable laws and rules is essential to advancing the marketing of environmentally friendly aviation fuel. To encourage investment and hasten the shift to sustainable aviation, tax breaks, subsidies, and emission reduction goals can be effective strategies. Airlines must use a specific amount of SAF by 2030 according to regulations in several European nations, and the U.S. Inflation Reduction Act (IRA) has provided them with an additional tax credit that is valid for the first two years of the program (PACE, 2023). In order to invest in biofuel production, carbon capture and sequestration, and the acquisition of SAF, a number of private sectors, international aviation organizations, and multinational enterprises have formed partnerships. One of such alliances is Eco-Skies Alliance which initiates to bring leading global corporations together with United Airlines to support sustainable flights. To implement the greener fuel, collaboration between governments, industry, and society is essential to overcome technological, economic, and social obstacles. To establish a more sustainable future for the aviation industry and ensure a maximum decrease in GHG emissions, airlines need a centralized, cloud-based data management system to monitor and analyze carbon emissions, establish sustainable fuel supply chains, and optimize fuel blend balance within Energy Efficiency and Renewable Energy (EERE) system.

8.7 CONCLUSION

Renewable gasoline, also known as green or drop-in gasoline, is a fuel synthesized from biomass sources using a range of biological, thermal, and chemical methods. This eco-friendly fuel mirrors petroleum gasoline chemically and can seamlessly operate within existing engines and infrastructure. Researchers are investigating diverse techniques for producing renewable gasoline, including traditional hydrotreating, which involves the reaction of lipids with hydrogen under specific conditions with a catalyst, a method currently employed in commercial plants. Additionally, biological sugar upgrading, catalytic conversion of sugars, gasification, pyrolysis, and hydrothermal processing are being explored. Each process presents unique pathways to convert biomass into hydrocarbon fuels. Renewable gasoline offers numerous advantages, such as compatibility with engines and infrastructure, bolstering energy security through domestic production, reduced emissions by capturing carbon dioxide during feedstock growth, and enhanced flexibility in product variety from various feedstocks and production methods.

Sustainable aviation fuels represent a crucial solution for the aviation sector to address climate change and meet increasing fuel demands. While alternative jet fuels show promise, global initiatives are underway to explore diverse fuel sources and technologies for widespread aviation use. However, ensuring the sustainability of alternative jet fuels requires the development of comprehensive sustainability metrics considering renewability, economic viability, environmental impact, energy security, and life cycle GHG balance. Integrating refinery operations can enhance the economics of drop-in biofuel production, with biocrudes potentially serving as intermediate commodities for further processing in refineries. While both conventional and advanced routes to drop-in biofuel production have been demonstrated, ongoing research is essential to understand the behavior of different biobased feedstocks in various reactors and their impact on product characteristics. Furthermore, elucidating the fate of renewable carbon distributed across different product fractions during refinery operations is crucial. Techno-economic assessments of different feedstock and reactor co-processing combinations are necessary to determine the economic feasibility of refinery integration.

REFERENCES

Abdellatief, T. M. M., Ershov, M. A., Makhmudova, A. E., Kapustin, V. M., Makhova, U. A., Klimov, N. A., Chernysheva, E. A., Ali Abdelkareem, M., Mustafa, A., & Olabi, A. G. (2024). Novel variants conceptional technology to produce eco-friendly sustainable high octane-gasoline biofuel based on renewable gasoline component. *Fuel*, *366*, 131400. https://doi.org/10.1016/j.fuel.2024.131400

Aboelela, D., Saleh, H., Attia, A. M., Elhenawy, Y., Majozi, T., & Bassyouni, M. (2023). Recent Advances in Biomass Pyrolysis Processes for Bioenergy Production: Optimization of Operating Conditions. *Sustainability*, *15*(14), Article 14. https://doi.org/10.3390/su151411238

Afonso, F., Sohst, M., Diogo, C. M. A., Rodrigues, S. S., Ferreira, A., Ribeiro, I., Marques, R., Rego, F. F. C., Sohouli, A., Portugal-Pereira, J., Policarpo, H., Soares, B., Ferreira, B., Fernandes, E. C., Lau, F., & Suleman, A. (2023). Strategies towards a more sustainable aviation: A systematic review. *Progress in Aerospace Sciences*, *137*, 100878. https://doi.org/10.1016/j.paerosci.2022.100878

Agarwal, P., Soni, R., Kaur, P., Madan, A., Mishra, R., Pandey, J., Singh, S., & Singh, G. (2022). Cyanobacteria as a Promising Alternative for Sustainable Environment: Synthesis of Biofuel and Biodegradable Plastics. *Frontiers in Microbiology*, *13*, 939347. https://doi.org/10.3389/fmicb.2022.939347

Bauen, A., Bitossi, N., German, L., Harris, A., & Leow, K. (2020). *Sustainable Aviation Fuels: Status, challenges and prospects of drop-in liquid fuels, hydrogen and electrification in aviation*. *64*(3), 263–278. https://doi.org/10.1595/205651320X15816756012040

Ben, H., Mu, W., Deng, Y., & Ragauskas, A. J. (2013). Production of renewable gasoline from aqueous phase hydrogenation of lignin pyrolysis oil. *Fuel*, *103*, 1148–1153. https://doi.org/10.1016/j.fuel.2012.08.039

Benavides, A., Benjumea, P., Cortés, F. B., & Ruiz, M. A. (2021). Chemical composition and low-temperature fluidity properties of jet fuels. *Processes*, *9*(7), Article 7. https://doi.org/10.3390/pr9071184

Bernabei, M., Reda, R., Galiero, R., & Bocchinfuso, G. (2003). Determination of total and polycyclic aromatic hydrocarbons in aviation jet fuel. *Journal of Chromatography A*, *985*(1), 197–203. https://doi.org/10.1016/S0021-9673(02)01826-5

Bhadha, J. H., Lang, T. A., Gomez, S. M., Daroub, S. H., & Giurcanu, M. C. (2015). Effect of water lettuce and filamentous algae on phosphorus loads in farm canals in the Everglades Agricultural Area. *J. Aquat. Plant Manage.*, 10.

Bohre, A., Dutta, S., Saha, B., & Abu-Omar, M. M. (2015). Upgrading Furfurals to Drop-in Biofuels: An Overview. *ACS Sustainable Chemistry & Engineering*, *3*(7), 1263–1277. https://doi.org/10.1021/acssuschemeng.5b00271

Bohutskyi, P., & Bouwer, E. (2013). *Biogas Production from Algae and Cyanobacteria Through Anaerobic Digestion: A Review, Analysis, and Research Needs* (J. W. Lee, Ed.; pp. 873–975). New York: Springer. https://doi.org/10.1007/978-1-4614-3348-4_36

Boldt, K. (1977). *Significance of Tests for Petroleum Products*. ASTM International.

Bowles, D. J. (2007). *Micro- and Macro-algae: Utility for Industrial Applications: Outputs from the EPOBIO Project, September 2007*. CPL Press.

Cabrera-Jiménez, R., Mateo-Sanz, J. M., Gavaldà, J., Jiménez, L., & Pozo, C. (2022). Comparing biofuels through the lens of sustainability: A data envelopment analysis approach. *Applied Energy*, *307*, 118201. https://doi.org/10.1016/j.apenergy.2021.118201

Chaiwong, K., Kiatsiriroat, T., Vorayos, N., & Thararax, C. (2013). Study of bio-oil and bio-char production from algae by slow pyrolysis. *Biomass and Bioenergy*, *56*, 600–606. https://doi.org/10.1016/j.biombioe.2013.05.035

Chang, S. H. (2018). Bio-oil derived from palm empty fruit bunches: Fast pyrolysis, liquefaction and future prospects. *Biomass and Bioenergy*, *119*, 263–276. https://doi.org/10.1016/j.biombioe.2018.09.033

Chavez-Rodriguez, M. F., & Nebra, S. A. (2010). Assessing GHG emissions, ecological footprint, and water linkage for different fuels. *Environmental Science & Technology*, *44*(24), 9252–9257. https://doi.org/10.1021/es101187h

Chisti, Y. (2016). Large-Scale Production of Algal Biomass: Raceway Ponds. In F. Bux & Y. Chisti (Eds.), *Algae Biotechnology: Products and Processes* (pp. 21–40). Springer International Publishing. https://doi.org/10.1007/978-3-319-12334-9_2

Cookson, D. J., Lloyd, C. P., & Smith, B. E. (1987). Investigation of the chemical basis of kerosene (jet fuel) specification properties. *Energy & Fuels*, *1*(5), 438–447. https://doi.org/10.1021/ef00005a011

Demirbaş, A. (2000). Mechanisms of liquefaction and pyrolysis reactions of biomass. *Energy Conversion and Management*, *41*(6), 633–646. https://doi.org/10.1016/S0196-8904(99)00130-2

Demirbas, A., & Arin, G. (2002). An overview of biomass pyrolysis. *Energy Sources*, *24*(5), 471–482. https://doi.org/10.1080/00908310252889979

Demirbas, A., & Demirbas, M. F. (2012). *Algae Energy*. London: Springer. https://link.springer.com/book/10.1007/978-1-84996-050-2

Diederichs, G. W., Ali Mandegari, M., Farzad, S., & Görgens, J. F. (2016). Techno-economic comparison of biojet fuel production from lignocellulose, vegetable oil and sugar cane juice. *Bioresource Technology*, *216*, 331–339. https://doi.org/10.1016/j.biortech.2016.05.090

Dutta, K., Daverey, A., & Lin, J.-G. (2014). Evolution retrospective for alternative fuels: First to fourth generation. *Renewable Energy*, *69*, 114–122. https://doi.org/10.1016/j.renene.2014.02.044

EA IEA. (2011). *Technology Roadmap—Biofuels for Transport – Analysis*. https://www.iea.org/reports/technology-roadmap-biofuels-for-transport

Eco-Skies Alliance | United Airlines. (n.d.). Fly - United Airlines. Retrieved March 29, 2024, from https://www.united.com/en/us/fly/company/responsibility/eco-skies-alliance.html

EERE. (n.d.). *Sustainable Aviation Fuel Grand Challenge*. Energy.Gov. Retrieved March 29, 2024, from https://www.energy.gov/eere/bioenergy/sustainable-aviation-fuel-grand-challenge

Emmanouilidou, E., Mitkidou, S., Agapiou, A., & Kokkinos, N. C. (2023). Solid waste biomass as a potential feedstock for producing sustainable aviation fuel: A systematic review. *Renewable Energy*, *206*(C), 897–907. https://ideas.repec.org//a/eee/renene/v206y2023icp897-907.html

Fairley, P. (2011). Introduction: Next generation biofuels. *Nature*, *474*(7352), S2–S5. https://doi.org/10.1038/474S02a

Fatih Demirbas, M. (2009). Biorefineries for biofuel upgrading: A critical review. *Applied Energy*, *86*, S151–S161. https://doi.org/10.1016/j.apenergy.2009.04.043

Fattah, I. M. R., Kalam, M. A., Masjuki, H. H., & Wakil, M. A. (2014). Biodiesel production, characterization, engine performance, and emission characteristics of Malaysian Alexandrian laurel oil. *RSC Advances*, *4*(34), 17787–17796. https://doi.org/10.1039/C3RA47954D

Gaughan, R., Seto, S., & Wilson, G. (2019). *Chapter 8 | Aviation Fuels*. https://doi.org/10.1520/MNL3720170010

Geleynse, S., Brandt, K., Garcia-Perez, M., Wolcott, M., & Zhang, X. (2018). The alcohol-to-jet conversion pathway for drop-in biofuels: Techno-economic evaluation. *ChemSusChem*, *11*(21), 3728–3741. https://doi.org/10.1002/cssc.201801690

Gomez, L. D., Steele-King, C. G., & McQueen-Mason, S. J. (2008). Sustainable liquid biofuels from biomass: The writing's on the walls. *New Phytologist*, *178*(3), 473–485. https://doi.org/10.1111/j.1469-8137.2008.02422.x

Jahirul, M. I., Rasul, M. G., Chowdhury, A. A., & Ashwath, N. (2012). Biofuels production through biomass pyrolysis—A technological review. *Energies*, *5*(12), Article 12. https://doi.org/10.3390/en5124952

Johnson, C., Moriarty, K., Alleman, T., & Santini, D. (2021). *History of Ethanol Fuel Adoption in the United States: Policy, Economics, and Logistics* (NREL/TP-5400-76260). National Renewable Energy Lab. (NREL), Golden, CO (United States). https://doi.org/10.2172/1832224

Jokiniemi, T., & Ahokas, J. (2013). A review of production and use of first generation biodiesel in agriculture. *Agronomy Research*, *11*(1), 239–248.

Kallio, P., Pásztor, A., Akhtar, M. K., & Jones, P. R. (2014). Renewable jet fuel. *Current Opinion in Biotechnology*, *26*, 50–55. https://doi.org/10.1016/j.copbio.2013.09.006

Karatzos, S., McMillan, J. D., & Saddler, J. N. (2014). *The potential and challenges of drop-in biofuels: A report by IEA Bioenergy Task 39* [Monograph]. https://library.wur.nl/WebQuery/titel/2098061

Karatzos, S., van Dyk, J. S., McMillan, J. D., & Saddler, J. (2017). Drop-in biofuel production via conventional (lipid/fatty acid) and advanced (biomass) routes. Part I. *Biofuels, Bioproducts and Biorefining*, *11*(2), 344–362. https://doi.org/10.1002/bbb.1746

Kim, H., Kim, D., Park, Y.-K., & Jeon, J.-K. (2018). Synthesis of jet fuel through the oligomerization of butenes on zeolite catalysts. *Research on Chemical Intermediates*, *44*(6), 3823–3833. https://doi.org/10.1007/s11164-018-3385-1

Kim, H., Lee, J., Kim, Y., Ha, J.-M., Park, Y.-K., Vlachos, D. G., Suh, Y.-W., & Jae, J. (2024). Continuous flow upgrading of lignin pyrolysis oils to drop-in bio-hydrocarbon fuels over noble metal catalysts. *Chemical Engineering Journal*, *481*, 148328. https://doi.org/10.1016/j.cej.2023.148328

Kumar, G., Shobana, S., Chen, W.-H., Bach, Q.-V., Kim, S.-H., Atabani, A. E., & Chang, J.-S. (2017). A review of thermochemical conversion of microalgal biomass for biofuels: Chemistry and processes. *Green Chemistry*, *19*(1), 44–67. https://doi.org/10.1039/C6GC01937D

Kumar, H. D., Yadava, P. K., & Gaur, J. P. (1981). Electrical flocculation of the unicellular green alga *Chlorella vulgaris* Beijerinck. *Aquatic Botany*, *11*, 187–195. https://doi.org/10.1016/0304-3770(81)90059-0

Kumar, N., Sonthalia, A., Pali, H. S., & Sidharth. (2020). Next-Generation Biofuels—Opportunities and Challenges. In A. K. Gupta, A. De, S. K. Aggarwal, A. Kushari, & A. Runchal (Eds.), *Innovations in Sustainable Energy and Cleaner Environment* (pp. 171–191). Springer. https://doi.org/10.1007/978-981-13-9012-8_8

Langholtz, M. (2024). *U.S. Department of Energy. 2024. 2023 Billion-Ton Report: An Assessment of U.S. Renewable Carbon Resources*. Oak Ridge National Laboratory. https://www.osti.gov/servlets/purl/2316165/

Lardon, L., Hélias, A., Sialve, B., Steyer, J.-P., & Bernard, O. (2009). Life-Cycle Assessment of Biodiesel Production from Microalgae. *Environmental Science & Technology*, *43*(17), 6475–6481. https://doi.org/10.1021/es900705j

Lee, D., Nam, H., Wang, S., Kim, H., Kim, J. H., Won, Y., Hwang, B. W., Kim, Y. D., Nam, H., Lee, K.-H., & Ryu, H.-J. (2021). Characteristics of fractionated drop-in liquid fuel of plastic wastes from a commercial pyrolysis plant. *Waste Management*, *126*, 411–422. https://doi.org/10.1016/j.wasman.2021.03.020

Lee, R. A., & Lavoie, J.-M. (2013). From first- to third-generation biofuels: Challenges of producing a commodity from a biomass of increasing complexity. *Animal Frontiers*, *3*(2), 6–11. https://doi.org/10.2527/af.2013-0010

Lü, J., Sheahan, C., & Fu, P. (2011). Metabolic engineering of algae for fourth generation biofuels production. *Energy & Environmental Science*, *4*(7), 2451–2466. https://doi.org/10.1039/C0EE00593B

Mabee, W. E., Gregg, D. J., Arato, C., Berlin, A., Bura, R., Gilkes, N., Mirochnik, O., Pan, X., Pye, E. K., & Saddler, J. N. (2006). Updates on Softwood-to-Ethanol Process Development. In J. D. McMillan, W. S. Adney, J. R. Mielenz, & K. T. Klasson (Eds.), *Twenty-Seventh Symposium on Biotechnology for Fuels and Chemicals* (pp. 55–70). Humana Press. https://doi.org/10.1007/978-1-59745-268-7_5

Malode, S. J., Prabhu, K. K., Mascarenhas, R. J., Shetti, N. P., & Aminabhavi, T. M. (2021). Recent advances and viability in biofuel production. *Energy Conversion and Management: X*, *10*, 100070. https://doi.org/10.1016/j.ecmx.2020.100070

Manikandan, S., Subbaiya, R., Biruntha, M., Krishnan, R. Y., Muthusamy, G., & Karmegam, N. (2022). Recent development patterns, utilization and prospective of biofuel production: Emerging nanotechnological intervention for environmental sustainability – A review. *Fuel*, *314*, 122757. https://doi.org/10.1016/j.fuel.2021.122757

Mascal, M., & Dutta, S. (2020). Synthesis of highly-branched alkanes for renewable gasoline. *Fuel Processing Technology*, *197*, 106192. https://doi.org/10.1016/j.fuproc.2019.106192

Melvin, J. (2023, June 21). *US EPA 'lowers ambitions' to chagrin of biofuel producers in final RFS rule for 2023-2025*. https://www.spglobal.com/commodityinsights/en/market-insights/latest-news/oil/062123-us-epa-finalizes-biofuel-blending-mandates-for-next-three-years

Mohamed, H. O., Abed, O., Zambrano, N., Castaño, P., & Hita, I. (2022). A zeolite-based cascade system to produce jet fuel from ethylene oligomerization. *Industrial & Engineering Chemistry Research*, *61*(43), 15880–15892. https://doi.org/10.1021/acs.iecr.2c02303

Mohan, D., Pittman, C. U. Jr., & Steele, P. H. (2006). Pyrolysis of wood/biomass for bio-oil: A critical review. *Energy & Fuels*, *20*(3), 848–889. https://doi.org/10.1021/ef0502397

Naik, S. N., Goud, V. V., Rout, P. K., & Dalai, A. K. (2010). Production of first and second generation biofuels: A comprehensive review. *Renewable and Sustainable Energy Reviews*, *14*(2), 578–597. https://doi.org/10.1016/j.rser.2009.10.003

National Research Council, D. on E. (2011). *Renewable Fuel Standard: Potential Economic and Environmental Effects of U.S. Biofuel Policy*. The National Academies Press. https://doi.org/10.17226/13105

Nigam, P. S., & Singh, A. (2011). Production of liquid biofuels from renewable resources. *Progress in Energy and Combustion Science*, *37*(1), 52–68. https://doi.org/10.1016/j.pecs.2010.01.003

PACE. (2023). *A Guide to Inflation Reduction Act in the Aviation Sector*. https://www.pace-esg.com/academy/a-guide-to-inflation-reduction-act-usa-for-the-aviation-sector/

Panwar, N. L., Kothari, R., & Tyagi, V. V. (2012). Thermo chemical conversion of biomass – Eco friendly energy routes. *Renewable and Sustainable Energy Reviews*, *16*(4), 1801–1816. https://doi.org/10.1016/j.rser.2012.01.024

Parmar, A., Singh, N. K., Pandey, A., Gnansounou, E., & Madamwar, D. (2011). Cyanobacteria and microalgae: A positive prospect for biofuels. *Bioresource Technology*, *102*(22), 10163–10172. https://doi.org/10.1016/j.biortech.2011.08.030

Paul, S., Bera, S., Dasgupta, R., Mondal, S., & Roy, S. (2021). Review on the recent structural advances in open and closed systems for carbon capture through algae. *Energy Nexus*, *4*, 100032. https://doi.org/10.1016/j.nexus.2021.100032

Pires, A. P. P., Han, Y., Kramlich, J., & Garcia-Pérez, M. (2018). *Chemical Composition and Fuel Properties of Alternative Jet Fuels. BioRes_13_2_2632*. https://rosap.ntl.bts.gov/view/dot/50415

Probstein, R. F., & Hicks, R. E. (2006). *Synthetic Fuels*. Courier Corporation.

Quintana, N., Van der Kooy, F., Van de Rhee, M. D., Voshol, G. P., & Verpoorte, R. (2011). Renewable energy from Cyanobacteria: Energy production optimization by metabolic pathway engineering. *Applied Microbiology and Biotechnology*, *91*(3), 471–490. https://doi.org/10.1007/s00253-011-3394-0

Ritchie, H., & Roser, M. (2024). Climate change and flying: What share of global CO2 emissions come from aviation? *Our World in Data*. https://ourworldindata.org/co2-emissions-from-aviation

Robles-Medina, A., González-Moreno, P. A., Esteban-Cerdán, L., & Molina-Grima, E. (2009). Biocatalysis: Towards ever greener biodiesel production. *Biotechnology Advances*, *27*(4), 398–408. https://doi.org/10.1016/j.biotechadv.2008.10.008

Sarisky-Reed, V. (2022). *Ethanol vs. Petroleum-Based Fuel Carbon Emissions*. Energy.Gov. https://www.energy.gov/eere/bioenergy/articles/ethanol-vs-petroleum-based-fuel-carbon-emissions

Schäppi, R., Rutz, D., Dähler, F., Muroyama, A., Haueter, P., Lilliestam, J., Patt, A., Furler, P., & Steinfeld, A. (2022). Drop-in fuels from sunlight and air. *Nature*, *601*(7891), 63–68. https://doi.org/10.1038/s41586-021-04174-y

Scheelhaase, J., Maertens, S., & Grimme, W. (2019). Synthetic fuels in aviation – Current barriers and potential political measures. *Transportation Research Procedia*, *43*, 21–30. https://doi.org/10.1016/j.trpro.2019.12.015

Sekomo, C. B., Rousseau, D. P. L., Saleh, S. A., & Lens, P. N. L. (2012). Heavy metal removal in duckweed and algae ponds as a polishing step for textile wastewater treatment. *Ecological Engineering*, *44*, 102–110. https://doi.org/10.1016/j.ecoleng.2012.03.003

Sembiring, K. C., Aunillah, A., Minami, E., & Saka, S. (2018). Renewable gasoline production from oleic acid by oxidative cleavage followed by decarboxylation. *Renewable Energy*, *122*, 602–607. https://doi.org/10.1016/j.renene.2018.01.107

Shell Global. (n.d.). *Civil Aviation Fuel | Jet Fuel Specifications | Shell Global*. Retrieved March 29, 2024, from https://www.shell.com/business-customers/aviation/aviation-fuel/civil-jet-fuel-grades.html

Shin, Y. S., Choi, H. I., Choi, J. W., Lee, J. S., Sung, Y. J., & Sim, S. J. (2018). Multilateral approach on enhancing economic viability of lipid production from microalgae: A review. *Bioresource Technology*, *258*, 335–344. https://doi.org/10.1016/j.biortech.2018.03.002

Sims, R. E. H., Mabee, W., Saddler, J. N., & Taylor, M. (2010). An overview of second generation biofuel technologies. *Bioresource Technology*, *101*(6), 1570–1580. https://doi.org/10.1016/j.biortech.2009.11.046

Sims, R., Taylor, M., & Mabee, W. (2008). *From 1st- to 2nd-Generation Biofuel Technologies – Analysis.* https://www.iea.org/reports/from-1st-to-2nd-generation-biofuel-technologies

Statista. (2024, February 2). *U.S. hydrogen production capacity in refineries*. Statista. https://www.statista.com/statistics/1447071/hydrogen-production-capacity-in-refineries-united-states/

Su, H., & Lin, J. (2023). Biosynthesis pathways of expanding carbon chains for producing advanced biofuels. *Biotechnology for Biofuels and Bioproducts*, *16*(1), 109. https://doi.org/10.1186/s13068-023-02340-0

Teoh, R., Schumann, U., Voigt, C., Schripp, T., Shapiro, M., Engberg, Z., Molloy, J., Koudis, G., & Stettler, M. E. J. (2022). Targeted Use of Sustainable Aviation Fuel to Maximize Climate Benefits. *Environmental Science & Technology*, *56*(23), 17246–17255. https://doi.org/10.1021/acs.est.2c05781

Thanigaivel, S., Vickram, S., Dey, N., Gulothungan, G., Subbaiya, R., Govarthanan, M., Karmegam, N., & Kim, W. (2022). The urge of algal biomass-based fuels for environmental sustainability against a steady tide of biofuel conflict analysis: Is third-generation algal biorefinery a boon? *Fuel*, *317*, 123494. https://doi.org/10.1016/j.fuel.2022.123494

Undavalli, V., Gbadamosi Olatunde, O. B., Boylu, R., Wei, C., Haeker, J., Hamilton, J., & Khandelwal, B. (2023). Recent advancements in sustainable aviation fuels. *Progress in Aerospace Sciences*, *136*, 100876. https://doi.org/10.1016/j.paerosci.2022.100876

US Department of Energy. (n.d.). *Alternative Fuels Data Center: Renewable Gasoline*. Alternative Fuels Data Center: Renewable Gasoline. Retrieved March 29, 2024, from https://afdc.energy.gov/fuels/emerging_hydrocarbon.html

U.S. EIA. (2019, November 6). *EIA projects energy consumption in air transportation to increase through 2050—U.S. Energy Information Administration (EIA)*. https://www.eia.gov/todayinenergy/detail.php?id=41913

Wang, W.-C., & Tao, L. (2016). Bio-jet fuel conversion technologies. *Renewable and Sustainable Energy Reviews*, *53*, 801–822. https://doi.org/10.1016/j.rser.2015.09.016

Watson, M. J., Machado, P. G., da Silva, A. V., Saltar, Y., Ribeiro, C. O., Nascimento, C. A. O., & Dowling, A. W. (2024). Sustainable aviation fuel technologies, costs, emissions, policies, and markets: A critical review. *Journal of Cleaner Production*, *449*, 141472. https://doi.org/10.1016/j.jclepro.2024.141472

Xie, S., Li, Z., Luo, S., & Zhang, W. (2024). Bioethanol to jet fuel: Current status, challenges, and perspectives. *Renewable and Sustainable Energy Reviews*, *192*, 114240. https://doi.org/10.1016/j.rser.2023.114240

Zinsmeister, J., Storch, M., Melder, J., Richter, S., Gaiser, N., Schlichting, S., Naumann, C., Schünemann, E., Aigner, M., Oßwald, P., & Köhler, M. (2023). Soot formation of renewable gasoline: From fuel chemistry to particulate emissions from engines. *Fuel*, *348*, 128109. https://doi.org/10.1016/j.fuel.2023.128109

Ziolkowska, J. R., & Simon, L. (2014). Recent developments and prospects for algae-based fuels in the US. *Renewable and Sustainable Energy Reviews*, *29*(C), 847–853. https://ideas.repec.org//a/eee/rensus/v29y2014icp847-853.html

9 Bioelectricity Generation
Powering the Future with Microbial Fuel Cells

Akansha Mohanty, Siddhika Ajmera, Yash Misra, Ranjeet Kumar Mishra, D. Jaya Prasanna Kumar, and Ravi Sankannavar

9.1 INTRODUCTION

The globe currently faces severe energy shortages, primarily because of the exhaustible and environmentally detrimental fossil fuels (DENG et al., 2014). Exploring and developing sustainable and ecologically sound alternative energy sources is thus growing (Beegle and Borole, 2018). A standout candidate among these alternatives is bioenergy, which comes from biological or natural sources and is renewable due to biomass growth, as opposed to fossil fuels, which have finite supplies (Taha et al., 2016). Microbial fuel cells (MFCs) have become the cutting edge of bioenergy research amid growing environmental concerns and energy demands. Within this rapidly developing industry, MFCs can address the needs of both energy production and environmental improvement at the same time. It becomes clear that the effective use of organic substrates is essential to maximum MFC performance. Microbes use many mechanisms in the membrane fuel cell to convert chemical energy from the oxidation of organic compounds into ATP (Prathiba et al., 2022). These reactions involve the moving of electrons to a terminal electron acceptor (TEA), which produces an electrical current. The cathode and anode compartments of conventional MFCs are divided by a cationic membrane.

The anode compartment is home to microbes that metabolize organic substances that serve as electron donors, including glucose. These organic compounds produce protons and electrons during their metabolism. After that, electrons are moved to the anode surface. While electrons migrate along the electrical circuit from the anode to the cathode, protons move via the electrolyte and the cationic membrane. By employing soluble electron acceptors like acidic permanganate, hexacyanoferrate, or oxygen, electrons and protons get used at the cathode (Pradhan and Joshi, 2022). To extract electrical energy, a load is positioned between the two electrode chambers (Chandrasekhar et al., 2015). The latest developments in MFC research emphasize the field's dual qualities of promise and complexity. Long-term stability is crucial when working with organic substrates that are intrinsically unstable (Fadzli et al., 2021). An exposition published may further explain the

 DOI: 10.1201/9781003585398-9

advancements obtained in increasing power output and lowering part and operating costs (Mahmoud et al., 2022). Still, several significant obstacles remain, primarily related to scalability, economic viability, and operational optimization, even in the face of these impressive advances. Expanding MFCs' economic viability is hindered by their low power output, high operating costs, and unstable power output. MFCs recommend a wide choice of applications, from effectively using bioenergy to cleaning wastewater, also to solving energy and waste management concerns (Tabassum et al., 2021). Their potential to offer complete solutions for producing clean energy can be boosted by investigating their seamless integration with additional renewable energy technologies. Although MFCs have great potential, there are obstacles to their actual use in terms of operational improvement, scalability, and cost-effectiveness (Choi, 2015). Within this framework, scholars highlight current developments expected to upgrade the stability, effectiveness, and financial sustainability of MFC research. Noteworthy inventions that convert solar energy into electricity, including unique hybrid systems that link live organisms with bio-battery tools, can completely transform the production of clean energy (Shlosberg et al., 2022a).

Furthermore, combining MFCs with bio-photoelectrochemical cells (BPECs) appears to have the potential to produce electricity from light. BPECs have the potential to produce environmentally benign, pollution-free energy when combined with a variety of genetic systems, including single-celled photosynthetic microorganisms, plants, and seaweeds. One example of a groundbreaking investigation at the nexus of energy and biology technology is using catalysts such as succulent plants in developing "bio-solar cells" built on the principles of photosynthesis (Shlosberg et al., 2022b). This work addresses the obstacles and presents progress in the field of MFCs. The objective is to clarify their capacity as agents for shifts in energy production and environmental balance. Researchers thoroughly investigated the complex mechanisms regulating electron transport and bioenergy production, considering many aspects influencing MFC efficiency. Close examination of the complicated relationships among organic matter, electrodes, and microbe goals to improve our knowledge regarding MFCs and their vital part in guiding us toward a more energy-rich and sustainable future. MFCs have great capability to be vital to the global movement toward ecologically friendly and sustainable energy solutions.

9.2 MICROBIAL FUEL CELLS AND THEIR MECHANISM

An MFC is made up of many components. These components are largely divided into two chambers: the anodic chamber with the anode and the cathodic chamber with the cathode. The anodic chamber contains the anode while the cathodic chamber consists of the cathode. A proton exchange membrane (PEM) is used to separate these two chambers as shown in Figure 9.1. The anodic chambers contain microbes that are supplied with substrates that degrade anaerobically to give electrons that move from anode to cathode through a circuit that is externally built. Protons are also generated in this process that are permeated through the exchange membrane. During

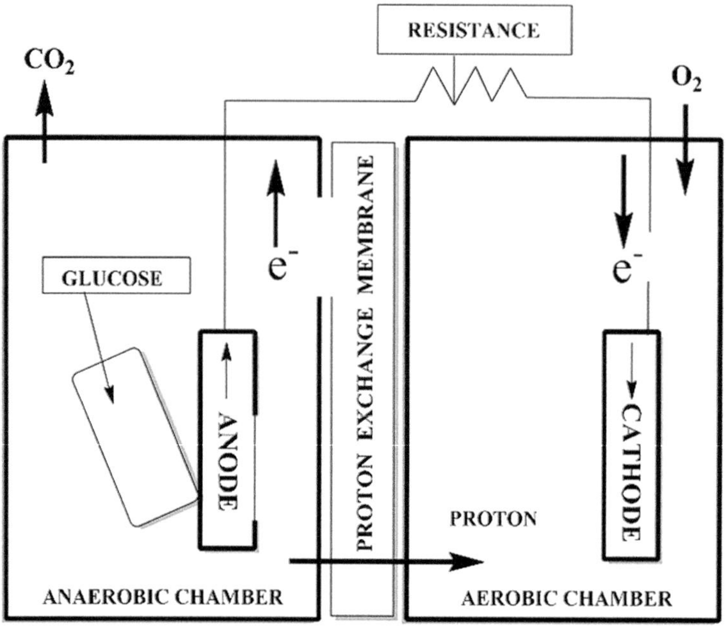

FIGURE 9.1 Diagram of a dual chamber MFCs for power generation.

this act of microbes in the anodic chamber, electrons along with protons move from the anode to the cathode to generate water by reacting with oxygen (Sharma and Li, 2010). Chemical energy can be converted into electrical energy with the help of MFCs. For this purpose, electrochemically active bacteria (EABs) carry out oxidation of several carbon sources as well as organic waste (Angenent et al., 2004; Lovley, 2008; Aelterman et al., 2008; Logan, 2009). In MFC, microorganisms catalyze the oxidation of several organic matter in the anode chamber and transfer electrons to an external circuit, ultimately leading to the reduction of an electron acceptor at the cathode. The overall process generates electricity while treating wastewater or organic substrates. Here are the reactions that typically occur at the anode and cathode.

Anodic Reaction (Oxidation)

At the anode, organic matter is oxidized by microorganisms, releasing electrons and protons. The primary reaction involves the breakdown of glucose by exoelectrogenic *Shewanella*. The overall reaction can be represented as:

$$C_6H_{12}O_6 + 6H_2O \rightarrow 6CO_2 + 24H^+ + 24e^- \quad (9.1)$$

In this equation, glucose is oxidized, and released electrons are transported through the external circuit to the cathode, whereas protons are transported through the PEM.

Cathodic Reaction (Reduction)

At the cathode, an electron acceptor is reduced, which is typically oxygen in aerobic MFCs or another suitable electron acceptor like ferricyanide in some cases. The reduction of oxygen is the most common cathodic reaction, represented as:

$$O_2 + 4H^+ + 4e^- \rightarrow 2H_2O \quad (9.2)$$

This reaction consumes the electrons and protons generated at the anode, and water is formed as the end product.

The electron transfer is a pivotal process in MFCs, particularly involving *Geobacter* and *Shewanella* sp. These microorganisms possess the ability to transport electrons beyond their cell membranes to insoluble electron acceptors, such as the anode surface in MFCs. This transfer occurs through distinct mechanisms: (a) Direct transfer: Certain bacteria, such as *Geobacter* develop conductive protein appendages known as pili or nanowires. These structures establish direct contact between bacterial cells and the electrode surface, facilitating the efficient transfer of electrons from the cell to the electrode. Furthermore, cytochromes also play a crucial role in facilitating the direct transfer of electrons from the microbial cell to the electrode surface by attaching to the anode surface. (b) Indirect electron transfer: In contrast, certain strains of microorganisms like *Shewanella* employ soluble redox mediators to transport electrons to the electrode surface. These mediators act as shuttles, facilitating electron transfer between microbial cells and the electrode, thereby enabling electron transport without direct physical contact.

Different materials can be used for the construction of these MFC chambers such as polycarbonate, plexiglass, and glass wherein the anode can be made up of carbon paper, carbon cloth, graphite felt, and graphite (Sangeetha and Muthukumar, 2013; Rhoads et al., 2005; Zhang et al., 2011; Wei et al., 2011; Zhou et al., 2011). The cathode is made up of materials like platinum-black or platinum that can sustain aerobic conditions. Some of the sustainable catholytes used in MFCs are permanganate (MnO^{4-}) or ferricyanide ($[Fe(CN_6)]^{3-}$) solutions (Pham et al., 2004; Wei et al., 2012). MFCs are working on the electron transfer mechanisms and Table 9.1 shows the bacterial electron transport mechanism in MFCs.

9.3 SUBSTRATES IN MFC

The microbes in the anodic chamber are given organic substrates to consume. These substrates are anaerobically degraded by the microbes during the process of bioelectricity generation. For a continuous generation of electricity, domestic wastewater can be used (Choi and Ahn, 2013). Min et al. (2005) employed swine wastewater as the substrate in a single chamber MFC to demonstrate the greatest power density generation (Choi and Ahn, 2013). An alternative for bioelectricity production can be oil wastewater (Choi and Liu, 2014; Yaqoob et al., 2020a). Waste sludge is another essential substrate that can be used for both hydrogen and bioelectricity generation (Choi and Liu, 2014; Ge et al., 2013). In single-chambered MFC, the microbes

TABLE 9.1
Bacterial Electron Transport Mechanism in Microbial Fuel Cells

Electron Transfer Mechanism	Description	Examples	References
Transmission mediated by naturally secreted redox mediators	Certain bacteria secrete soluble redox compounds that can move electrons from the cell to the anode. These materials can move electrons through the biofilm	*Pseudomonas*, *Bacillus*	(Tamirat et al., 2020)
Transfer mediated by artificially introduced redox mediators	Certain bacteria are unable to send electrons to the anode either directly or indirectly. On the other hand, artificial redox chemicals that can transport electrons from the cell to the anode can be added to stimulate them. These substances can improve MFCs efficiency by speeding up the rate of electron transport	*Escherichia*, *Enterococcus*	(Rosenberg et al., 2014)
Transferring directly via nanowires or conductive pili	Some bacteria can create pili, or nanowires, which are hair-like projections that can make direct contact with the anode and transport electrons. These highly conductive structures are made from protein subunits	*Geobacter*, *Shewanella*	(Rosenberg et al., 2014)

separated from the high Andean region are provided with substrates made of fruit and vegetable wastes (Ghangrekar and Shinde, 2006). Food waste leachate acquired from biohydrogen fermentation is reported to be one of the potential substrates for electricity generation (Ghangrekar and Shinde, 2006). A study shows some simple substrates used in MFCs are acetate, butyrate, glucose, and propionate (Ghangrekar and Shinde, 2006). These are in the order of acetate > butyrate > propionate when measured depending on the power density. When organic wastes break down acidogenically, a large number of volatile fatty acids originate. The microbes in the anodic chamber are given organic substrates to consume. These substrates are anaerobically degraded by the microbes during the process of bioelectricity generation. For a continuous generation of electricity, domestic wastewater can be used (Choi and Ahn, 2013). Min et al. (2005) employed swine wastewater as the substrate in a single chamber MFC to demonstrate the greatest power density generation (Min et al., 2005). An alternative for bioelectricity production can be oil wastewater (Jiang et al., 2013; Choi and Ahn, 2015). Waste sludge is another essential substrate that can be used for both hydrogen and bioelectricity generation (Choi and Ahn, 2015; Ge et al., 2013). In single-chambered MFC, the microbes separated from the high Andean region are provided with substrates made of fruit and vegetable wastes (Ghangrekar and Shinde, 2006). Food waste leachate acquired from biohydrogen fermentation is reported to be one of the potential substrates for electricity generation (Choi and Ahn, 2015). A study shows some simple substrates used in MFCs are acetate, butyrate, glucose, and propionate (Ahn and Logan, 2010). These are in the order of acetate > butyrate > propionate when measured, depending on the power density. When organic wastes break down acidogenically, a large number of volatile fatty acids are produced. These fatty acids, depending on how much they attract microbes, can affect how much electricity is generated (Choi and Ahn, 2015). After looking over the main MFC substrates and the many works related to them, it's important to investigate the microbe that powers the MFC technique and produces electricity. Electrons from the metabolism of organic substances can be transferred to the anode by microorganisms. Table 9.2 lists them along with their substrates and mechanisms of function. Effective resources for microorganisms for the MFC catalyst unit include organic-rich materials like soil, freshwater sediment, activated sludge, wastewater, and marine sediment. Anaerobic substrate digestion is allowed in complex mixed cultures; however, mixed cultures are more commonly produced. These fatty acids, depending on how much they attract microbes, can affect how much electricity is generated (Choi and Ahn, 2015).

9.4 PROTON EXCHANGE MEMBRANES (PEM)

It is crucial to maintain electro-neutrality between the two chambers in MFC technology. This balance is achieved by the PEM, which allows protons to move across the membrane, ensuring efficient operation (Rozendal et al., 2006). One of the main components of the MFC setup is the PEM as it separates the anode and cathode by enabling the transport of protons from the anode to the cathode to maintain the current (Rozendal et al., 2006). An ideal PEM would hold the following characteristics: (i) increased proton conductivity, (ii) cost-effectiveness, (iii) increased mechanical strength, (iv) good segregational properties, (v) electronically resistive, and (vi) endurance against heat and chemicals. The surface of the PEM is very crucial for the generation of power. It was stated that the power output is affected by the surface area of PEM as it decreases if the surface is smaller than the electrodes as it experiences internal resistance (Rozendal et al., 2006). In MFC technology, Ultrex (Rabaey et al., 2005; Taşkan et al., 2014), salt bridge (Sridevi et al., 2020), and Nafion (Sridevi et al., 2020) are present in the PEM used. Nafion, which is made by Dupont, is a widely used copolymer in MFC technology. It's a type of sulfonated tetrafluoroethylene with a hydrophobic fluorocarbon backbone (–CF_2–CF_2–) connected to hydrophilic sulfonate groups (SO^{3-}) (Sridevi et al., 2020). However, the application of Nafion can also be difficult at times when cathodic pH increases while anodic pH decreases resulting in the lowering of working efficiency of the MFCs (Sridevi et al., 2020). These complications are caused due to the accumulation of additional cations, other than the protons, that transport through the membrane to reach the cathodic chamber and amplify the conductivity in that chamber (Rozendal et al., 2006). Chitosan membranes cross-linked with sulfuric acid are one of the most popular choices for PEM, and they have been analyzed using a variety of criteria. Chitosan is generated from

TABLE 9.2
The Used Substrate with Their Properties in MFCs for Power Generation

Properties	Glucose	Acetic	Malate	Glycerol	Butyrate	Citrate
Molecular formula	$C_6H_{12}O_6$	$C_2H_4O_2$	$C_6H_8O_5$	$C_3H_8O_3$	$C_4H_8O_2$	$C_6H_8O_7$
Solubility in water	Readily dissolved in water	Fully miscible	Soluble in water	Miscible in water	Insoluble in water	Soluble in water and insoluble in alcohol
Density	1.54 g/cm^3	1.049 g/cm^3	–	1.26 g/cm^3	0.96 g/m^3	1.66 g/cm^3
Acidity pKa	10–12	4.76	3.40–5.20	–	4.5–7.0	2.92–5.21
Boiling point	100°C	118.10°C	306.40°C	290°C	163.50°C	310°C
Appearance	Colorless solution	Colorless liquid or crystal	–	Colorless hygroscopic liquid	Oily and colorless liquid	White, crystalline powder
Power density output (W/m^3)	3600	506	–	–	305	–
Concentration (mg/L)	500–3000	800	–	–	1000	–

chitin, a form of carbohydrate (polysaccharide) present in several organisms. Sulfonated polyimide membranes are also evaluated for their use as PEM in MFCs (Rozendal et al., 2006). One of the core molecules in the construction of PEM is styrene and its derivatives. These are produced on a commercial scale; Dias Analytics produces sulfonated styrene-ethylene-butylene-styrene (SEBS), and Ballard produces BAM (Ballard Advanced Materials) (Rozendal et al., 2006). The efficiency of any PEM is majorly influenced by two factors, its water absorption capacity and conductivity. The water absorptivity of the membrane affects the proton movement across the fuel cell, while also influencing the mechanical characteristics of the membrane. Both of these are a function of the concentration of ion conducting units (Rozendal et al., 2006). Cation exchange membranes (CEM) act as separators and have anionic groups like $C_6H_4O^-$, PO^{3-}, and COO^-. These groups help positive ions move across the membranes while stopping the negative ions from passing through. Anion exchange membranes (AEM) are membranes that have cationic groups (NR_2H^+, NH^{3+}, NHR^{2+}, SR^{2+}, PR^{3+}, and NR^{3+}) attached to them. These groups help the negatively charged groups pass through the membranes, while at the same time, they block the anions from getting through (Rozendal et al., 2006). Multiple MFCs were making use of various membranes consisting of CEM, AEM, and three kinds of ultrafiltration membranes with different molecular cut-offs (Rozendal et al., 2006). They were then tested for their abilities in the generation of power density and Coulombic efficiencies (CE). MFCs consisting of AEM produced the highest power density and Coulombic efficiency when compared to the others. This was due to the proton charge transfer facilitated by phosphate anions and lower internal resistance (Rozendal et al., 2006). This resulted in the realization that the AEM matrix's positively charged groups not only had an advantage over CEM by preventing excess cations from moving through the chambers and thus increasing the production of electricity but also made it easier for protons (phosphate anions) to travel over the membrane. As demonstrated in Figure 9.2, the bipolar membranes are made up of two separate monopolar membranes (CEM and AEM) that are placed together with a space in between them. The movement of ions across the bipolar membrane happens because of a water-splitting reaction that occurs at the interface of the AEM and CEM. This reaction produces protons and hydroxide ions, which then travel through the CEM and AEM, respectively (Rozendal et al., 2006). To examine and enhance the MFC assembly efficiency, bipolar membranes were combined with ferric ion reduction in graphite cathode by making use of ferric iron sulfate and ferric iron chloride as catholytes (ter Heijne et al., 2006). A combination of ionic salt (NaOH or KCl) and agar is poured into a cylindrical cast which is then solidified and utilized to form a salt bridge. After the solidification process, it is positioned between the two MFC chambers, thereby acting as a PEM and enabling proton transfer. One of the simplest forms of PEM that is utilized in MFCs is salt bridge. It was noted that the salt bridge produced the maximum power density (84.99 mW/m^3) when the salt concentration was only 5% (Sevda and Sreekrishnan, 2012). Variation in the salt bridge agarose concentration in MFCs was tested to conclude that 10% of agarose concentration acts as the ideal amount for the generation of voltage and current (Sridevi et al., 2020).

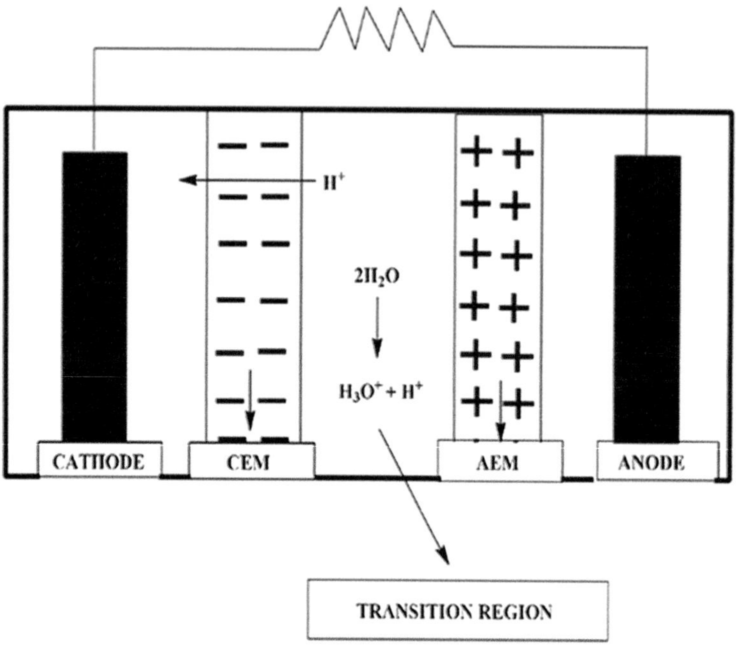

FIGURE 9.2 Systematic structure of a bipolar membrane.

9.5 MEDIATORS

Electrons generated through anaerobic degradation necessitate a mediator to be transported to the electrode (Fultz and Durst, 1982). A mediator is a compound with low redox potential that is fed to the growth media at a specific concentration to help generate electrons from the metabolic reactions of the microbes and transport them to the anode electrode. Features of an ideal MFC mediator are (Park and Zeikus, 2000): (i) it should possess stability in its reduced as well as oxidized form. (ii) It should be able to connect with NADH alongside having a high negative E^{0-} value. (iii) It should be capable enough to form a reversible redox couple at the electrode. (iv) Finally, it should possess solubility in aqueous systems. The electron transportation to the anode after its generation is aided via an exogenous electron carrier compound (methylene blue, neutral red, thionine, potassium ferricyanide) (Babanova et al., 2011; Delaney et al., 2008; Park and Zeikus, 2000; Rabaey et al., 2004).

9.6 CATHODE AND ANODE PERFORMANCE

The cathode is the electrode where water is produced due to the integration of protons with oxygen and electrons. Cathodes can be Pt-coated carbons that are made up of expensive carbon cloth; therefore, other materials like graphite granules, graphite fiber, and granular activated carbon are also used. The cathode's surface area is crucial when it comes to generating power in MFCs. The surface of the cathodes can be coated with catalysts such as cobalt tetramethoxyphenyl porphyrin (CoTMPP) which can help in achieving a power of 18 W/m^3 (Zuo et al., 2007). On the other hand, choosing materials for an MFC arrangement that maximizes power generation and Columbic efficiency while also minimizing costs is a significant task. A few of the materials utilized in the cathode are carbon paper, felt, brush, fiber, and graphite of different kinds. Pt is also frequently used as the cathode catalyst, while other polymer binders, including *naphiron* (*perfluorosulfonic acid*), have also been tested (Rahimnejad et al., 2015). Table 9.2 comprises the number of the MFC electrodes that provide the most power, current, and voltage. Anode is the electrode where microbes supplied with substrates are present. Materials used for anodes are of different forms such as rods, plates, granules, and fibers (Figure 9.3). The surface area plays a significant role in anodes as well. A three-dimensional anode having more surface area would prove to be better for power generation (Di Lorenzo et al., 2010). One way to increase the surface area is to use graphite granules packed into a bed. A viable environmentally friendly way to produce value bioenergy from natural waste and sustainability is to use natural recyclable sources. The most used natural precursor for carbon-based anodes is plant waste. This is referred to as anode material obtained from natural waste. One common method to produce porous carbon material is to directly carbonize plant precursors at super high temperatures, around 1050°C (Antolini, 2016). Plant heteroatoms serve as natural dopants throughout the preparation process and later as self-doped material. Water evaporation during the carbonization process (at 1050 °C) causes the carbon substance to create a porous structure (Antolini, 2016). Table 9.4 shows a few anodes that are utilized for the generation of power.

FIGURE 9.3 Some of the materials used in MFC construction for power generation (adopted from Obileke et al. (2021) and permission from Elsevier (Obileke et al., 2021)).

TABLE 9.3
Some of the Anode Materials Are Used in MFCs for Power Generation

Anode Material	Synthesis Method	Organic Substrate	Source of Inoculum	Power Density	References
Chestnut shell	Carbonization and activation	Acetate	Pre-acclimated bacteria	759 ± 38 mW/m^2	(Chen et al., 2016)
Silk cocoon	Simple carbonization	Acetate	Activated anaerobic sludge	5.0 mW/m^2	(Lu et al., 2017)
Chestnut shell	Carbonization and activation	Acetate	Municipal waste	23.6 W/m^3	(Chen et al., 2018)
Neem wood	Simple carbonization	Simple glucose	Pre-acclimated bacteria	256 ± 25 mW/m^2	(Senthilkumar et al., 2019)
Pinecone	Simple carbonization	Acetate	Pre-acclimated bacteria	10.88 W/m^3	(Wang et al., 2019)
Coffee	Carbonization	Acetate	Domestic waste	3927 mW/m**2**	(Hung et al., 2019)
Cellulose-wastes	Carbonization and modification	Sweet potato waste	Synthetic wastewater	0.11 mW/m^2	(Yaqoob et al., 2020b)

TABLE 9.4
Several of the MFC Electrodes Provide the Most Power, Current, and Voltage

Cathode	Max Current Density	Max Power Density	Max Voltage	References
Air-cathode with graphite	1210 mA/m^2	283 mW/m^2	440 mV	(Rahimnejad et al., 2011)
Activated carbon fiber felt (ACFF)	1.67×10^{-3} mA/m^2	315 mW/m^2 (0.7 W/m^3)	679 mV	(Deng et al., 2010)
Plain carbon	67 mW/m^2 (0.1 W/m^3)	1.5 mA/m^2	598 mV	(Deng et al., 2010)
Carbon felt	6×10^{-3} mA/m^2	77 mW/m^2 (0.2 W/m^3)	575 mV	(Deng et al., 2010)
Biocathode	41.78 A/m^3	19.53 W/m^3	432 mV	(Chen et al., 2008)
Graphite felt	3145 mA/m^2	539 mW/m^2	742.3 mV	(ter Heijne et al., 2006)
Air-cathode with carbon cloth	363 A/m^3	50 W/m^3	710 mV	(You et al., 2008)
Parallel sheets of carbon paper secured by carbon fiber coated with Pt	13.16 A/m^3	7.29 W/m^3	553 mV	(Fornero et al., 2008)

9.7 TYPES OF MFCS

9.7.1 Mediator-less MFCs

To eradicate the potential poisoning of the microbial flora that are usually employed in MFCs, mediator-less MFCs are developed as some mediators being inorganic could be toxic to the microbial flora. The electron transfer is initiated from the substrate to the electrode employed by a class of microbes known as EAB in two main ways (Aelterman et al., 2008): directly through special proteins like filaments, pili, and c-type cytochromes that are attached to their membranes (Ahn and Logan, 2010) and indirectly through mediators (Wang and Ren, 2013). In a study conducted by Bond and Lovley (2003), it was observed that the *Geobacter* genus can create networks of filaments (pili), which are highly conductive (Bond and Lovley, 2003). These pili, along with a protein called c-type cytochrome, facilitate the transfer of electrons. Another genus called *Shewanella* can generate bacterial nanowires which can be utilized as a mediator for electron transfer from a bacterial cell to a solid phase located at distances (Gorby et al., 2006). The highest power density equivalent to 0.56 W/m^2 was generated when artificial wastewater was used in mediator-less MFCs (Rane et al., 2022). A mediator-less MFC was built to study and enhance various operational parameters such as resistance, pH, electrolyte used, and dissolved oxygen (DO) concentration in the cathode compartment (Gil et al., 2003). It was noted that for an effective mediator-less MFC, lower resistance, pH of 7, buffer with NaCl solution, and increased aeration rate were ideal.

9.7.2 Membrane-less MFCs

The transfer of protons generated via the microbial degradation to the cathodic chamber is aided by the means of a PEM. In membrane-less MFCs, this usage of PEM is eliminated, creating a transmembrane potential difference leading to resistance to the flow of ions in electrolytes (Feregrino-Rivas et al., 2023). Usually, for wastewater treatment, membrane-less MFCs are preferred owing to the fouling caused by the membrane behaving like an electronic insulator due to suspended solids and soluble contaminants (Jang et al., 2004). Without a membrane, the loss of electrical resistance in the electrolyte can be reduced by minimizing the distance between the electrodes (Feregrino-Rivas et al., 2023). Ghangrekar and Shinde (2006) examined the same by constructing membrane-less MFC utilizing wastewater and noted the power density production corresponding to various electrode distances (Ghangrekar and Shinde, 2006). At an electrode spacing of 20 cm, the maximum power densities of 10.9 and 10.16 mW/m^2 were attained. Jang et al. (2004) developed MFCs which were deprived of both a mediator and membrane. These MFCs use wastewater for generating current and removing COD (chemical oxygen demand) (Jang et al., 2004). However, its efficiency was compromised because the MFC's design allowed oxygen to flow to the anode from the cathode. Jadhav et al. constructed dual-chambered membrane-less MFCs to treat synthetic wastewater, accomplishing a maximum power density and power of 0.52 mW/m^2 and 3.8 mA/m^2, respectively (NIESSEN et al., 2004).

9.7.3 Catalyst Coated Electrode MFCs

There are numerous ways to enhance the efficiency of an electrode. One such technique is coating the electrode surface with certain chemicals. Conductive polymers like polyaniline can help transfer electrons to the electrode in these chemicals. Niessen et al. (2004) investigated fluorinated aniline polymer's effectiveness as an electrode modifier (NIESSEN et al., 2004). They also found that the polymers heightened the catalytic activity of Pt in producing hydrogen and protected it against poisoning. Another polymer polytetrafluoroethylene (PTFE) combined with 30% graphite anode generated a power density equivalent to 760 mW/m^2 when used in MFCs (NIESSEN et al., 2004). To enhance the MFC efficiency, the possible use of carbon nanotubes/polyaniline composite structure was demonstrated to further modify the anode (NIESSEN et al., 2004). Also, when the anode was pretreated with ammonia, there was an increase in the power density output due to the presence of functional groups on the surface of the anode (NIESSEN et al., 2004). Likewise, oxygen is reduced on the cathode surface to speed up the process; a catalyst like ferricyanide and Pt is commonly used in the cathodic chambers of MFCs. This helps overcome the slow kinetics of oxygen reduction on plain graphite (Rahimnejad et al., 2015).

9.7.4 Sediment-Type MFCs

This MFC design revolves around the idea of having an electrode (anode) in a sediment under anaerobic conditions along with both the organic substrate and the microbial community. The anode is then linked with a cathode which is situated in the aerobic water. Thomas et al. (2013) created sediment-type MFCs and linked them to a wireless telecommunications gadget as part of a test (Thomas et al., 2013). The MFCs managed to sustain power to a device for a considerable duration autonomously, without relying on an external source. In this kind of MFC, there is scope for alterations. It can be modified by considering factors such as the type of the microbial catalyst used and the process of generating and transferring electrons (Thomas et al., 2013).

9.8 ADVANCES IN MFCS

With the application of numerous novel tactics, there has been a notable advancement in improving MFC performance (Thomas et al., 2013). This section examines significant progressions in the technology of MFCs, including developing new electrode materials, using synthetic biology techniques to increase the efficiency of electron transfer, integrating MFCs into wastewater treatment systems, and expanding MFCs for practical uses.

9.8.1 Electrodes

Creating novel electrode materials that promote the best possible electron transmission between microbes and the electrode surface is essential to the further advancement of MFC technology. While materials based on carbon such as carbon cloth or graphite were used in earlier MFCs, recent research indicates that the performance of MFCs can be significantly impacted by alternate materials (Thomas et al., 2013). Tables 9.5 and 9.6 provide an overview of these new electrode materials and how they affect MFC efficiency. These innovative materials can provide higher power densities and current densities because of their increased surface area, improved biocompatibility, and lower electrical resistance. MFCs may be enhanced by the utilization of these cutting-edge electrode materials, potentially leading to more efficient and sustainable energy conversion.

9.8.2 Integration with Wastewater Treatment Systems

Integrating MFCs with a system of wastewater treatment is among the most favorable uses. By using this method, organic wastes can be treated in addition to producing bioelectricity (Cavelius et al., 2023). The organic matter in wastewater fuels microbial populations on the anode of MFCs, which operate as bio-electrochemical systems (Cavelius et al., 2023). In addition to producing electricity by consuming organic materials, microorganisms decompose wastewater pollutants. This combination produces a win-win scenario with less environmental contamination and a sustainable energy supply. Various commercial and pilot-scale applications are already in progress because of the tremendous advancements gained in the design of MFCs for efficient wastewater treatment (Cavelius et al., 2023).

9.9 APPLICATIONS OF MFCS

Biofuel cells powered by carbohydrates can address the limited supply of fossil fuels and environmental concerns. One liter of concentrated carbohydrate solution can power a car for 20–30 miles, so a car with a 50-liter tank can go more than 1000 km before refueling (Cavelius et al., 2023). This is good for the environment since it lowers the dangers related to moving volatile fuels and the potential for fires to start after an accident. Compared to conventional fuel cells, biofuel cells provide many benefits, such as high efficiency, low maintenance, and low cost. Their uses include the production of bioenergy, bioremediation, and wastewater treatment. Biosensors that can identify glucose and lactate and the possible use of oxygen and nearby fuel sources as power sources for implanted medical devices are examples of successful applications. By oxidizing substances in wastewater streams, MFCs remove organic contaminants and produce electricity. For instance, a town's wastewater can produce up to 2.3 MW of electricity. However, 0.5 MW is a more reasonable estimate (Cavelius et al., 2023). Up to 80% less chemical oxygen is needed in wastewater when using MFCs, and the electricity they generate can be used to power additional wastewater treatment procedures (Cavelius et al., 2023). Additionally, new approaches to attaching bacterial enzymes to electrodes have been developed, such as a biofuel cell system for ammonia production, which suggests advancements in the field of more effective technology of biofuel cells. The idea of biofuel-powered robots has been studied; one such robot called "slugbot" harvests slugs and stores their digestive juices in a MFC to generate electricity. This process recharges the robot's battery and opens the door for future autonomous robots that can produce power independently

TABLE 9.5
The Effects of Electrode Materials on MFCs Performance

Electrode Material	Impact on MFCs Performance	Advantages	Disadvantages	References
Conductive polymers (CPs)	Enhanced biofilm development, power production, and current density	Cheap, simple to make and alter, and highly biocompatible	Poor mechanical strength, low stability, and fouling susceptibility	(Namsheer and Rout, 2021)
Carbon nanotubes (CNTs)	Increased biofilm growth, power production, and current density	High electrical conductivity, biocompatibility, and surface area	Expensive, challenging to produce and handle	(Cavelius et al., 2023)
Metal oxides (MOs)	Varying electron transport, power production, and current density based on the kind and makeup of MOs	Excellent catalytic activity, great stability, and low cost	Poor biocompatibility, low electrical conductivity, and potential metal ion leaching	(Cavelius et al., 2023)

TABLE 9.6
Electrode Materials Were Selected to be Utilized with Their Performance of Power Density Generation in MFCs

Electrode Material	Configuration	Type of Reactor Configuration	Source of Inoculation	Size of Electrode Material (cm)	Power Density (W/m^3)	References
Carbon brush	Brush	Single chamber	Bacteria from an active MFC	4	2400	–
Carbon (granular)	Packed	Single chamber	Domestic wastewater	--	5	--
Carbon (cloth)	Plane	Single chamber	Bacteria from an active MFC	7	46	(Cavelius et al., 2023)
Carbon paper	Plane	Double chamber	Primary clarifier overflow	22.50	600	(Cavelius et al., 2023)
Metal plate	Plane	Artificial marine	Marine sediment	0.12	23	–

(Cavelius et al., 2023). Microbial fuel cells, or MFCs, are a cutting-edge technology with a wide range of uses. However, they are essential for treating wastewater and simultaneously generating power. This allows MFCs to reduce pollution while generating energy. Their contribution to biofuel synthesis from organic substrates also offers a viable path toward developing renewable energy sources. Bioremediation helps remove poisons from soil or water, while biosensing skills allow for quickly identifying pollutants or pathogens. Additionally, MFCs demonstrate their promise in a variety of sectors by aiding in desalination efforts, CO_2 sequestration, marine sediment power production, microbial electrosynthesis, phototrophic systems, and biomedical applications. However, issues including limited power output, scale constraints, and conversion efficiency optimization continue to be critical areas for MFC technology research and real-world application (Choi, 2015).

9.9.1 Production of Bioelectricity Using MFCs

Electrical power can also be generated efficiently in MFCs by using photosynthetic bacteria. Photosynthetic bacteria remove carbon dioxide from the atmosphere along with generating power, which makes it beneficial for the environment as well (Rosenbaum et al., 2010). Strains *Anaebaena and Nostoc* are some of the biocatalysts that are used earlier in MFCs (Tanaka et al., 2007; Yagishita et al., 1998, 1997). An alternative idea for the generation of electricity via MFCs is to use the synergistic relationship between the heterotrophic and photosynthetic bacteria which function symbiotically. The heterotrophic bacteria utilize all the organic matter produced due to photosynthesis. Another way is to use strains of *Pseudomonas aeruginosa* combined with NAD co-factor; it helps boost the metabolic rate and enhances the bacteria's potential for producing biofuels. The leading groups responsible for the electric power generation are anaerobic acidogenesis of cattle dung disclosed *Ochrobactrum pseudogrignonense*, *Pseudomonas luteola*, and *Clostridium* sp. (Zhao et al., 2012). To generate both bioelectricity and biofuel together, researchers have utilized a specific type of algae called *Leptolyngbya* sp. JPMTW1 (Zhao et al., 2012). MFCs are designed with various microbial populations, including brewery wastewater, sediments from marine and lake environments, and natural microbial communities (Feng et al., 2008; Logan, 2005; Maity et al., 2014).

Microorganisms and solid-state electrodes interact to produce bioelectricity in the MFC context. The basic process is that microorganisms oxidize organic materials at the anode, which releases protons and electrons. Following their journey through an external circuit, these electrons arrive at the cathode, combining with protons and a TEA (often oxygen or another oxidant) to create water. The external circuit's electron flow produces an electric current with a variety of applications. Critical elements impact the performance of MFCs. The effectiveness of electron transport is strongly impacted by anode metabolism and microbial diversity. The selection of electrode material influences the rates of electron transfer and microbial adherence. The kind and quantity of organic matter supplied have a direct impact on electron production and overall energy yield (Chandrasekhar et al., 2015). In MFCs, microbial performance is impacted by ecological factors such as temperature and pH. Ensuring products and substrates are effectively transported to and from electrode surfaces is essential. System stability and power density must be balanced in MFCs since external resistance impacts current and power output (Chandrasekhar et al., 2015). The effectiveness of MFCs in the production of sustainable bioelectricity is determined by these elements taken together.

The primary characteristic of MFCs is the utilization of organic carbohydrate substrates obtained from sources such as industrial, agricultural, and municipal wastes to generate bioelectricity. MFCs have the extra advantage of converting fuel molecules directly into electricity without producing heat. This eliminates the limitations of the Carnot cycle, which affects the efficiency of thermal energy conversion. As a result, MFCs can achieve a higher conversion efficiency of over 70 (Feregrino-Rivas et al., 2023). While MFCs currently can't be considered an economical approach for power generation, in recent years, they have progressed toward being more effective in power generation. By utilizing substrates such as glucose and domestic wastewater, power yield of 250–500 and 10–50 mW/m^2 have been generated, respectively. Making use of glucose and a mixed consortium of microbial communities as substrate (Feregrino-Rivas et al., 2023). Stacked MFCs are another option for increasing electricity generation while also serving a variety of other purposes. In a study conducted by Aelterman et al., they connected six individual MFC units in a stacked arrangement and achieved an average power generation of 258 W/m^3 (Feregrino-Rivas et al., 2023). Although the power generation by MFCs falls short of other fuel cells, such as methanol drivel MFCs, the difference in substrate usage adds an incentive feature to the system (Feregrino-Rivas et al., 2023). Also, the utilization of substrates (organics or nutrients) is eliminated by the creation of self-sustained phototrophic MFCs which also provide a continuous power output under illumination; thus, if upgraded, MFCs could become a great sustainable energy choice (Feregrino-Rivas et al., 2023). Rosenbaum et al. assembled an MFC that made use of the metabolic activity of *Rhodobacter sphaeroids*. This microorganism was able to oxidize photo-biological hydrogen in the MFC, resulting in the production of electricity (Feregrino-Rivas et al., 2023). Hence, MFCs are a technology with the potential to be used as a sustainable energy source. MFC technology can be implemented toward creating bio-batteries. These bio-batteries could be used to recharge low-voltage appliances and gadgets. Various reformations are being implemented into the core and basic design of MFCs, giving the foundation for various upcoming ideas and applications. Table 9.7 lists the latest research that has demonstrated excellent power generation efficiency using MFCs and a variety of waste sources.

9.10 LIMITATION

MFCs have tremendous promise as long-term bioenergy solutions since they employ microbial processes to remediate organic waste and produce electricity. Not withstanding their potential, MFCs have several issues and restrictions that need to be resolved to increase their efficiency and wide range of applications (Sonawane et al., 2022). The main challenges MFCs face are covered in this section, along with possible solutions.

TABLE 9.7
An Overview of the Generation of Bioelectricity with Various MFC Setups

Configuration Type	Electrode Materials		Membrane Type	Substrate	Working Volume (mL)	Operation (Days)	Max. Power Generation	Reference
	Anode	Cathode						
T-MFC	Graphite rod	Carbon cloth coated Pt (200 cm^2)	Nanocomposite	Sewage wastewater	300	3 weeks	138 mW/m^3	(Feregrino-Rivas et al., 2023)
DC-MFC	Carbon fiber	Carbon fiber	SPEEK-goethite	–	–	–	73.7 mW/m^3	(Feregrino-Rivas et al., 2023)
SC-MFC	Carbon felt (16 cm^2)	Carbon felt (31 cm^2)	Clay ware	Synthetic wastewater	150	30	995.73 mW/m^3	(Feregrino-Rivas et al., 2023)
SMFC	Carbon-polymer composite	Carbon cloth (3 × 3 cm^2)	–	Sediment from wastewater	100	30	1056.6 W/m^3	(Feregrino-Rivas et al., 2023)
MFC	Carbon fiber brushes	Carbon fiber brushes	Nafion 117	Glucose, MB, and yeast	800	-	5.55 W/m^3	(Feregrino-Rivas et al., 2023)
S-MFC	Graphite felt (7 × 7 × 0.4 cm^3)	Carbon cloth coated-Pt, plain carbon cloth, and graphite felt	–	Soil	–	–50	87.3 mW/m^3	(Feregrino-Rivas et al., 2023)
DC-MFC	Graphite filter	Stainless steel mesh	Carbon ceramic composite	Wastewater	5.3 (cm^3)	–	0.699 W/m^3	(Feregrino-Rivas et al., 2023)
SC-MFC	Wired stainless steel 60 mesh	Wired stainless steel 60 mesh	Cylindrical terra cotta pots	Textile effluent	–	–	21–42 mW/m^3	(Feregrino-Rivas et al., 2023)
SC-MFC	Graphite brush	Graphite-based nanomaterials	–	Anaerobic mud	–	30	2203 mW/m^3	(Feregrino-Rivas et al., 2023)
SC-SMFC	Unidirectional carbon fiber (total area 81 cm^2)	Unidirectional carbon fiber (total area 40.5 cm^2)	–	Marine and fluvial sediments	2000	30	70 mW/cm^3	(Feregrino-Rivas et al., 2023)

9.10.1 Low Power Throughput

A notable obstacle MFCs face is their reduced power output in contrast to conservative energy sources. The MFCs have very low densities of power to meet mainstream power generation demands or be widely deployed. Numerous factors, such as inadequate electrode materials, ineffective electron transfer pathways, and constraints on microbial metabolic activity, contribute to this limitation (Sonawane et al., 2022). Researchers are experimenting with several strategies to boost MFCs power production. A potential approach to enhance the metabolic activity and electron-transfer efficiency of MFCs involves optimizing the microbial populations. Genetic engineering methods can improve the capacity of microbial strains to transfer electrons to the electrode surface (DENG et al., 2014). Moreover, electrode material developments, like conductive polymers, nanomaterials, or composite materials, can greatly improve the efficiency of MFCs and electron transport kinetics. Furthermore, hybrid systems that overcome the drawbacks of MFCs low power output and provide dependable energy production may be created by combining MFCs with other renewable energy technologies, such as solar panels or wind turbines.

9.10.2 Commercialization and Scale-up

The affordability of the MFC technology is one more obstacle to its broad acceptance. Because the cost of production and operation of MFCs can be high, traditional energy sources may not be as competitive with them commercially. Researchers are investigating scalable manufacturing techniques and substituting inexpensive electrode materials to increase affordability. MFCs can add value by treating organic waste and producing electricity simultaneously when integrated into already-existing wastewater treatment plants, increasing the technology's economic appeal. Additionally, government funding for development research and incentives can be vital in lowering expenses and advancing the commercialization of MFC technology. The development of MFCs for widespread use poses considerable scientific and practical obstacles despite their promising performance in the lab. Controlling increasing reactor contents, ensuring sufficient material flow, and ensuring consistent electrode shape are all part of this. Handling rising reactor capacities, maximizing chemical mass transfer, and preserving electrode uniformity are all difficult. Optimizing power output and system efficiency in large MFC systems is greatly influenced by reactor design and engineering. To enhance mixing, maximize reactor stability, and guarantee uniform dispersion of microbial populations, design and engineering considerations are essential (DENG et al., 2014). Large-scale MFC deployment is also more doable thanks to innovative strategies like stackable and modular MFC designs, which offer scalability and more straightforward installation (Figure 9.4). In summary, even with the challenges and limitations that MFCs encounter, these can be overcome with ongoing study and technological advancements. MFCs can be established as a sustainable and feasible bioenergy source, potentially revolutionizing the renewable energy market, by tackling issues including cost-effectiveness, power output, large-scale integration, and microbial stability.

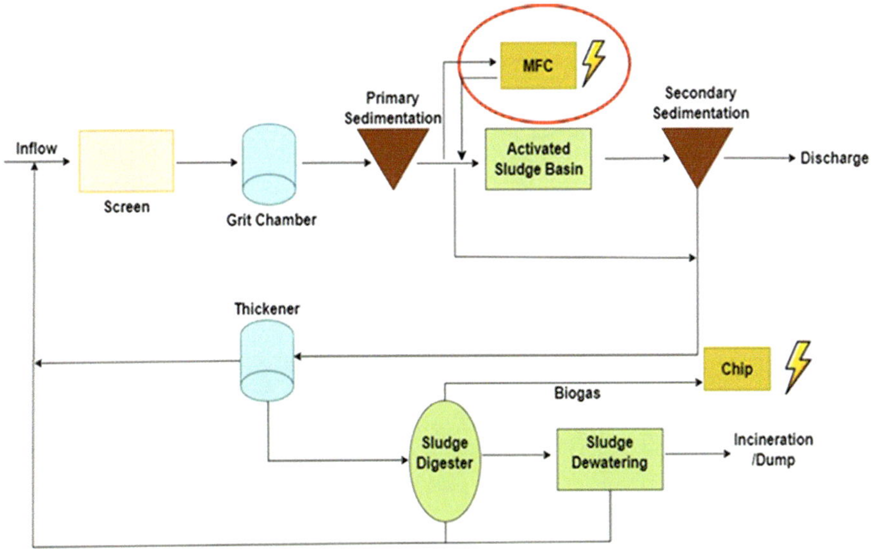

FIGURE 9.4 A 45-liter pilot MFC device integrated into a full-scale wastewater treatment plant. (redrawn as per Hiegemann et al,; DENG et al., 2014).

9.10.3 Genetic Engineering

Plant biomass yield must be increased through genetic engineering to synthesize bioenergy precursors. Energy crop performance can be significantly enhanced by modifying genetic targets researchers have identified. A primary objective is to enhance the efficiency of photosynthesis. By overexpressing genes related to assimilation and carbon fixation, scientists have effectively raised energy crops biomass yield. Because the processes involved in making biofuels require changes in gene expression to produce lignin and cellulose, energy crops with lower recalcitrance are also better suited to these processes. Furthermore, new biochemical pathways have been introduced into plants, enabling direct synthesis of bioenergy precursors thanks to advancements in synthetic biology. Researchers have engineered plants that can manufacture commercial bioenergy products without requiring costly downstream processing by putting the genes responsible for producing biofuels like ethanol or hydrocarbons into the plants. Certain genetic targets have been identified by researchers and modifying them can greatly improve the productivity of energy crops. Table 9.8 illustrates genetic targets and their influence on biomass and bioenergy.

9.10.4 Plant-Microbe Interaction

The rhizosphere and the soil surrounding a plant's roots are shaped by interactions between plants and microbes, which affects nutrient uptake, growth promotion, and disease resistance. These interactions can potentially improve the MFCs and energy

TABLE 9.8
The Impact of Genetic Targeting on Biomass and Biofuels

Genetic Target	Description	Impact on Bioenergy and Biomass	References
Photosynthesis genes	The light responses and carbon fixation processes of photosynthesis are mediated by genes	By improving these genes, one can boost carbon uptake and photosynthetic efficiency, which can result in a greater buildup of biomass	(DENG et al., 2014)
Transcription factors	Genes that control how other genes are expressed in different biological processes	It is possible to regulate the expression of numerous downstream genes linked to biomass and bioenergy properties by overexpressing or silencing these genes	(DENG et al., 2014)
Cell wall biosynthesis genes	Genes related to the production of wall proteins, lignin, hemicellulose, cellulose, and pectin	The composition and structure of the cell wall can be changed by modifying these genes, which may have an impact on the efficiency of saccharification and the digestibility of biomass	(DENG et al., 2014)
Hormone biosynthesis and signaling genes	Genes related to the production and sensing of plant hormones, including salicylic acid, ethylene, gibberellin, auxin, cytokinin, abscisic acid, brassinosteroid, and jasmonic acid	Numerous aspects of plant growth and development, including cell division, elongation, differentiation, senescence, flowering, and stress responses, can be influenced by altering these genes. This could have an impact on the amount and caliber of biomass generated	(DENG et al., 2014)

crop performance in MFC situations. To encourage the growth of energy crops, for instance, mycorrhizal fungi and bacteria that boost plant growth might be specifically chosen and added (DENG et al., 2014). As demonstrated in Figure 9.5, these bacteria enhance the accessibility of nutrients, augment water absorption, and shield plants from diverse stressors, culminating in a rise in the total production of biomass. Moreover, certain plant-microbe partnerships involve reciprocal nutrient exchanges. Bacteria receive organic carbon from plants in return for vital nutrients like phosphorus and nitrogen. Enhancing these mutualistic connections has the potential to increase the effectiveness of the nutrient cycle, leading to increased yields of energy crops and improved circumstances for MFC activities (DENG et al., 2014). The combination of plant molecular biology and MFC technology has great promise for enhancing the generation of sustainable bioenergy. Through the utilization of photosynthesis, genetic engineering, and the interactions between microbes and plants, scientists want to maximize the effectiveness of MFCs and energy crops, paving the way for a more environmentally friendly and sustainable future (Figure 9.6). To fully realize this synergy's potential, collaborations between plant biologists and bioenergy engineers are essential.

9.11 SEPARATORS AND BIOSENSORS

Clayware and ceramic have been utilized as MFC separators for some time. Ceramic membranes are viable options due to their low cost and mechanical, chemical, and

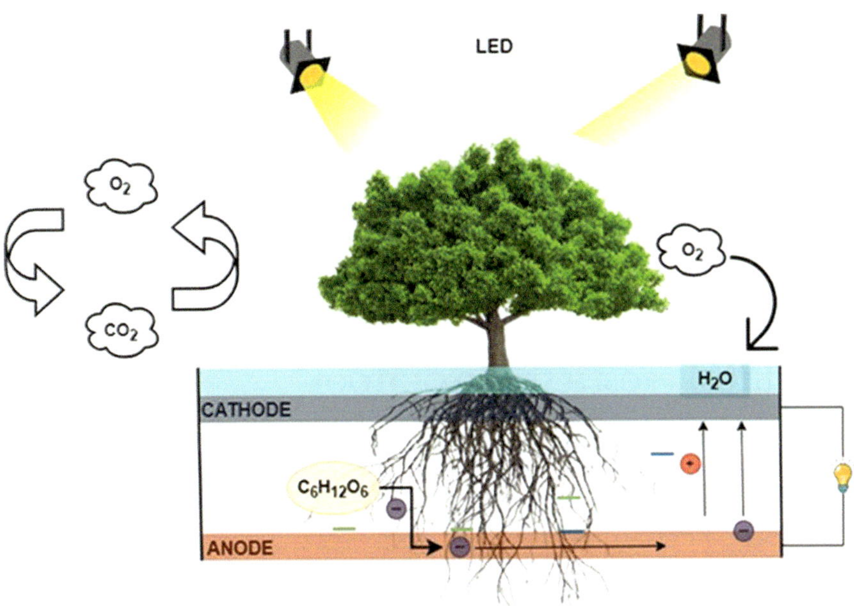

FIGURE 9.5 Illustration of a microbe-plant interaction-based MFC (redrawn as per Chicas et al.; DENG et al., 2014).

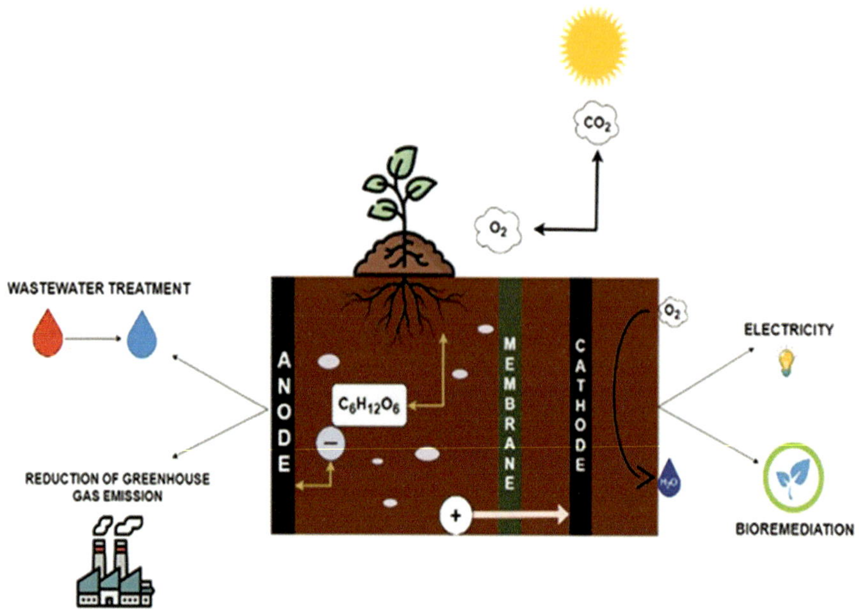

FIGURE 9.6 Plant-microbe relations using interactions to build an MFC (redrawn as per Deng et al., 2014; DENG et al., 2014).

thermal stability. They are an excellent choice for managing membrane fouling and operating MFCs at high temperatures. Nonetheless, further investigation will focus on enhancing clay material properties to boost microbial colonization (Srivastava et al., 2020). For in-situ environmental monitoring, MFC-based biosensors are self-sufficient devices. The two MFC-based biosensors most frequently utilized are toxicity and water quality. This commercial use of MFC technology is promising. However, significant research is required to enhance the functionality of these biosensors. Future studies could focus on improving data-driven evaluation models, matching detection algorithms, lowering the detection limit for organic matter monitoring, and improving sensitivity (Srivastava et al., 2020).

9.12 FUTURE PROSPECTS AND OUTLOOK

MFCs have attracted much attention lately due to research conducted on a lab scale, but little is known about their effective commercial deployment. Investigating nanocomposites as coating materials has been a focal point of efforts to enhance their performance through anode changes. Numerous research works have demonstrated the efficiency of nanocomposites, mainly metal or metal oxide blends on the anode surface, in improving conductivity, electron transmission, and ohmic losses while increasing bacterial adherence (Srivastava et al., 2020). Despite a wealth of research on the coexistence of electricity production and wastewater treatment, there are currently no thorough reports on the practical implementation of MFC-built wetlands or continuous flow systems (Srivastava et al., 2020). Aeration expenses may be decreased by adding photosynthetic microorganisms to MFCs, as they exhibit promise in producing oxygen and capturing CO_2. Several obstacles hamper commercialization despite significant advances. The fundamental obstacle to scaling up MFCs is their high internal resistance, which is made worse by problems including low and unstable voltage production, high internal resistance, and potential losses from ohmic, metabolic, mass transfer, and activation losses, among other sources. Although there are suggestions for improving the effectiveness of electrode changes, there are issues with their higher cost and stability over prolonged use. Membrane biofouling also reduces longevity and increases operating expenses by requiring frequent membrane replacements. Resilience against corrosion, instability, and biofouling is necessary to ensure MFC's lifespan (Gajda et al., 2018).

9.13 CONCLUSIONS

Microorganisms act as biocatalysts in MFCs, bio-electrochemical tools that turn chemical energy from biological substrates into electrical energy. Research in this field has increased dramatically in the last few years, mainly concentrating on improving reactor designs, electrode materials, and membranes to maximize voltage production. Because it is still challenging to scale up MFCs for practical usage, there is a tendency toward smaller units and the modular stacking of several units for practical uses. An increasing field of research is using phototrophic microorganisms to help cathodes reduce the need for energy-intensive mechanical aeration. Research efforts are increasingly focused on producing bioelectricity and bioremediation simultaneously. MFC technology has promising uses in many areas such as biosensors, bioremediation, and biohydrogen production. Continuous attempts are made to improve the efficiency and performance of MFCs but scaling them up for widespread commercial use is still a challenge. Durability requires components that balance performance and cost and resist biofouling and corrosion.

ACKNOWLEDGMENTS

The authors thanks Department of Chemical Engineering, Manipal Institute of Technology Manipal, Ramaiah Institute of Technology Bangalore, and Indian Institute of Technology Goa for providing all the necessary facilities.

REFERENCES

Aelterman, P., Freguia, S., Keller, J., Verstraete, W., Rabaey, K., 2008. The anode potential regulates bacterial activity in microbial fuel cells. *Appl. Microbiol. Biotechnol.* 78, 409–418. https://doi.org/10.1007/s00253-007-1327-8

Ahn, Y., Logan, B.E., 2010. Effectiveness of domestic wastewater treatment using microbial fuel cells at ambient and mesophilic temperatures. *Bioresour. Technol.* 101, 469–475. https://doi.org/10.1016/j.biortech.2009.07.039

Angenent, L.T., Karim, K., Al-Dahhan, M.H., Wrenn, B.A., Domíguez-Espinosa, R., 2004. Production of bioenergy and biochemicals from industrial and agricultural wastewater. *Trends Biotechnol.* 22, 477–485. https://doi.org/10.1016/j.tibtech.2004.07.001

Antolini, E., 2016. Nitrogen-doped carbons by sustainable N- and C-containing natural resources as nonprecious catalysts and catalyst supports for low temperature fuel cells. *Renew. Sustain. Energy Rev.* 58, 34–51. https://doi.org/10.1016/j.rser.2015.12.330

Babanova, S., Hubenova, Y., Mitov, M., 2011. Influence of artificial mediators on yeast-based fuel cell performance. *J. Biosci. Bioeng.* 112, 379–387. https://doi.org/10.1016/j.jbiosc.2011.06.008

Beegle, J.R., Borole, A.P., 2018. Energy production from waste: Evaluation of anaerobic digestion and bioelectrochemical systems based on energy efficiency and economic factors. *Renew. Sustain. Energy Rev.* 96, 343–351. https://doi.org/10.1016/j.rser.2018.07.057

Bond, D.R., Lovley, D.R., 2003. Electricity production by Geobacter sulfurreducens attached to electrodes. *Appl. Environ. Microbiol.* 69, 1548–1555. https://doi.org/10.1128/AEM.69.3.1548-1555.2003

Cavelius, P., Engelhart-Straub, S., Mehlmer, N., Lercher, J., Awad, D., Brück, T., 2023. The potential of biofuels from first to fourth generation. *PLOS Biol.* 21, e3002063. https://doi.org/10.1371/journal.pbio.3002063

Chandrasekhar, K., Lee, Y.-J., Lee, D.-W., 2015. Biohydrogen production: Strategies to improve process efficiency through microbial routes. *Int. J. Mol. Sci.* 16, 8266–8293. https://doi.org/10.3390/ijms16048266

Chen, G.-W., Choi, S.-J., Lee, T.-H., Lee, G.-Y., Cha, J.-H., Kim, C.-W., 2008. Application of biocathode in microbial fuel cells: cell performance and microbial community. *Appl. Microbiol. Biotechnol.* 79, 379–388. https://doi.org/10.1007/s00253-008-1451-0

Chen, Q., Pu, W., Hou, H., Hu, J., Liu, B., Li, J., Cheng, K., Huang, L., Yuan, X., Yang, C., Yang, J., 2018. Activated microporous-mesoporous carbon derived from chestnut shell as a sustainable anode material for high performance microbial fuel cells. *Bioresour. Technol.* 249, 567–573. https://doi.org/10.1016/j.biortech.2017.09.086

Chen, S., Tang, J., Jing, X., Liu, Y., Yuan, Y., Zhou, S., 2016. A hierarchically structured urchin-like anode derived from chestnut shells for microbial energy harvesting. *Electrochim. Acta* 212, 883–889. https://doi.org/10.1016/j.electacta.2016.07.077

Choi, J., Ahn, Y., 2015. Enhanced bioelectricity harvesting in microbial fuel cells treating food waste leachate produced from biohydrogen fermentation. *Bioresour. Technol.* 183, 53–60. https://doi.org/10.1016/j.biortech.2015.01.109

Choi, J., Ahn, Y., 2013. Continuous electricity generation in stacked air cathode microbial fuel cell treating domestic wastewater. *J. Environ. Manage.* 130, 146–152. https://doi.org/10.1016/j.jenvman.2013.08.065

Choi, J., Liu, Y., 2014. Power generation and oil sands process-affected water treatment in microbial fuel cells. *Bioresour. Technol.* 169, 581–587. https://doi.org/10.1016/j.biortech.2014.07.029

Choi, S., 2015. Microscale microbial fuel cells: Advances and challenges. *Biosens. Bioelectron.* 69, 8–25. https://doi.org/10.1016/j.bios.2015.02.021

Delaney, G., Bennetto, H., Mason, J., Roller, S., Stirling, J., Thurston, C., 2008. Electron-transfer coupling in microbial fuel cells. 2. performance of fuel cells containing selected microorganism-mediator-substrate combinations. *J. Chem. Technol. Biotechnol. Biotechnol.* 34, 13–27. https://doi.org/10.1002/jctb.280340104

Deng, H., Wu, Y.-C., Zhang, F., Huang, Z.-C., Chen, Z., Xu, H.-J., Zhao, F., 2014. Factors affecting the performance of single-chamber soil microbial fuel cells for power generation. *Pedosphere* 24, 330–338. https://doi.org/10.1016/S1002-0160(14)60019-9

Deng, Q., Li, X., Zuo, J., Ling, A., Logan, B.E., 2010. Power generation using an activated carbon fiber felt cathode in an upflow microbial fuel cell. *J. Power Sources* 195, 1130–1135. https://doi.org/10.1016/j.jpowsour.2009.08.092

Di Lorenzo, M., Scott, K., Curtis, T.P., Head, I.M., 2010. Effect of increasing anode surface area on the performance of a single chamber microbial fuel cell. *Chem. Eng. J.* 156, 40–48. https://doi.org/10.1016/j.cej.2009.09.031

Fadzli, F.S., Bhawani, S.A., Adam Mohammad, R.E., 2021. Microbial fuel cell: Recent developments in organic substrate use and bacterial electrode interaction. *J. Chem.* 2021, 1–16. https://doi.org/10.1155/2021/4570388

Feng, Y., Wang, X., Logan, B.E., Lee, H., 2008. Brewery wastewater treatment using air-cathode microbial fuel cells. *Appl. Microbiol. Biotechnol.* 78, 873–880. https://doi.org/10.1007/s00253-008-1360-2

Feregrino-Rivas, M., Ramirez-Pereda, B., Estrada-Godoy, F., Cuesta-Zedeño, L.F., Rochín-Medina, J.J., Bustos-Terrones, Y.A., Gonzalez-Huitron, V.A., 2023. Performance of a sediment microbial fuel cell for bioenergy production: Comparison of fluvial and marine sediments. *Biomass and Bioenergy* 168, 106657. https://doi.org/10.1016/j.biombioe.2022.106657

Fornero, J.J., Rosenbaum, M., Cotta, M.A., Angenent, L.T., 2008. Microbial fuel cell performance with a pressurized cathode chamber. *Environ. Sci. Technol.* 42, 8578–8584. https://doi.org/10.1021/es8015292

Fultz, M., Durst, R., 1982. Mediator compounds for the electrochemical study of biological redox systems: A compilation. *Anal. Chim. Acta* 140, 1–18. https://doi.org/10.1016/S0003-2670(01)95447-9

Gajda, I., Greenman, J., Ieropoulos, I.A., 2018. Recent advancements in real-world microbial fuel cell applications. *Curr. Opin. Electrochem.* 11, 78–83. https://doi.org/10.1016/j.coelec.2018.09.006

Ge, Z., Zhang, F., Grimaud, J., Hurst, J., He, Z., 2013. Long-term investigation of microbial fuel cells treating primary sludge or digested sludge. *Bioresour. Technol.* 136, 509–514. https://doi.org/10.1016/j.biortech.2013.03.016

Ghangrekar, M.M. and Shinde, V.B., 2006. Wastewater Treatment in Microbial Fuel Cell and Electricity Generation: A Sustainable Approach. Int. Sustain. Dev. Res. Conf.

Gil, G.-C., Chang, I.-S., Kim, B.H., Kim, M., Jang, J.-K., Park, H.S., Kim, H.J., 2003. Operational parameters affecting the performannce of a mediator-less microbial fuel cell. *Biosens. Bioelectron.* 18, 327–334. https://doi.org/10.1016/S0956-5663(02)00110-0

Gorby, Y.A., Yanina, S., McLean, J.S., Rosso, K.M., Moyles, D., Dohnalkova, A., Beveridge, T.J., Chang, I.S., Kim, B.H., Kim, K.S., Culley, D.E., Reed, S.B., Romine, M.F., Saffarini, D.A., Hill, E.A., Shi, L., Elias, D.A., Kennedy, D.W., Pinchuk, G., Watanabe, K., Ishii, S., Logan, B., Nealson, K.H., Fredrickson, J.K., 2006. Electrically conductive bacterial nanowires produced by Shewanella oneidensis strain MR-1 and other microorganisms. *Proc. Natl. Acad. Sci.* 103, 11358–11363. https://doi.org/10.1073/pnas.0604517103

Hung, Y.-H., Liu, T.-Y., Chen, H.-Y., 2019. Renewable coffee waste-derived porous carbons as anode materials for high-performance sustainable microbial fuel cells. *ACS Sustain. Chem. Eng.* 7, 16991–16999. https://doi.org/10.1021/acssuschemeng.9b02405

Jang, J.K., Pham, T.H., Chang, I.S., Kang, K.H., Moon, H., Cho, K.S., Kim, B.H., 2004. Construction and operation of a novel mediator- and membrane-less microbial fuel cell. *Process Biochem.* 39, 1007–1012. https://doi.org/10.1016/S0032-9592(03)00203-6

Jiang, Y., Ulrich, A.C., Liu, Y., 2013. Coupling bioelectricity generation and oil sands tailings treatment using microbial fuel cells. *Bioresour. Technol.* 139, 349–354. https://doi.org/10.1016/j.biortech.2013.04.050

Namsheer, K., Rout, C.S., 2021. Conducting polymers: A comprehensive review on recent advances in synthesis, properties and applications. *RSC Adv.* 11, 5659–5697. https://doi.org/10.1039/D0RA07800J

Logan, B.E., 2009. Exoelectrogenic bacteria that power microbial fuel cells. *Nat. Rev. Microbiol.* 7, 375–381. https://doi.org/10.1038/nrmicro2113

Logan, B.E., 2005. Simultaneous wastewater treatment and biological electricity generation. *Water Sci. Technol. J. Int. Assoc. Water Pollut. Res.* 52, 31–37.

Lovley, D.R., 2008. The microbe electric: conversion of organic matter to electricity. *Curr. Opin. Biotechnol.* 19, 564–571. https://doi.org/10.1016/j.copbio.2008.10.005

Lu, M., Qian, Y., Yang, C., Huang, X., Li, H., Xie, X., Huang, L., Huang, W., 2017. Nitrogen-enriched pseudographitic anode derived from silk cocoon with tunable flexibility for microbial fuel cells. *Nano Energy* 32, 382–388. https://doi.org/10.1016/j.nanoen.2016.12.046

Mahmoud, R.H., Gomaa, O.M., Hassan, R.Y.A., 2022. Bio-electrochemical frameworks governing microbial fuel cell performance: technical bottlenecks and proposed solutions. *RSC Adv.* 12, 5749–5764. https://doi.org/10.1039/D1RA08487A

Maity, J., Hou, C.-P., Majumder, D., Bundschuh, J., Kulp, T., Chen, C.-Y., Chuang, L.-T., Chen, N., Jean, J.-S., Yang, T.-C., Chen, C.-C., 2014. The production of biofuel and bioelectricity associated with wastewater treatment by green algae. *Energy*. https://doi.org/10.1016/j.energy.2014.06.023

Min, B., Kim, J., Oh, S., Regan, J.M., Logan, B.E., 2005. Electricity generation from swine wastewater using microbial fuel cells. *Water Res.* 39, 4961–4968. https://doi.org/10.1016/j.watres.2005.09.039

Niessen, J., Schroder, U., Scholz, F., 2004. Exploiting complex carbohydrates for microbial electricity generation? A bacterial fuel cell operating on starch. *Electrochem. Commun.* 6, 955–958. https://doi.org/10.1016/j.elecom.2004.07.010

Obileke, K., Onyeaka, H., Meyer, E.L., Nwokolo, N., 2021. Microbial fuel cells, a renewable energy technology for bio-electricity generation: A mini-review. *Electrochem. Commun.* 125, 107003. https://doi.org/10.1016/j.elecom.2021.107003

Park, D.H., Zeikus, J.G., 2000. Electricity generation in microbial fuel cells using neutral red as an electronophore. *Appl. Environ. Microbiol.* 66, 1292–1297. https://doi.org/10.1128/AEM.66.4.1292-1297.2000

Pham, T.H., Jang, J.K., Chang, I.S., Kim, B., 2004. Improvement of cathode reaction of a mediatorless microbial fuel cell. *J. Microbiol. Biotechnol.* 14, 324–329.

Pradhan, S., Joshi, J., 2022. A review on microbial fuel cell performance for energy generation. *Nepal J. Sci. Technol.* 21, 101–106. https://doi.org/10.3126/njst.v21i1.49919

Prathiba, S., Kumar, P.S., Vo, D.-V.N., 2022. Recent advancements in microbial fuel cells: A review on its electron transfer mechanisms, microbial community, types of substrates and design for bio-electrochemical treatment. *Chemosphere* 286, 131856. https://doi.org/10.1016/j.chemosphere.2021.131856

Rabaey, K., Boon, N., Siciliano, S.D., Verhaege, M., Verstraete, W., 2004. Biofuel cells select for microbial consortia that self-mediate electron transfer. *Appl. Environ. Microbiol.* 70, 5373–5382. https://doi.org/10.1128/AEM.70.9.5373-5382.2004

Rabaey, K., Clauwaert, P., Aelterman, P., Verstraete, W., 2005. Tubular Microbial Fuel Cells for Efficient Electricity Generation. *Environ. Sci. Technol.* 39, 8077–8082. https://doi.org/10.1021/es050986i

Rahimnejad, M., Adhami, A., Darvari, S., Zirepour, A., Oh, S.-E., 2015. Microbial fuel cell as new technology for bioelectricity generation: A review. *Alexandria Eng. J.* 54, 745–756. https://doi.org/10.1016/j.aej.2015.03.031

Rahimnejad, M., Ghoreyshi, A.A., Najafpour, G., Jafary, T., 2011. Power generation from organic substrate in batch and continuous flow microbial fuel cell operations. *Appl. Energy* 88, 3999–4004. https://doi.org/10.1016/j.apenergy.2011.04.017

Rane, N.V., Kumari, A., Holkar, C., Pinjari, D. V., Pandit, A.B., 2022. Microbial fuel cell technology for wastewater treatment, in: *Sustainable Water Treatment*. Wiley, pp. 271–323. https://doi.org/10.1002/9781119480075.ch7

Rhoads, A., Beyenal, H., Lewandowski, Z., 2005. Microbial fuel cell using anaerobic respiration as an anodic reaction and biomineralized manganese as a cathodic reactant. *Environ. Sci. Technol.* 39, 4666–4671. https://doi.org/10.1021/es048386r

Rosenbaum, M., He, Z., Angenent, L.T., 2010. Light energy to bioelectricity: photosynthetic microbial fuel cells. *Curr. Opin. Biotechnol.* 21, 259–264. https://doi.org/10.1016/j.copbio.2010.03.010

Rosenberg, E., Long, E.F., Loy, S., Stackebrandt, E., Thompson, F., 2014. *The Prokaryotes*. Berlin, Heidelberg: Springer. https://doi.org/10.1007/978-3-642-39044-9

Rozendal, R.A., Hamelers, H.V.M., Buisman, C.J.N., 2006. Effects of membrane cation transport on pH and microbial fuel cell performance. *Environ. Sci. Technol.* 40, 5206–5211. https://doi.org/10.1021/es060387r

Sangeetha, T., Muthukumar, M., 2013. Influence of electrode material and electrode distance on bioelectricity production from sago-processing wastewater using microbial fuel cell. *Environ. Prog. Sustain. Energy* 32, 390–395. https://doi.org/10.1002/ep.11603

Senthilkumar, N., Pannipara, M., Al-Sehemi, A.G., Gnana Kumar, G., 2019. PEDOT/NiFe 2 O 4 nanocomposites on biochar as a free-standing anode for high-performance and durable microbial fuel cells. *New J. Chem.* 43, 7743–7750. https://doi.org/10.1039/C9NJ00638A

Sevda, S., Sreekrishnan, T., 2012. Effect of salt concentration and mediators in salt bridge microbial fuel cell for electricity generation from synthetic wastewater. *J. Environ. Sci. Health. A. Tox. Hazard. Subst. Environ. Eng.* 47, 878–886. https://doi.org/10.1080/10934529.2012.665004

Sharma, Y., Li, B., 2010. The variation of power generation with organic substrates in single-chamber microbial fuel cells (SCMFCs). *Bioresour. Technol.* 101, 1844–1850. https://doi.org/10.1016/j.biortech.2009.10.040

Shlosberg, Y., Schuster, G., Adir, N., 2022a. Harnessing photosynthesis to produce electricity using cyanobacteria, green algae, seaweeds and plants. *Front. Plant Sci.* 13, 955843. https://doi.org/10.3389/fpls.2022.955843

Shlosberg, Y., Schuster, G., Adir, N., 2022b. Self-enclosed bio-photoelectrochemical cell in succulent plants. *ACS Appl. Mater. Interfaces* 14, 53761–53766. https://doi.org/10.1021/acsami.2c15123

Sonawane, J.M., Mahadevan, R., Pandey, A., Greener, J., 2022. Recent progress in microbial fuel cells using substrates from diverse sources. *Heliyon* 8, e12353. https://doi.org/10.1016/j.heliyon.2022.e12353

Sridevi, V., Basheerunnisa, S., Venkat Rao, P., Rushi Varma, G., Pavan Srivatsav, D., 2020. Performance of salt bridge in microbial fuel cell at various agarose concentrations for treating sugar industrial effluent. *Int. J. Adv. Sci. Eng.* 06, 1513–1519. https://doi.org/10.29294/IJASE.6.4.2020.1513-1519

Srivastava, P., Yadav, A.K., Garaniya, V., Lewis, T., Abbassi, R., Khan, S.J., 2020. Electrode dependent anaerobic ammonium oxidation in microbial fuel cell integrated hybrid constructed wetlands: A new process. *Sci. Total Environ.* 698, 134248. https://doi.org/10.1016/j.scitotenv.2019.134248

Tabassum, N., Islam, N., Ahmed, S., 2021. Progress in microbial fuel cells for sustainable management of industrial effluents. *Process Biochem.* 106, 20–41. https://doi.org/10.1016/j.procbio.2021.03.032

Taha, M., Foda, M., Shahsavari, E., Aburto-Medina, A., Adetutu, E., Ball, A., 2016. Commercial feasibility of lignocellulose biodegradation: possibilities and challenges. *Curr. Opin. Biotechnol.* 38, 190–197. https://doi.org/10.1016/j.copbio.2016.02.012

Tamirat, A.G., Guan, X., Liu, J., Luo, J., Xia, Y., 2020. Redox mediators as charge agents for changing electrochemical reactions. *Chem. Soc. Rev.* 49, 7454–7478. https://doi.org/10.1039/D0CS00489H

Tanaka, K., Kashiwagi, N., Ogawa, T., 2007. Effects of light on the electrical output of bioelectrochemical fuel-cells containing Anabaena variabilis M-2: Mechanism of the post-illumination burst. *J. Chem. Technol. Biotechnol.* 42, 235–240. https://doi.org/10.1002/jctb.280420307

Taşkan, E., Ozkaya, B., Hasar, H., 2014. Effect of Different Mediator Concentrations on Power Generation in MFC Using Ti-TiO2 Electrode. *Int. J. Energy Sci.* 4, 9. https://doi.org/10.14355/ijes.2014.0401.02

ter Heijne, A., Hamelers, H.V.M., de Wilde, V., Rozendal, R.A., Buisman, C.J.N., 2006. A bipolar membrane combined with ferric iron reduction as an efficient cathode system in microbial fuel cells. *Environ. Sci. Technol.* 40, 5200–5205. https://doi.org/10.1021/es0608545

Thomas, Y.R.J., Picot, M., Carer, A., Berder, O., Sentieys, O., Barrière, F., 2013. A single sediment-microbial fuel cell powering a wireless telecommunication system. *J. Power Sources* 241, 703–708. https://doi.org/10.1016/j.jpowsour.2013.05.016

Wang, H., Ren, Z.J., 2013. A comprehensive review of microbial electrochemical systems as a platform technology. *Biotechnol. Adv.* 31, 1796–1807. https://doi.org/10.1016/j.biotechadv.2013.10.001

Wang, R., Liu, D., Yan, M., Zhang, L., Chang, W., Sun, Z., Liu, S., Guo, C., 2019. Three-dimensional high performance free-standing anode by one-step carbonization of pinecone in microbial fuel cells. *Bioresour. Technol.* 292, 121956. https://doi.org/10.1016/j.biortech.2019.121956

Wei, J., Liang, P., Cao, X., Huang, X., 2011. Use of inexpensive semicoke and activated carbon as biocathode in microbial fuel cells. *Bioresour. Technol.* 102, 10431–10435. https://doi.org/10.1016/j.biortech.2011.08.088

Wei, L., Han, H., Shen, J., 2012. Effects of cathodic electron acceptors and potassium ferricyanide concentrations on the performance of microbial fuel cell. *Int. J. Hydrogen Energy* 37, 12980–12986. https://doi.org/10.1016/j.ijhydene.2012.05.068

Yagishita, T., Sawayama, S., Tsukahara, K.-I., Ogi, T., 1998. Performance of photosynthetic electrochemical cells using immobilized Anabaena variabilis M-3 in discharge/culture cycles. *J. Ferment. Bioeng.* 85, 546–549. https://doi.org/10.1016/S0922-338X(98)80106-2

Yagishita, T., Sawayama, S., Tsukahara, K., Ogi, T., 1997. Behavior of glucose degradation in Synechocystis sp. M-203 in bioelectrochemical fuel cells. *Bioelectrochemistry Bioenerg.* 43, 177–180. https://doi.org/10.1016/S0302-4598(96)05145-8

Yaqoob, A.A., Mohamad Ibrahim, M.N., Rafatullah, M., Chua, Y.S., Ahmad, A., Umar, K., 2020a. Recent Advances in Anodes for Microbial Fuel Cells: An Overview. *Materials (Basel).* 13, 2078. https://doi.org/10.3390/ma13092078

Yaqoob, A.A., Mohamad Ibrahim, M.N., Umar, K., Bhawani, S.A., Khan, A., Asiri, A.M., Khan, M.R., Azam, M., AlAmmari, A.M., 2020b. Cellulose derived graphene/polyaniline nanocomposite anode for energy generation and bioremediation of toxic metals via benthic microbial fuel cells. *Polymers (Basel).* 13, 135. https://doi.org/10.3390/polym13010135

You, S., Zhao, Q., Zhang, J., Liu, H., Jiang, J., Zhao, S., 2008. Increased sustainable electricity generation in up-flow air-cathode microbial fuel cells. *Biosens. Bioelectron.* 23, 1157–1160. https://doi.org/10.1016/j.bios.2007.10.010

Zhang, Y., Mo, G., Li, X., Zhang, W., Zhang, J., Ye, J., Huang, X., Yu, C., 2011. A graphene modified anode to improve the performance of microbial fuel cells. *J. Power Sources* 196, 5402–5407. https://doi.org/10.1016/j.jpowsour.2011.02.067

Zhao, G., Ma, F., Wei, L., Chua, H., Chang, C.-C., Zhang, X.-J., 2012. Electricity generation from cattle dung using microbial fuel cell technology during anaerobic acidogenesis and the development of microbial populations. *Waste Manag.* 32, 1651–1658. https://doi.org/10.1016/j.wasman.2012.04.013

Zhou, M., Chi, M., Luo, J., He, H., Jin, T., 2011. An overview of electrode materials in microbial fuel cells. *J. Power Sources* 196, 4427–4435. https://doi.org/10.1016/j.jpowsour.2011.01.012

Zuo, Y., Cheng, S., Call, D., Logan, B.E., 2007. Tubular membrane cathodes for scalable power generation in microbial fuel cells. *Environ. Sci. Technol.* 41, 3347–3353. https://doi.org/10.1021/es0627601

10 Bio-Oil Production

Unlocking the Liquid Fuel Potential of Biomass

Ahmad Nawaz, Hayat Ali Haddad, Ayah Stif, and Shaikh Abdur Razzak

10.1 INTRODUCTION

One of the implications of the world's population expansion and rapid industrialization is an increase in consumption rates, which influences subsistence, products, and power. This inevitably strong participation in consumer-oriented activity generates a large quantity of trash, necessitating thorough waste management measures. Special care is being given to managing various types of waste in a responsible and sustainable manner, which includes food waste, solid municipal garbage, and agricultural byproducts (Harirchi et al., 2022). Presently, a significant worry revolves around the fact that nearly one-third of the global annual food supply is either discarded or wasted, with negative consequences for both the planet and the health of humans (Dou and Toth, 2021). Further, serving as a pivotal component and propellant for global economic advancement, the global per capita energy demand is on the rise. Presently, fossil fuel reservoirs, including coal, petroleum, natural gas, and other variables, account for approximately 90% of the world's energy requirements. The persistent use of petroleum and coal is resulting in diminishing reserves, escalating prices, and the establishment of numerous environmental pollutants. This has spurred scientists to actively pursue alternative sources of energy that are endless, cost-effective, sustainable, clean, and environmentally friendly (Aysu and Durak, 2015; Nawaz et al., 2023). Biomass is garnering attention due to its resilience as a source of energy in the context of fossil fuels, its convenience, and its environment friendly nature. It encompasses natural substances produced by plants during photosynthesis, extending to include organic waste like garbage and sewage, as well as plant life in terrestrial and aquatic environments. Current transformation methods employ just a tiny portion of the biomass (which sometimes competes with food) and provide unclear CO_2 benefits. Lignocellulosic biomass is utilized for second-generation biofuels; however, it requires more investment and produces highly costly fuels.

Organic substances may be converted into fuel using three methods: thermochemical transformation, biochemical conversion, and physical or mechanical transformation. Combustion, gasification, torrefaction, liquefaction, pyrolysis, and hydrothermal

DOI: 10.1201/9781003585398-10

treatment exemplify various thermochemical processes. Pyrolysis, a method within this category, converts biomass into higher-value products, including bio-oil, gases, char, and other concentrated organic compounds. Functioning at temperatures between 300°C and 700°C, this method produces bio-oil that includes key constituents such as hydrocarbons, phenols, alkanes, and various oxygenated compounds like ketones, esters, and ethers. Chemical processing can add worth to solid char by improving its physicochemical qualities for usage as soil additives and adsorbent for ecological purposes (Nair and Vinu, 2016). Non-condensable gases, varying with the fuel type, typically consist of lighter gases like carbon dioxide (CO_2), carbon monoxide (CO), hydrogen (H_2), and small hydrocarbons ranging from C1 to C5, including methane, ethylene, propane, ethane, propylene, butanes, and butylenes. The two critical factors that affect the corresponding outputs of the goods are temperature and residence time. Pyrolysis is classed into three sorts based on the temperature of the technique and the rate of heating: sluggish pyrolysis, rapid pyrolysis, and flash pyrolysis. Bio-oil is divided into two types: an aqueous state comprising oxygenated organic substances with low molecular weight and a non-aqueous portion comprising heavy organic substances (mostly aromatics). The components of bio-oil embody molecules of numerous sizes, attributable to the depolymerization and disintegration processes of the three essential structural elements of biomass: cellulose, hemicellulose, and lignin. The aqueous phase has an excessive concentration of oxygen, which comes from the cellulose and hemicellulose additives of biomass. The water-insoluble phase is derived from the lignin portion of biomass and has a high viscosity and low oxygen concentration (Piskorz et al., 2000). Although bio-oil is a more appealing fuel than biomass, it has some unfavorable characteristics (corrosiveness, instabilities under preservation and heating conditions, insoluble with petroleum fuels, acidic nature, high viscosity, and low calorific value) which act as elementary obstacles to further usage and commercial operations. As a result, before bio-oil can be used, its quality must be improved (Shahbeik et al., 2023; Zhou et al., 2023). Leveraging heterogeneous catalysis in biomass pyrolysis stands out as a highly promising method to enhance the quality of bio-oil by reducing undesirable characteristics or increasing the synthesis of high-value compounds. However, pyrolysis oil's value competitiveness is reduced compared to fossil fuels due to increased production costs.

Biochemical conversion is the decomposition of biomass utilizing bacteria, microorganisms, and enzymes to create liquid fuels or gases. Anaerobic digestion and fermentation are the two most frequent ways. Several processes are involved in the biological process of producing ethanol from lignocellulosic substrate. The first step is to remove cellulose and hemicellulose by dissolving the lignin complex by delignification. The second stage involves depolymerization, which breaks down polymers into monomer units to produce sugar molecules. During the third and final phase, sugar molecules, including hexoses and pentoses, undergo fermentation to produce ethanol (Afraz et al., 2024).

This section provides an overview of the liquid fuel potential inherent in biomass. Further, microbial biomass transformation into bio-oil is discussed followed by its composition, properties, and effect of processing parameters. Applications of microbial lipid synthesis using biomass pyrolysis products, carbon efficient biofuel production, bio-oil as a raw material for biochemical processes, and fermentation of bio-oil to obtain bioethanol and polyhydroxyalkanoates are also discussed. The chapter finishes with a discussion of stable biochar in microbial petrol cells, a techno-economic analysis, and an examination of prospective future advances.

10.2 MICROBIAL BIOMASS TRANSFORMATION INTO BIO-OIL

The major natural source of proteins and biofuels is microbial biomass. Bacteria, yeasts, fungi, and algae are the most common microbial populations used for biomass generation. The selection of microbes is typically influenced by factors such as rapid growth rate, efficient conversion of substrate to biomass, the ability to reach high cell densities, effectiveness in breaking down complex substrates, affinity toward the substrate, and minimal nutritional needs. Additionally, factors such as raw material accessibility, nutrient availability (including energy values, protein levels, and amino acid balance), as well as technical considerations like culture type and nutritional toxicity, can all influence biomass output (Ravi et al., 2021).

To produce bio-oil efficiently, it is crucial to prepare the biomass feedstock for optimal heat transfer rates. However, the type of raw material plays a significant role in this process. During the production of bio-oil, the structural components of biomass, such as hemicellulose, cellulose, and lignin, are broken down. The proportion and composition of these elements in lignocellulosic biomass are vital in determining how effectively it can be converted into renewable fuels or useful chemicals (Singh et al., 2023). The elemental, proximate, and fiber analysis of the different types of biomass is presented in Table 10.1.

The content of hemicellulose, cellulose, and lignin varies in different types of agricultural waste. Rapeseed, sunflowers, herbs, sugarcane bagasse, coconut shells, rice husks, cotton, and maize stalks are popular agricultural leftovers used in bio-oil production. To decide the potential of biomass for bio-oil production with a higher calorific value, it's crucial to decide its elemental composition (carbon, oxygen, nitrogen, hydrogen, and sulfur), as well as its ash and moisture content. The composition and water content of bio-oil is influenced by the initial moisture content of biomass and the properties of pyrolysis products. A high water content in bio-oil is commonly seen as a drawback for its use as a gas source (Ahmad et al., 2020; Singh et al., 2023; Welker et al., 2015).

10.3 EFFECTS OF PROCESSING PARAMETERS, BIO-OIL COMPOSITION, AND PROPERTIES

Process variables, encompassing heating rate, temperature, catalyst type, and dosage play a crucial role as they significantly influence the physicochemical properties of bio-oil throughout the pyrolysis process. Essential considerations for pyrolysis also include the moisture content and particle size of the agricultural feedstock. Among

TABLE 10.1
Physicochemical Characterization of Different Types of Biomass in Weight Percent Basis (wt%)

Biomass Feedstock	MC	VM	FC	Ash	C	H	N	O	S	Hemi	Cell	Lig	Reference
Poplar wood sawdust (Raw)	9.6	75.54	11.15	3.7	45.5	6.26	1.04	47.2	–	16.73	44.75	30.72	(Gu et al., 2014)
Phyllanthus emblica	6.56	75.20	15.55	2.69	48.76	5.91	2.01	43.31	–	21.98	48.11	5.48	(Mishra and Mohanty, 2020)
Eucalyptus sawdust	6.12	74.38	18.79	0.81	49.75	5.77	0.17	44.26	0.03	24.74	33.89	20.77	(Chen et al., 2015)
Arhar stalk	4.67	77.33	16.33	1.67	36.19	3.64	0.42	59.35	–	18.0	24.0	19.0	(Kumar et al., 2019)
Banana peels	8.53	66.79	20.55	4.13	45.43	5.67	2.31	36.40	0.35	–	–	–	(Tahir et al., 2019)
Banana trunk	6.67	74.33	7.33	11.67	33.09	2.94	0.94	63.03	–	25.0	25.0	15.0	(Kumar et al., 2019)
Barley straw	6.9	80.3	4.8	9.8	41.4	6.2	0.6	51.7	0.01	–	–	–	(Naik et al., 2010)
Cashew nutshell	10.4	69.3	19.3	1.0	48.7	6.9	0.4	42.9	–	18.6	41.3	40.1	(Das and Ganesh, 2003)
Cherry seed shells	6.1	76.1	17.0	0.8	48.9	6.3	3.1	41.6	0.1	31.9	27.2	36.9	(Duman et al., 2011)
Chestnut shells	10.17	65.55	23.08	1.20	48.14	5.47	0.6	45.79	–	22.64	31.61	42.69	(Özsin and Pütün, 2017)

C. humicola	3.42	55.6	14.18	26.8	33.16	5.58	4.8	27.54	2.42	–	–	–	(Kirtania and Bhattacharya, 2013)
C. pilulifera	10.5	32.2	18.4	38.6	–	–	–	–	–	–	–	–	(Li et al., 2011)
Microalgae *Chlorella*	6.8	72.2	15.1	5.9	6.7	38.6	–	–		47.5	9.5	7.1	(Phukan et al., 2011)
Polysiphonia elongate	11.55	48.2	12.8	27.45	35.81	5.93	6.86	51.40	–	–	–	–	(Ceylan et al., 2014)
Reeds	5.89	72.12	13.52	8.47	42.78	5.17	1.33	50.51	0.21	30.68	43.05	20.34	(Li et al., 2019)
Wolffia arrhizal	4.76	72.6	–	10.4	35.55	6.36	5.25	35.87	1.16	–	–	–	(Ahmad et al., 2018)
Kitchen garbage	–	63.5	8.8	27.8	–	4.4	2.3	5.3	0.6	–	–	–	(Luo et al., 2010)

these factors, temperature stands out as a pivotal operational parameter in thermochemical processing, exerting a substantial impact on the characteristics and yield of bio-oil. The yield of bio-oil will increase with the upward thrust in temperature and reach a stage at around 500°C. However, the yield starts to decrease beyond this temperature due to the formation of gaseous compounds that occur during the breakdown of volatile tar (Stegen and Kaparaju, 2020). A study conducted on the composition of pyrolysis oil produced from the polar wood revealed that the formation of levoglucosan, a typical sugar originating from cellulose, initially increased as the temperature rose from 350°C to 550°C. However, this production began to decline when the temperature was further increased to 1000°C (Singh et al., 2023). This phenomenon could be attributed to a trade-off between the production and depolymerization of levoglucosan. Another possible explanation is the occurrence of secondary cracking of levoglucosan at elevated temperatures. A pyrolysis study involving rice husks in a fluidized bed reactor illustrated the correlation between temperature elevation and the increased generation of carbon monoxide. In another investigation focusing on palm tree empty fruit bunches, it was observed that raising the pyrolysis temperature from 400°C to 500°C led to an increase in bio-oil output from 36.8 to 46.1 wt%, followed by a decline after reaching 800°C (Isahak et al., 2012). *Trapa natans* peel was pyrolyzed in a fixed bed reactor and Figure 10.1 depicts the reactor configuration for bio-oil production (Nawaz et al., 2023). The optimal breakdown, maximizing both mass and heat transfer, of WCP biomass resulted in the highest bio-oil production at 600°C, reaching 42.2 ± 1.2%. Earlier studies have identified hemicellulose and cellulose as pivotal factors influencing bio-oil yield (Quan et al., 2016). The rate of heating significantly affects the biomass to bio-oil conversion. Swift heating maximizes bio-oil production, while slower heating tends to favor char formation. Nair and Vinu (2016) demonstrated the notable impact of the heating rate on this process. The complete feedstock breakdown led to a peak bio-oil yield of 42.2% at a heating rate of 25°C/min. An escalation in the heating rate resulted in a marginal decrease in oil output; however, the production of gases increased owing to the rapid degradation of WCP biomass. Moreover, at lower heating rates of 10 and 15°C/min, incomplete pyrolysis occurred, resulting in elevated biochar production and reduced bio-oil and non-condensable gas generation due to partial mass and heat transfer among feedstock elements. The elemental composition and physicochemical properties of agricultural residues serve as crucial determinants of biomass potential for bio-oil conversion. However, the selection of feedstock is predominantly influenced by the thermochemical method employed. In contrast to herbaceous feedstock, agricultural biomass with a high lignin concentration (woody biomass) proves advantageous for pyrolysis, yielding higher outputs while minimizing catalytic poisoning and equipment corrosion.

The residences of bio-oils produced through pyrolysis are appreciably distinctive from those of traditional mineral oils. The composition and characteristics of bio-oil are influenced by several factors, including the type of biomass used as feedstock, the moisture content in the feed, process parameters such as vapor residence time, temperature, and pressure, reactor type, recovery system design, and the scale of operation. All these factors contribute to the unique attributes exhibited by bio-oils,

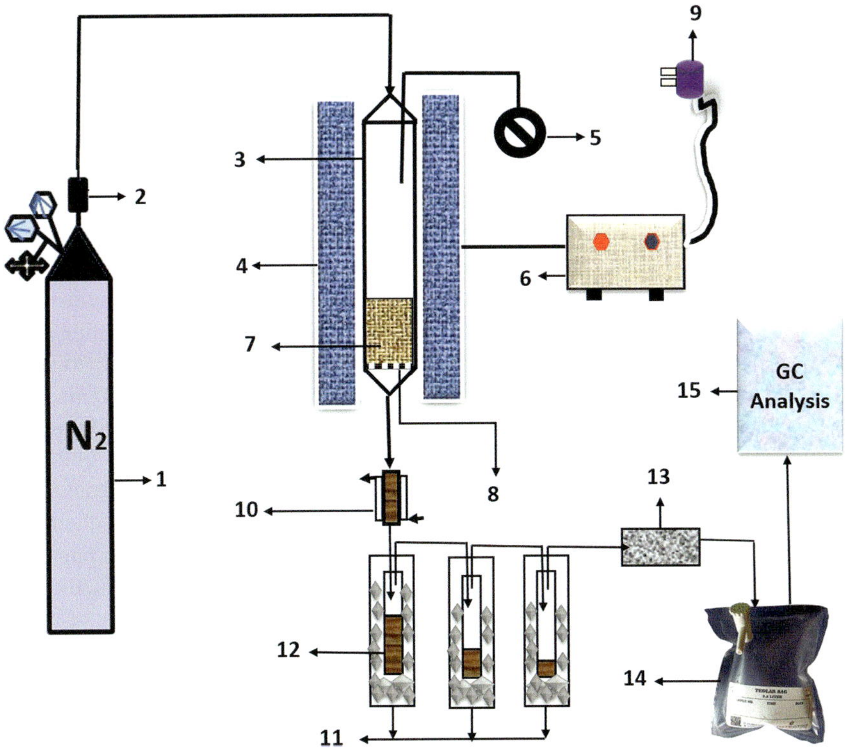

FIGURE 10.1 Experimental setup for pyrolysis as described in the reference (Nawaz et al., 2023).

highlighting the complex and multifaceted nature of the pyrolysis process and its impact on the final product (Solantausta et al., 1993).

Bio-oils exhibit elevated levels of moisture, suspended particles, char, oxygen, and acidic chemicals, resulting in modest heating and calorific values. In contrast to mineral oils that contain oxygen at ppm levels, bio-oils have a significantly higher oxygen content, ranging from 35% to 40%. This oxygen is present in various organic functional groups within the bio-oil. These functional groups contribute to the bio-oil's markedly polar character. The distinctive composition of bio-oils, with their unique combination of elements and organic functional groups, underscores the challenges and opportunities associated with their utilization as a renewable and sustainable energy source (Lehto et al., 2014). Due to the existence of polar hydrophilic functional groups, water forms associations with lignin oligomers present in biomass. The primary component in bio-oil is moisture, and its content can vary from 15% to 30%, contingent on the specific type of feedstock employed. This linkage between water and lignin oligomers, along with fluctuations in moisture levels, contributes to the distinctive composition of bio-oil, emphasizing the influence of feedstock type on its characteristics (Park et al., 2012; Xiao and Yang, 2013). Pyrolysis produces moisture as a by-product of dehydration reactions. However, storing pyrolysis oil at room temperature can cause it to age and form high molecular weight molecules and water due to various reactions such as esterification, a reaction between hydroxyl and carbonyl groups. The presence of char particles can further accelerate these aging processes, leading to an increase in the viscosity and molecular weight of pyrolysis oil. However, the addition of alcohols like ethanol and methanol to pyrolysis oil can help regulate these processes. This highlights the dynamic nature of pyrolysis oil and the possibility of controlled manipulation to improve its stability (Oasmaa and Peacocke, 2001). A key distinction between bio-oils and traditional hydrocarbon fuels is the presence of oxygen in bio-oils. This leads to a lower energy density, which is typically less than 50% of that of conventional fuels, and also results in bio-oils being immiscible with these conventional fuels. The literature documents the presence of up to 35–40% oxygen concentration in over 300 compounds within bio-oils. This high oxygen content contributes to the unique properties of bio-oils, influencing their energy density and compatibility with traditional hydrocarbon-based fuels. The challenge posed by oxygen content underscores the need for innovative approaches to enhance the utility and integration of bio-oils in existing energy systems (Czernik and Bridgwater, 2004). Bio-oils are made up of a variety of chemicals, such as carboxylic acids, hydroxyaldehydes, hydroxyketones, sugars, and phenolics. Bio-oil's viscosity can vary from 35 to 1000 cP at 40°C based on various factors such as the feedstock type, pre-processing and processing conditions, and the effectiveness of collecting components with low boiling points. The intricate composition and broad viscosity range highlight how crucial it is to take into account a variety of factors when producing and refining bio-oils (Czernik and Bridgwater, 2004). Due to chemical interactions among the elements within bio-oil, its viscosity tends to increase with storage time. This phenomenon underscores the impact of chemical changes on the physical properties of bio-oil over time, emphasizing the importance of monitoring and managing these interactions to maintain their stability and performance (Czernik et al., 1994). Bio-oil, which is produced from biomass, may contain carboxylic acids such as acetic acid and formic acid. These acids make the pH of bio-oil acidic, typically in the range of 2–3. This acidity can cause corrosion of equipment and storage containers, especially in high-temperature and moisture content conditions, respectively. Furthermore, bio-oil also contains varying levels of ash that come from the source feedstock. The ash is mainly composed of alkali metals such as sodium, potassium, and vanadium. These chemical properties make bio-oil corrosive and can lead to the presence of ash components, which pose a challenge for proper handling, storage, and processing. To avoid negative impacts on equipment and system performance, it's crucial to handle, store, and process bio-oil carefully, considering its corrosive and potentially ash-containing nature.

10.4 MICROBIAL SUBSTRATES FOR BIOFUEL PRODUCTION

The successful generation of biofuel from lignocellulosic waste materials depends on identifying and utilizing an appropriate fermentative microbial strain. This strain should have the ability to efficiently use both hexose and pentose-rich sugar moieties derived from lignocellulosic feedstock to produce biofuel. Additionally, it should be resilient in the presence of inhibitory chemicals that might be generated during the pretreatment phase. The microbial strain's ability to withstand inhibitory substances is crucial for ensuring a robust and efficient biofuel production process from lignocellulosic materials (Bilal et al., 2018). Bacterial fermentation processes often encounter difficulties in efficiently assimilating five-carbon sugars. Even in cases where fermentation is possible, the process is hindered by the production of unwanted by-products and end products. This challenge highlights the complexity of bacterial fermentation with regard to certain sugar types and emphasizes the need for strategies to enhance the efficiency and selectivity of these processes. Overcoming these obstacles is essential for optimizing bacterial fermentation and improving its overall effectiveness in biofuel production and other applications (Canilha et al., 2012). *Zymomonas mobilis* and *Saccharomyces cerevisiae* are well-known microbial strains capable of fermenting sucrose and hexose carbohydrates into ethanol. However, their progress is impeded by the accumulation of end products (Moysés et al., 2016). Additionally, in the presence of end products, pentose-fermenting platform microorganisms such as *Pichia stipitis*, *Candida shehatae*, and *Pachysolen tannophilus* are also inhibited (Chandel et al., 2010; Martiniano et al., 2013; Senatham et al., 2016). While filamentous fungi can withstand interfering chemicals, their extended development period, low productivity, and yields make them unsuitable for biofuel production. This highlights the ongoing challenges in selecting microbial strains for efficient and productive biofuel generation, taking into consideration their ability to withstand inhibitory substances and optimize overall process performance (Chandel et al., 2010). The target strain may be a naturally occurring cellulose-hydrolyzing microbe that is skilled in producing biofuels, or it may be a genetically modified candidate that carries the necessary genes for effective biofuel production. In lignocellulosic hydrolysates, this strain ought to demonstrate increased cell mass growth and biofuel synthesis efficiency. Optimizing biofuel production processes requires the selection of a robust and versatile microbial strain, especially in the complex environment of lignocellulosic feedstocks (P. Li, Fu, et al., 2019; X. Li, Xia, et al., 2019; McMillan and Beckham, 2017). This strain should exhibit high metabolic fluxes and follow a rapid route leading to a single fermentation product. Operating at higher temperatures and effectively managing bioreactors can enhance reaction efficiency, reduce culture broth viscosities, and decrease the risk of contamination during the manufacturing process. Additionally, a strain with reduced pH adaptation capacity can contribute to minimizing contamination from various interfering bacteria. These attributes collectively underscore the importance of selecting a microbial strain with optimal characteristics for biofuel production, emphasizing efficiency, robustness, and the ability to withstand challenging conditions in the production environment (McMillan and Beckham, 2017). All of these characteristics should be considered by researchers and metabolic scientists while establishing the best microbial strain for industrial-level production of biofuels (Figure 10.2).

Metabolic engineers, synthetic biologists, and transcription factors responsive to metabolites for industrial applications. Master regulator transcription factors (MRTFs) can be self-adapted by changing the architecture of the cell chassis. Using this method, Xu et al. (2014) developed a natural transcriptional regulator called FapR to control the *E. coli* fatty acid synthesis pathway. Effective gene expression regulation led to an implicit improvement in the production of fatty acids by

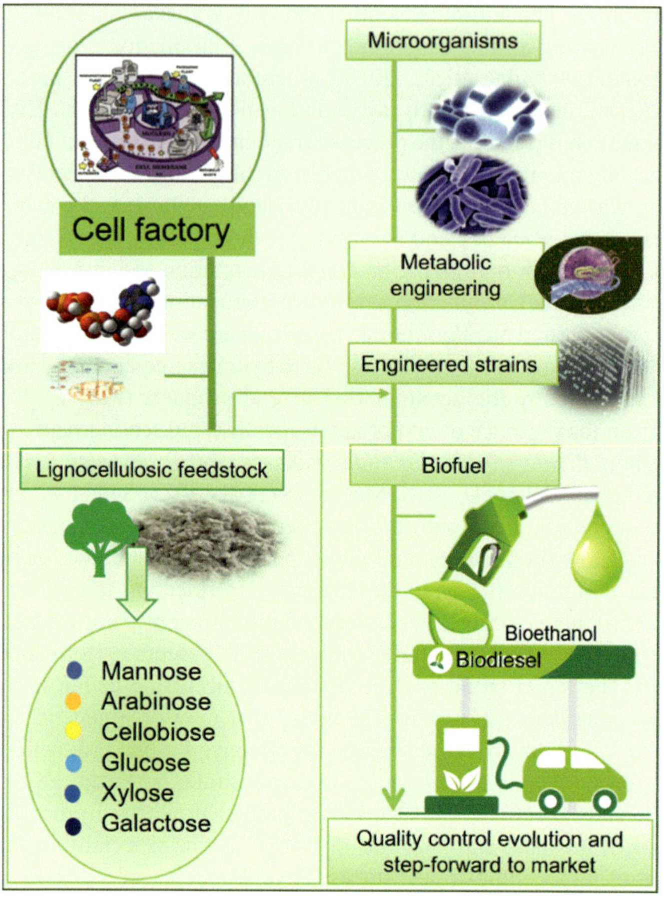

FIGURE 10.2 Specially designed microbial cell factory engineered to convert lignocellulosic biomass into biofuels (Lu et al., 2022).

promoting a balanced metabolism between cell development and product generation. The enzyme activity of a microbial strain is influenced by changes in post-translational proteins, translation, and gene expression. Even though an individual biosystem's enzymatic characteristics are unique, studying the characteristics of heterologous enzymes is helpful when introducing them to change metabolic fluxes (Xu, 2018).

10.5 MICROBIAL LIPID SYNTHESIS USING BIOMASS PYROLYSIS PRODUCTS

Pyrolysis produces liquid and gaseous by-products that can be used as ideal carbon sources for microbial fermentations. Although solid by-products cannot be used for this purpose, they are valuable due to their absorption qualities. However, fermenting bio-oil using microorganisms is a complex process because of the oil's chemical diversity. Often, the specific toxic effects of bio-oil components on microorganisms are either ignored or deduced from studies on model organisms. While some bio-oil components are toxic to certain microbes, others can withstand and even utilize these components in their metabolic pathways. Anaerobic bacteria are particularly efficient in this regard, as they have the ability to ferment non-condensable gases (Neumann et al., 2015). Three important compounds that are frequently present in the light phase of bio-oil and can cause damage to many fungi are hydroxyacetone, hydroxyacetaldehyde, and phenolic compounds (Dörsam et al., 2016). On the other hand, acetic acid increases the synthesis of lipids in conditions where nitrogen is scarce. As a result, due to the microbial uptake of components found in bio-oil, separation and detoxification procedures must be implemented after utilization (Neumann et al., 2015). Modern oxidation techniques, like catalytic wet air oxidation, offer a straightforward method because they can convert bio-oil's light phase directly into acetic acid that is readily fermented in a single step (He et al., 2020).

Shafizadeh and Stevenson (1982) were the first researchers who investigated *S. cerevisiae* to produce ethanol. Further Lian et al. (2010) were the first to introduce the application of anhydrosugars obtained from pyrolysis via the microbial fermentation processes. Their study looked at the detrimental effects of different bio-oil constituents on the fermentation process and discovered that furans and alkanes were equally as harmful as carboxylic acids and phenols. A particular study used the oleaginous yeasts *Rhodotorula glutinis* and *Cryptococcus curvatus* to produce lipids from pyrolytic anhydrosugars. Initially, a liquid-to-liquid extraction procedure was used to separate the bio-oil's anhydrosugar-rich aqueous phase. Sulfuric acid was then applied to break down the sugars. Lastly, the procedure involved filtering through activated carbon to remove dangerous materials. The production of lipids was demonstrated by combining glucose with upgraded pyrolyzate (Luque et al., 2016). The upgraded pyrolyzate was obtained by chemically hydrolyzing anhydrosugars and then going through a three-step detoxification process. The effects of nitrogen limitation in media with glucose as the sole carbon source were further investigated in this study. They noticed that the dry cell weights in media with low carbon-to-nitrogen (C/N) ratios were constant, averaging 12 g/L in both cases. However, lipid productivity significantly increased in media with high C/N ratios (about 0.020 and 0.068 g/(L h), respectively). The idea that a deficiency in nitrogen leads to fat accumulation was supported by this outcome. Reductions of over two and ten times, respectively, in dry cell weight and lipid production were observed when enhanced pyrolyzate was combined with a glucose medium with a high C/N ratio. It was found that in low-C/N ratio environments, the yield was higher, reaching up to 17 grams of dry cell per liter and 0.027 g/(L h). These findings indicate that nitrogen limitation should be circumvented when glucose and improved pyrolyzate are used together as a carbon source. The fats made using pure high-quality biodiesel require both a low cetane number (CN) and cold flow plugging point (CFPP), which can be achieved when glucose is utilized in a low C/N medium. Comparatively, under the same conditions, biodiesel made by blending enhanced pyrolyzate, which mostly contained oleic acid, displayed a lower CFPP but a slightly higher CN. In spite of this, the results obtained may still be appropriate for low-emission biodiesel, especially in warmer regions.

The primary short-chain organic acid generated by biomass pyrolysis is acetic acid. Additionally, propionic acid and formic acid in smaller amounts are produced (Pinheiro Pires et al., 2019). It is crucial to differentiate these from volatile fatty acids, which are carboxylic acids with a carbon chain length ranging from C2 to C6. These acids are produced by anaerobic digestion of organic waste that contains little lignin (Kim et al., 2018). Lian et al. (2012) studied the accumulation of lipids in oleaginous yeast that was obtained from bio-oil fractions rich in C1–C4 chemicals. In the beginning, they fermented single commercial model molecules in bio-oil aqueous phase to distinguish between appropriate and hazardous components. They created a method that separated the bio-oil's aqueous phase using fractional condensation and then enhanced it by progressively removing its hazardous components. Following fermentation with *C. curvatus*, an aqueous phase containing high quantities of formate and acetate was formed, which was subsequently diluted accordingly. Xu et al. (2017) employed *Y. lipolytica* MTYL065 to explore lipid accumulation in oleaginous yeast produced from C1–C4 chemical-rich bio-oil fractions. In the beginning, they fermented single commercial model molecules in bio-oil aqueous phase to distinguish between appropriate and hazardous components. They used a tangential flow filtration system for cell recycling and modified the nitrogen and acetate feeds based on respiratory quotient monitoring. This method helped in creating high-density microbial cultures with a mere 3% acetate concentration. As a result, they were able to produce 115 grams of lipid per liter, yielding 0.16 grams per gram and achieving a productivity of 0.8 grams per liter per hour. Furthermore, this procedure produced lipid yields that were almost at the predicted maximum of 0.16 in addition to effectively utilizing acetate.

10.6 FERMENTATION OF BIO-OIL TO OBTAIN BIOETHANOL AND POLYHYDROXYALKANOATES

Pyrolysis oil has various applications in microbiological procedures. The pyrolysis of specific sugars, like levoglucosan, the fermentation of raw or crude pyrolysis oil, and the fermentation of sugars that have been acid-hydrolyzed, like glucose, are some examples of these. The elimination of preprocessing steps is one advantage of directly fermenting pyrolysis oil. It's important to keep in mind that at high concentrations, pyrolysis oil can be harmful to bacterial cultures. Most microbes cannot use levoglucosan directly, but some are able to metabolize it. Plant species include *Cryptococcus*, while fungal species include *Aspergillus niger*, *Penicillium*, *Chrysogenum*, *Phanerochaete chrysosporium*, and *Candida utilis* (Jiang et al., 2019).

Prosen et al. (1993) conducted a study on 12 strains, including eight yeast and four fungal strains, to assess their capability to utilize pyrolysis oil for carbon and energy. They experimented with three distinct fractions of pyrolysis oil: the aqueous fraction treated with activated carbon, the fraction hydrolyzed using 2% v/v sulfuric acid, and the aqueous fraction obtained by blending pyrolysis oil with water at a 1:2 ratio. The investigation discovered that the initial aqueous fraction did not support the growth of the fungal strains. On the other hand, five yeast strains were successful in growing on each of the three pyrolytic fractions. Concurrently, no *S. salmonicolor* or *C. albidus* growth was observed in any of the pyrolysate samples. The components of the pyrolysis oil can be metabolized by yeast and fungus strains, as per this study. However, due to water-soluble aromatics derived from lignin, aqueous pyrolysis oil is more detrimental to fungal strains than yeast strains. Remarkably, it was discovered that some strains of yeast were unable to break down the hydrolyzed and detoxified pyrolysate. This suggests that a critical step in the microbial conversion process of pyrolysis oils is selecting the appropriate microbial strain or pre-adapting them. A thorough summary of the different microbial products that can be obtained from the various pyrolysis oil fractions is given in Table 10.2.

One innovative method for creating bio-ethanol could be through the fermentation of bio-oil. First, treated lignocellulosic biomass is pyrolyzed quickly to produce bio-oil rich in pyrolytic sugars. Subsequently, the bio-oil's sugars can undergo direct bio-oil fermentation, which yields bio-ethanol, or indirect bio-oil fermentation, which first converts the sugars into glucose before fermenting them. The operational and capital costs associated with the indirect conversion of bio-oil to ethanol were assessed. Their results indicated that the biochemical conversion of lignocellulosic biomass into ethanol is not as cost-competitive as fast pyrolysis followed by indirect fermentation, which costs $0.42 per liter of ethanol ($2.38 per gallon of gasoline equivalent) (So and Brown, 1999). The amount of ethanol generated from pyrolysis oil is significantly influenced by the type of biomass used. For instance, ethanol yields from pyrolysis sugar solutions derived from poplar and pine biomasses varied significantly under the same processing conditions, with poplar yielding only approximately 4 mg/L and pine yielding about 70 mg/L (Chi et al., 2019). The principle motive could be the better awareness of fragrant hydrocarbons, like anisoles, within the poplar pyrolytic sugar solutions. In a special examination (Yu and Zhang, 2003), *S. cerevisiae* produced 40.2 g/L ethanol yields after being pre-tailored to pyrolysis sugar hydrolysates for 12 trials. This was 47% better than the consequences from the prior assessments. Chi et al. (2013) discovered that pyrolytic sugar concentrations greater than 0.25 wt% dramatically suppressed the synthesis of ethanol. Similar outcomes were mentioned by Chi et al. (2019), who reported that no ethanol generation was observed when the pyrolytic sugar content exceeded 0.25 wt%. The strain of *E. coli* KO11 + *lgk* was employed in both experiments. In a different study, Wang et al. (2012) found that yeast growth was completely inhibited by pyrolysis sugar hydrolysates in YPD medium at concentrations greater than 10% (v/v). When the concentration was higher than 5% (v/v), significant inhibition was also observed. Pyrolysis sugar fermentation was significantly impacted by the number of yeast cells present in the fermentation medium.

Polyhydroxyalkanoates (PHA), a type of polyester used in bioplastic production, are of significant interest in terms of environmental and industrial relevance. Moita Fidalgo et al. (2014) conducted a study on microbial PHA creation using pyrolytic sugars derived from chicken bedding pyrolysis oil. They achieved a maximum concentration of 9.8% PHA using unprocessed pyrolysis oil. The monomer breakdown was 75% hydroxybutyrate (HB) and 25% hydroxyvalerate (HV). Additionally, through a two-step process, PHA can be produced from pyrolysis oil. First, the oil is used as a feed for mixed microbial cultures to produce volatile fatty acids (VFA),

TABLE 10.2
Compounds Derived from Fermentation of Biomass-Derived Bio-Oil Constituents

Biomass-Pyrolysis Oil Fraction	Strain	Productivity/Yield	Product	Reference
Pine-pyrolytic sugars	*E. coli* KO11 + *lgk*	70 mg/L	Ethanol	(Chi et al., 2019)
Loblolly pine hydrolyzed pyrolysis oil aqueous phase	*S. pastorianus* ATCC 234	0.4 g ethanol/g glucose (80% of theoretical yield)	Ethanol	(Wang et al., 2012)
Waste cotton-acid hydrolyzed pyrolysate	*S. cerevisiae* 2.399	0.43 g ethanol/g glucose	Ethanol	(Yu and Zhang, 2003)
Red oak- pyrolytic sugars	*E. coli* LJS1	37 mg/L	Styrene	(Lian et al., 2016)
Chicken beds pyrolysis oil		0.098 g of PHA/g of cell dry weight	PHA	(Moita Fidalgo et al., 2014)
Douglas fir-pyrolytic levoglucosan aqueous phase	*R. glutinis*	0.78 g/L	Lipids (mainly palmitic and oleic acids)	(Lian et al., 2013)
Rice husk-aqueous fraction of pyrolysis oil	*E. coli* MG 165	2.42 g/L	Succinic acid	(Wang et al., 2013)
Demineralized pinewood-pyrolytic sugars	*R. diobovatum*	26.9 mg/(L h)	Lipids	(Luque et al., 2016)
Cotton cellulose pyrolytic sugars	*A. niger* CBX-209	1.9 g/L	Citric acid	(Zhuang et al., 2001)

which are subsequently fermented to produce PHA. Lian et al. (2016) did the production of styrene using pure levoglucosan as well as pyrolytic sugars extracted from red oak pyrolysis oil, which contained 4 g/L levoglucosan. The results showed that the fermentation medium containing 5 g/L of levoglucosan, glucose, and pyrolytic sugars produced similar amounts of styrene. Because the LGK enzyme was missing, the *E coli* strain NST74—which was used to produce styrene—was unable to metabolize levoglucosan. The same study used *E coli* LJS1, a genetically modified strain, to address this.

10.7 UTILIZATION OF BIO-OIL

The use of bio-oil in boilers instead of heavy fossil fuels has a large positive environmental impact. It results in reduced NO_x, SO_x, and CO_2 emissions, which helps to reduce air pollution. This benefit may remove the need for additional procedures to reduce SO_x and NO_x emissions in boiler systems (Hou et al., 2016). One instance is the co-combustion of 2.5% bio-oil with heavy fuel oil, which led to reductions in NO and SO_2 levels of about 2.6% and 7.9%, respectively. This decrease is explained by biomass's low levels of sulfur, especially sulfur species (Lehto et al., 2014), and nitrogen (leaf excluded). Table 10.2 compares the general characteristics of conventional fossil fuels and bio-oil.

It is clear from analyzing Table 10.2 data that bio-oil doesn't have the same fuel qualities as heavy fossil fuels. One important distinction is that bio-oil has a higher water content (20–30%) than heavy fossil fuels (0.32 wt%). When compared to heavy fossil fuels which have a heating value of 40.63–42.39 MJ/kg, bio-oil has a significantly lower heating value of 13–19 MJ/kg. This suggests that it is impossible to directly burn bio-oil in burners made for heavy fossil fuels. Additionally, the pumping and storage systems for bio-oil in boiler applications need to be modified because it has a lower pour point, flash point, and viscosity compared to conventional heavy fuels. Sweden's Arsta District Heating Plant holds the distinction of being the first in Europe to use bio-oil for large-scale heating. Originally designed to run on fossil fuels, a 9 MW boiler has been modified to run on various kinds of bio-oils. Based on their experience, bio-oil can successfully take the place of fossil fuels in large-scale boiler systems and still achieve acceptable combustion performance (Lehto et al., 2014). Fortum, a firm, utilized bio-oil instead of light fuel oil in a 400 kW boiler experiment. In each experiment, they burnt more than 12 m^3 of bio-oil and solely utilized bio-oil as boiler fuel for over 1500 cycles. Their goal was to maintain less than 0.1% weight of solids and 0.03% weight of sand in the boiler wall. According to their findings, NO_x emissions have decreased to an acceptable level (between 100 and 150 ppm) (Lehto et al., 2014). Moreover, they found that a unique bio-oil burner system was needed, which came at a higher cost than standard burners for fossil fuels (Lehto et al., 2014). In Finland, Fortum may burn over 40 tons of biofuel in a different heating facility with a 1.5 MW heat capacity (Lehto et al., 2014). Valtion Teknillinen Tutkimuskeskus (VTT) in Finland investigated different types of bio-oils in a boiler that could produce up to 4 MW of heat. They discovered that employing the best possible nozzle design, air to bio-oil flow rate ratio, and bio-oil pumping can greatly increase boiler efficiency. Furthermore, the properties of bio-oil such as viscosity, water and solid matter content, and so forth have a significant influence on combustion efficiency (Chiaramonti et al., 2014). The lack of any additives was the most amazing aspect of these experiments. Moreover, different bio-oils had different degrees of flammability and release. When compared to mineral oil combustion, bio-oil ignition produced a larger flame and lasted longer. The amount of release increases with bio-oil's moisture content and variability (Chiaramonti et al., 2014; Lehto et al., 2014).

Based on the findings of previous research, it is reasonable to assume that pyrolysis oil may be utilized as a fuel for huge engines and boilers. However, in order to use bio-oil as fuel, conventional burner systems must be modified due to their distinct properties from fossil fuels. For instance, adjustments must be made to the air-to-fuel ratio. Additionally, materials that will not corrode when exposed to the acidic elements in the bio-oil should be used for the engine, burner, and other parts that come into contact with the bio-oil. Reduce the amount of solid particles in the bio-oil and/or design a filtration device to collect ash to prevent clogging in the combustion chamber. In recent years, several companies have developed a novel burner technology for the thorough purification of bio-oil derived from pyrolysis, one of which is clean combustion. However, because of the high setup costs, they cannot be used in large-scale applications. Thus, there needs to be more research done in this field. Broadly speaking, bio-oil can be viewed as a possible fuel for boilers; nonetheless, modifications to the structural and mechanical designs of existing large engines are required. To prevent blockage in the burners' nozzles and sprayers, a mixture of bio-oil and fossil fuels such as coal, diesel, and natural gas was used (Chiaramonti et al., 2014). Additionally, co-firing with fossil fuels may increase engine efficiency while lowering the cost of burner system and engine body changes in comparison to direct combustion of bio-oil (Chiaramonti et al., 2014; Lehto et al., 2014). Diesel and wood pyrolysis produced bio-oil, which Ormrod Diesels fed into a diesel engine (Chiaramonti et al., 2003; Czernik and Bridgwater, 2004). There won't be any detrimental effects on the engine's performance after 400 hours of operation. Nevertheless, deposits gathered in the injectors and feeding pump. The Italian scientists looked into blending diesel and 50% bio-oil. They discovered that the bio-oil had poor miscibility with diesel and partially damaged the injectors. The engine's combustion effectiveness did not change (Czernik and Bridgwater, 2004). Bio-oil could co-fire with other fossil fuels at low co-feeding volumes of 5% without visibly damaging the furnace. However, the burner system will need to be modified, which could be costly, due to the high co-feeding levels (>5%). Further research is needed in a long-term application to optimize the co-feeding ratio while considering modification costs. It is also necessary to develop a new economic burner to support higher co-feeding ratios (Hu and Gholizadeh, 2020).

10.8 UTILIZING SOLID BIOCHAR IN MICROBIAL FUEL CELL SYSTEMS

Because of its superior qualities, biochar offers a reliable substitute for conventional electrodes in the production of microbial fuel cells (MFCs) that are incredibly efficient. It is perfect for electrodes in MFC systems due to its large surface area and high porosity (You et al., 2021). Moreover, 40–50% of the overall output cost of most commercially available MFC systems is attributed to the cost of the electrodes used in the system. Therefore, it is strongly recommended to utilize biochar, a material rich in carbon obtained by heating waste biomass to very high temperatures, a process that can occur in the presence or absence of oxygen (Zhang et al., 2009). Biochar is commonly employed in MFCs as a practical and cost-effective alternative to traditional anode materials such as graphite and granular activated carbon. Furthermore, the cost of power output using biochar ranges from $17 to $35 per watt, significantly cheaper – by about 90% – than the costs associated with graphite ($392 per watt) and activated carbon ($402 per watt) (Qian et al., 2015). In a different study, biochar produced by carbonizing mills and forestry wastes was used as an anode in an MFC. The anode made of biochar produced more electrons with less input than anodes made of graphite or activated carbon (Huggins et al., 2014). The results of the study showed that when temperatures were raised, the pore density increased, increasing the anode's surface area and facilitating power generation. The goal of the research was to use straightforward carbonization techniques to create conductive electrode materials using three distinct organic materials: maize stem, king mushroom, and wild mushroom. Using biofilm electroactivity and an electrochemical redox probe ($[Fe(CN)_6]^{3-/4-}$), the relationship between the electrode's structure and reactivity was tested. The electrochemical and bioelectrochemical traits of the carbonized electrodes have been assessed using techniques including inductance, cyclic voltammetry (CV), and chronoamperometry. Key parameters like electron-transfer fee, price-transfer resistance, oxidative energy output, and electroactive components were investigated. The results showed that the electrocatalytic current produced by these electrodes was up to eight times higher than that of a standard graphite electrode. (Chakraborty et al., 2020). It was also shown that the power output of MFCs is inversely related to the size of the biochar particle. The negatively charged surface of biochar might inhibit the adhesion of biofilms. The hydrodynamic properties of MFC anodes have recently improved with the addition of nanomaterials. Enhanced surface area, enhanced electron transport, and enhanced electroactive biofilm development are some of these improvements. Additionally, the greater microbe-electrode contact increases the production of direct current. Adopting nanomaterial-based anodes has several disadvantages, though, including challenging production methods and quality degradation. Biochar is an excellent source of anode materials in many ways (Yaqoob et al., 2021). To enhance the application of biochar as an anode material, it's crucial to adopt a novel viewpoint by thoroughly investigating the electrophysiological characteristics of the electrode.

In the meantime, cathode material plays a significant role in the design and functionality of the system, and cathode efficacy is essential for MFC power generation (Huggins et al., 2014). Due to its increased porosity and surface area, biochar can function effectively as a cathode catalyst, aiding the movement of H_2 ions and O_2. This significantly improves the oxygen reduction reaction (ORR) (Ma et al., 2020). Numerous investigations have established the good-sized electrocatalytic hobby of biochar which is generated at excessive temperatures or following chemical activation. Moreover, biochar produced at decreased temperature has less electroactivity because of the presence of unpyrolyzed organic additives (Lee et al., 2019). Among its many noteworthy physical and chemical characteristics are its large surface area, high cation-exchange capacity, and porosity. Biochar is an easily accessible carbon-based material. These characteristics improve the connection between significant nutrient cycles and microbial growth (Allam et al., 2020). Currently, available affordable biochar produced from agricultural waste exhibits ORR activity. Compared to metal electrocatalysts, carbon electrocatalysts are more affordable and environmentally friendly (Chakraborty et al., 2020). In order to create PEMs that are both affordable and highly efficient for use in MFCs, numerous studies have been carried out recently. A novel proton exchange membrane (PEM) named G-5 has been developed for use in MFCs as a separator. Its primary components are 5% goethite and natural clay. Utilizing the G-5 membrane led to enhancements in chemical oxygen demand (COD) removal and the coulombic efficiency of the MFC by 22% and 9%, respectively, in comparison to Nafion-117. Additionally, this innovation resulted in a five-fold reduction in cost (Das et al., 2020). A synthetic membrane constructed from ceramic and charcoal exhibits a lower oxygen diffusion coefficient compared to Nafion membranes. In terms of cost, a PEM made from charcoal and ceramic materials is projected to be more economical, estimated at around $\$45/m^2$, as opposed to Nafion-117's cost (Das et al., 2020). While biochar, even after being subjected to sulfonation, doesn't reach the performance levels of Nafion, its significantly reduced cost is a notable advantage (Neethu et al., 2019). Biochar's suitability for PEM applications is attributed to its high porosity, plentiful surface-active sites, and strong cation-exchange properties.

10.9 TECHNO-ECONOMIC ANALYSIS

It is necessary to comprehend the metabolic activity of numerous microbial species in a synergistic setting. Microbial strains that have undergone genetic modification may simultaneously increase product production and make processes more feasible. Currently, there is a lot of interest in the fermentation of ethanol from oil sugar fractions obtained during pyrolysis. Shemfe et al. (2015) estimated that at a biomass feeding rate of 72 MT/day, the plant would eventually produce biofuel at a cost of £6.25 per gallon. According to Ringer et al. (2006), it was concluded that the ultimate cost of producing bio-oil was $7.62/GJ. It was stated that this cost exceeded that of fossil fuels like gasoline, diesel, and so forth. According to Wright et al. (2010), the cost per gallon of producing bio-oil was estimated to be $2.11. Corn stover was their diet, and they fed 2000 MT of it every day. In summary, research predicts that the final production costs for biofuels and bio-oil would be higher than the current pricing of fossil fuels such as heavy fuel oil, diesel, and petrol. Lignocellulosic waste is the type of agricultural waste that is most frequently used to produce bioethanol. Massive volumes of lignocellulosic wastes are produced worldwide. The main feedstocks being researched for the production of bioethanol are sugarcane bagasse, rice straws, wheat straws, and maize stover (Medina Jr, 2020). From $1.5 to $150 per gallon, pyrolysis oil can be used to produce ethanol. Changes in the specified process parameters are the cause of the significant variation in production costs. The preprocessing of sugars used in pyrolysis and the chemicals introduced to the fermentation medium also significantly affected the cost of producing ethanol. Hydrolyzed levoglucosan fermentation raised the cost of producing ethanol by 26% in comparison to direct levoglucosan fermentation (Claypool and Simmons, 2016). The pyroligneous fraction, which has a heating value of 46 MJ/kg, may sell for $2.90 per gallon, according to reports. It is more economically feasible to produce compounds from pyrolysis oil than it is to produce electrical energy (Medina Jr, 2020).

10.10 FUTURE PERSPECTIVES

Bio-oil has attracted a lot of attention as a possible renewable energy source and is being used in engines, gas turbines, boiler fuels, and the chemical industry. Pyrolysis is the method of choice for commercial applications due to its ability to yield a significant amount of high-grade bio-oil. Using hydrothermal liquefaction (HTL), significant amounts of organic waste can be converted into profitable oil in an inventive way. In contrast to pyrolysis, the field of HTL is still infancy. Additional research is necessary in a number of areas, such as reaction kinetics, energy efficiency, production rate and quality, and process scalability. However, bio-oil isn't always suited for direct use; it has excessive oxygen and moisture contents, a low heating value, and high viscosity, making it less exceptional than fossil fuels. Several technological developments, including emulsification, hydrodeoxygenation, catalytic cracking, hot vapor filtration, solvent addition, esterification, supercritical fluid, etc. have been implemented to raise the bio-oil quality. It's significant to highlight that the majority of current research has focused on conventional methods like solvent extraction and adsorption. The existing techniques do not effectively remove inhibitory compounds from pyrolysis oil fractions, making them either expensive or ineffective. Lowering the microbial toxicity of the pyrolysis oils may be possible through the use of two-step fermentation or co-fermentation with different strains, allowing for multiple or co-production. Further research could focus on various combinations of microorganisms that consume various types of pyrolysis oil chemicals simultaneously or sequentially. Understanding the metabolic activity of multiple microbial species in a synergistic environment is essential. Genetically modified microbial strains have the potential to both improve process viability and boost product output. Currently, there is a lot of interest in the fermentation of ethanol from oil sugar fractions obtained during pyrolysis. Research is necessary for the development of a variety of new chemicals. For instance, certain microbial strains have the ability to metabolize substrates rich in acetate and produce a range of biochemicals. In this regard, the aqueous component rich in acetic acid, which results from pyrolysis, could be significant.

10.11 CONCLUSIONS

Pyrolysis can be used in conjunction with a variety of microbial processes, including algal cultivation, MFCs, fermentation, and anaerobic digestion. Compared to traditional biochemical conversion processes, pyrolysis integrated microbial technologies typically have lower product yields and production rates. Moreover, improving the interplay between chemical and biological conversion mechanisms could improve pyrolysis's economic feasibility. Still, pyrolysis oil integration is difficult in a microbiologically based economy. The complex and varied makeup of pyrolysis oil, as well as its detrimental effects on microbial growth and resistance to analytical techniques, provides this difficulty. Biochar is a reliable replacement for regular electrodes in the manufacturing of rather green MFCs due to its splendid functions. It is appropriate for electrodes in MFC systems because of its high porosity and huge surface area. Additionally, pretreatment plans and process engineering procedures must be implemented in order for pyrolysis oil to be used as a substitute feedstock in a biobased economy. Finally, a range of microbial products can be produced by biomass pyrolysis in conjunction with microbial conversion activities. However, various practical issues must be addressed before this approach may be used on a commercial level.

REFERENCES

Afraz, M., Muhammad, F., Nisar, J., Shah, A., Munir, S., Ali, G., Ahmad, A., 2024. Production of value added products from biomass waste by pyrolysis: An updated review. *Waste Manag. Bull.* 1, 30–40. https://doi.org/10.1016/j.wmb.2023.08.004

Ahmad, M.S., Mehmood, M.A., Liu, C.G., Tawab, A., Bai, F.W., Sakdaronnarong, C., Xu, J., Rahimuddin, S.A., Gull, M., 2018. Bioenergy potential of Wolffia arrhiza appraised through pyrolysis, kinetics, thermodynamics parameters and TG-FTIR-MS study of the evolved gases. *Bioresour. Technol.* 253, 297–303. https://doi.org/10.1016/j.biortech.2018.01.033

Ahmad, S.F.K., Md Ali, U.F., Isa, K.M., 2020. Compilation of liquefaction and pyrolysis method used for bio-oil production from various biomass: A review. *Environ. Eng. Res.* 25, 18–28. https://doi.org/10.4491/eer.2018.419

Allam, F., Elnouby, M., El-Khatib, K.M., El-Badan, D.E., Sabry, S.A., 2020. Water hyacinth (Eichhornia crassipes) biochar as an alternative cathode electrocatalyst in an air-cathode single chamber microbial fuel cell. *Int. J. Hydrogen Energy* 45, 5911–5927. https://doi.org/10.1016/j.ijhydene.2019.09.164

Aysu, T., Durak, H., 2015. Thermochemical conversion of Datura stramonium L. by supercritical liquefaction and pyrolysis processes. *J. Supercrit. Fluids* 102, 98–114. https://doi.org/10.1016/j.supflu.2015.04.008

Bilal, M., Nawaz, Z.M., Iqbal, M.N.H., Hou, J., Mahboob, S., Al-Ghanim, A.K., Cheng, H., 2018. Engineering Ligninolytic Consortium for Bioconversion of Lignocelluloses to Ethanol and Chemicals. *Protein Pept. Lett.* http://doi.org/10.2174/0929866525666180122105835

Canilha, L., Chandel, A.K., Suzane dos Santos Milessi, T., Antunes, F.A.F., Luiz da Costa Freitas, W., das Graças Almeida Felipe, M., da Silva, S.S., 2012. Bioconversion of Sugarcane Biomass into Ethanol: An Overview about Composition, Pretreatment Methods, Detoxification of Hydrolysates, Enzymatic Saccharification, and Ethanol Fermentation. *J. Biomed. Biotechnol.* 2012, 989572. https://doi.org/10.1155/2012/989572

Ceylan, S., Topcu, Y., Ceylan, Z., 2014. Thermal behaviour and kinetics of alga Polysiphonia elongata biomass during pyrolysis. *Bioresour. Technol.* 171, 193–198. https://doi.org/10.1016/j.biortech.2014.08.064

Chakraborty, I., Das, S., Dubey, B.K., Ghangrekar, M.M., 2020. Novel low cost proton exchange membrane made from sulphonated biochar for application in microbial fuel cells. *Mater. Chem. Phys.* 239, 122025. https://doi.org/10.1016/j.matchemphys.2019.122025

Chakraborty, I., Sathe, S.M., Dubey, B.K., Ghangrekar, M.M., 2020. Waste-derived biochar: Applications and future perspective in microbial fuel cells. *Bioresour. Technol.* 312, 123587. https://doi.org/10.1016/j.biortech.2020.123587

Chandel, A.K., Singh, O. V., Chandrasekhar, G., Rao, L.V., Narasu, M.L., 2010. Key drivers influencing the commercialization of ethanol-based biorefineries. *J. Commer. Biotechnol.* 16, 239–257. https://doi.org/10.1057/jcb.2010.5

Chen, Z., Zhu, Q., Wang, X., Xiao, B., Liu, S., 2015. Pyrolysis behaviors and kinetic studies on Eucalyptus residues using thermogravimetric analysis. *Energy Convers. Manag.* 105, 251–259. https://doi.org/10.1016/j.enconman.2015.07.077

Chi, Z., Rover, M., Jun, E., Deaton, M., Johnston, P., Brown, R.C., Wen, Z., Jarboe, L.R., 2013. Overliming detoxification of pyrolytic sugar syrup for direct fermentation of levoglucosan to ethanol. *Bioresour. Technol.* 150, 220–227. https://doi.org/10.1016/j.biortech.2013.09.138

Chi, Z., Zhao, X., Daugaard, T., Dalluge, D., Rover, M., Johnston, P., Salazar, A.M., Santoscoy, M.C., Smith, R., Brown, R.C., Wen, Z., Zabotina, O.A., Jarboe, L.R., 2019. Comparison of product distribution, content and fermentability of biomass in a hybrid thermochemical/biological processing platform. *Biomass and Bioenergy* 120, 107–116. https://doi.org/10.1016/j.biombioe.2018.11.006

Chiaramonti, D., Bonini, M., Fratini, E., Tondi, G., Gartner, K., Bridgwater, A. V., Grimm, H.P., Soldaini, I., Webster, A., Baglioni, P., 2003. Development of emulsions from biomass pyrolysis liquid and diesel and their use in engines—Part 1: emulsion production. *Biomass and Bioenergy* 25, 85–99. https://doi.org/10.1016/S0961-9534(02)00183-6

Chiaramonti, D., Prussi, M., Buffi, M., Tacconi, D., 2014. Sustainable bio kerosene: Process routes and industrial demonstration activities in aviation biofuels. *Appl. Energy* 136, 767–774. https://doi.org/10.1016/j.apenergy.2014.08.065

Claypool, J.T., Simmons, C.W., 2016. Hybrid thermochemical/biological processing: The economic hurdles and opportunities for biofuel production from bio-oil. *Renew. Energy* 96, 450–457. https://doi.org/10.1016/j.renene.2016.04.095

Czernik, S., Bridgwater, A. V., 2004. Overview of applications of biomass fast pyrolysis oil. *Energy and Fuels* 18, 590–598. https://doi.org/10.1021/ef034067u

Czernik, S., Johnson, D.K., Black, S., 1994. Stability of wood fast pyrolysis oil. *Biomass and Bioenergy* 7, 187–192. https://doi.org/10.1016/0961-9534(94)00058-2

Das, I., Das, S., Dixit, R., Ghangrekar, M.M., 2020. Goethite supplemented natural clay ceramic as an alternative proton exchange membrane and its application in microbial fuel cell. *Ionics (Kiel).* 26, 3061–3072. https://doi.org/10.1007/s11581-020-03472-1

Das, P., Ganesh, A., 2003. Bio-oil from pyrolysis of cashew nut shell—a near fuel. *Biomass and Bioenergy* 25, 113–117. https://doi.org/10.1016/S0961-9534(02)00182-4

Dörsam, S., Kirchhoff, J., Bigalke, M., Dahmen, N., Syldatk, C., Ochsenreither, K., 2016. Evaluation of pyrolysis oil as carbon source for fungal fermentation. *Front. Microbiol.* 7. https://doi.org/10.3389/fmicb.2016.02059

Dou, Z., Toth, J.D., 2021. Global primary data on consumer food waste: Rate and characteristics – A review. *Resour. Conserv. Recycl.* 168, 105332. https://doi.org/10.1016/j.resconrec.2020.105332

Duman, G., Okutucu, C., Ucar, S., Stahl, R., Yanik, J., 2011. The slow and fast pyrolysis of cherry seed. *Bioresour. Technol.* 102, 1869–1878. https://doi.org/10.1016/j.biortech.2010.07.051

Gu, X., Liu, C., Jiang, X., Ma, X., Li, L., Cheng, K., Li, Z., 2014. Thermal behavior and kinetics of the pyrolysis of the raw/steam exploded poplar wood sawdust. *J. Anal. Appl. Pyrolysis* 106, 177–186. https://doi.org/10.1016/j.jaap.2014.01.018

Harirchi, S., Wainaina, S., Sar, T., Nojoumi, S.A., Parchami, M., Parchami, M., Varjani, S., Khanal, S.K., Wong, J., Awasthi, M.K., Taherzadeh, M.J., 2022. Microbiological insights into anaerobic digestion for biogas, hydrogen or volatile fatty acids (VFAs): A review. *Bioengineered* 13, 6521–6557. https://doi.org/10.1080/21655979.2022.2035986

He, S., Bijl, A., Barana, P.K., Lefferts, L., Kersten, S.R.A., Brem, G., 2020. Recycling strategy for bioaqueous phase via catalytic wet air oxidation to biobased acetic acid solution. *ACS Sustain. Chem. Eng.* 8, 14694–14699. https://doi.org/10.1021/acssuschemeng.0c05946

Hou, S.-S., Huang, W.-C., Rizal, F.M., Lin, T.-H., 2016. Co-firing of fast pyrolysis bio-oil and heavy fuel oil in a 300-kWth furnace. *Appl. Sci.* https://doi.org/10.3390/app6110326

Hu, X., Gholizadeh, M., 2020. Progress of the applications of bio-oil. *Renew. Sustain. Energy Rev.* 134, 110124. https://doi.org/10.1016/J.RSER.2020.110124

Huggins, T., Wang, H., Kearns, J., Jenkins, P., Ren, Z.J., 2014. Biochar as a sustainable electrode material for electricity production in microbial fuel cells. *Bioresour. Technol.* 157, 114–119. https://doi.org/10.1016/j.biortech.2014.01.058

Isahak, W.N.R.W., Hisham, M.W.M., Yarmo, M.A., Yun Hin, T., 2012. A review on bio-oil production from biomass by using pyrolysis method. *Renew. Sustain. Energy Rev.* 16, 5910–5923. https://doi.org/10.1016/j.rser.2012.05.039

Jiang, L.-Q., Fang, Z., Zhao, Z.-L., Zheng, A.-Q., Wang, X.-B., Li, H.-B., 2019. Levoglucosan and its hydrolysates via fast pyrolysis of lignocellulose for microbial biofuels: A state-of-the-art review. *Renew. Sustain. Energy Rev.* 105, 215–229. https://doi.org/10.1016/j.rser.2019.01.055

Kim, N.-J., Lim, S.-J., Chang, H.N., 2018. Volatile fatty acid platform: Concept and application, in: *Emerging Areas in Bioengineering*, pp. 173–190. https://doi.org/10.1002/9783527803293.ch10

Kirtania, K., Bhattacharya, S., 2013. Pyrolysis kinetics and reactivity of algae–coal blends. *Biomass and Bioenergy* 55, 291–298. https://doi.org/10.1016/j.biombioe.2013.02.019

Kumar, M., Upadhyay, S.N., Mishra, P.K., 2019. A comparative study of thermochemical characteristics of lignocellulosic biomasses. *Bioresour. Technol. Reports* 8, 100186. https://doi.org/10.1016/j.biteb.2019.100186

Lee, J., Sarmah, A.K., Kwon, E.E., 2019. Chapter 1 - Production and Formation of Biochar, in: Ok, Y.S., Tsang, D.C.W., Bolan, N., Novak, J.M.B.T.-B. from B. and W. (Eds.), Elsevier, pp. 3–18. https://doi.org/10.1016/B978-0-12-811729-3.00001-7

Lehto, J., Oasmaa, A., Solantausta, Y., Kytö, M., Chiaramonti, D., 2014. Review of fuel oil quality and combustion of fast pyrolysis bio-oils from lignocellulosic biomass. *Appl. Energy* 116, 178–190. https://doi.org/10.1016/j.apenergy.2013.11.040

Li, D., Chen, L., Zhang, X., Ye, N., Xing, F., 2011. Pyrolytic characteristics and kinetic studies of three kinds of red algae. *Biomass and Bioenergy* 35, 1765–1772. https://doi.org/10.1016/j.biombioe.2011.01.011

Li, J., Qiao, Y., Zong, P., Qin, S., Wang, C., Tian, Y., 2019. Fast pyrolysis characteristics of two typical coastal zone biomass fuels by thermal gravimetric analyzer and down tube reactor. *Bioresour. Technol.* 283, 96–105. https://doi.org/10.1016/j.biortech.2019.02.097

Li, P., Fu, X., Zhang, L., Li, S., 2019. CRISPR/Cas-based screening of a gene activation library in Saccharomyces cerevisiae identifies a crucial role of OLE1 in thermotolerance. *Microb. Biotechnol.* 12, 1154–1163. https://doi.org/10.1111/1751-7915.13333

Li, X., Xia, J., Zhu, X., Bilal, M., Tan, Z., Shi, H., 2019. Construction and characterization of bifunctional cellulases: Caldicellulosiruptor-sourced endoglucanase, CBM, and exoglucanase for efficient degradation of lignocellulose. *Biochem. Eng. J.* 151, 107363. https://doi.org/10.1016/j.bej.2019.107363

Lian, J., Chen, S., Zhou, S., Wang, Z., O'Fallon, J., Li, C.-Z., Garcia-Perez, M., 2010. Separation, hydrolysis and fermentation of pyrolytic sugars to produce ethanol and lipids. *Bioresour. Technol.* 101, 9688–9699. https://doi.org/10.1016/j.biortech.2010.07.071

Lian, J., Garcia-Perez, M., Chen, S., 2013. Fermentation of levoglucosan with oleaginous yeasts for lipid production. *Bioresour. Technol.* 133, 183–189. https://doi.org/10.1016/j.biortech.2013.01.031

Lian, J., Garcia-Perez, M., Coates, R., Wu, H., Chen, S., 2012. Yeast fermentation of carboxylic acids obtained from pyrolytic aqueous phases for lipid production. *Bioresour. Technol.* 118, 177–186. https://doi.org/10.1016/j.biortech.2012.05.010

Lian, J., McKenna, R., Rover, M.R., Nielsen, D.R., Wen, Z., Jarboe, L.R., 2016. Production of biorenewable styrene: utilization of biomass-derived sugars and insights into toxicity. *J. Ind. Microbiol. Biotechnol.* 43, 595–604. https://doi.org/10.1007/s10295-016-1734-x

Lu, H., Yadav, V., Zhong, M., Bilal, M., Taherzadeh, M.J., Iqbal, H.M.N., 2022. Bioengineered microbial platforms for biomass-derived biofuel production – A review. *Chemosphere* 288, 132528. https://doi.org/10.1016/j.chemosphere.2021.132528

Luo, S., Xiao, B., Hu, Z., Liu, S., 2010. Effect of particle size on pyrolysis of single-component municipal solid waste in fixed bed reactor. *Int. J. Hydrogen Energy* 35, 93–97. https://doi.org/10.1016/j.ijhydene.2009.10.048

Luque, L., Orr, V.C.A., Chen, S., Westerhof, R., Oudenhoven, S., van Rossum, G., Kersten, S., Berruti, F., Rehmann, L., 2016. Lipid accumulation from pinewood pyrolysates by Rhodosporidium diobovatum and Chlorella vulgaris for biodiesel production. *Bioresour. Technol.* 214, 660–669. https://doi.org/10.1016/j.biortech.2016.05.030

Ma, L.-L., Liu, W.-J., Hu, X., Lam, P.K.S., Zeng, J.R., Yu, H.-Q., 2020. Ionothermal carbonization of biomass to construct sp2/sp3 carbon interface in N-doped biochar as efficient oxygen reduction electrocatalysts. *Chem. Eng. J.* 400, 125969. https://doi.org/10.1016/j.cej.2020.125969

Martiniano, S.E., Chandel, A.K., Soares, L.C.S.R., Pagnocca, F.C., da Silva, S.S., 2013. Evaluation of novel xylose-fermenting yeast strains from Brazilian forests for hemicellulosic ethanol production from sugarcane bagasse. *3 Biotech* 3, 345–352. https://doi.org/10.1007/s13205-013-0145-1

McMillan, J.D., Beckham, G.T., 2017. Thinking big: towards ideal strains and processes for large-scale aerobic biofuels production. *Microb. Biotechnol.* 10, 40–42. https://doi.org/10.1111/1751-7915.12471

Medina, J.D.C., Jr, A.I.M., 2020. Ethanol Production, Current Facts, Future Scenarios, and Techno-Economic Assessment of Different Biorefinery Configurations, in: Inambao, F. (Ed.), . IntechOpen, Rijeka, p. Ch. 2. https://doi.org/10.5772/intechopen.95081

Mishra, R.K., Mohanty, K., 2020. Kinetic analysis and pyrolysis behaviour of waste biomass towards its bioenergy potential. *Bioresour. Technol.* 311, 123480. https://doi.org/10.1016/j.biortech.2020.123480

Moita Fidalgo, R., Ortigueira, J., Freches, A., Pelica, J., Gonçalves, M., Mendes, B., Lemos, P.C., 2014. Bio-oil upgrading strategies to improve PHA production from selected aerobic mixed cultures. *N. Biotechnol.* 31, 297–307. https://doi.org/10.1016/j.nbt.2013.10.009

Moysés, D.N., Reis, V.C., Almeida, J.R., Moraes, L.M., Torres, F.A., 2016. Xylose Fermentation by Saccharomyces cerevisiae: Challenges and Prospects. *Int. J. Mol. Sci.* https://doi.org/10.3390/ijms17030207

Naik, S., Goud, V. V., Rout, P.K., Jacobson, K., Dalai, A.K., 2010. Characterization of Canadian biomass for alternative renewable biofuel. *Renew. Energy* 35, 1624–1631. https://doi.org/10.1016/j.renene.2009.08.033

Nair, V., Vinu, R., 2016. Peroxide-assisted microwave activation of pyrolysis char for adsorption of dyes from wastewater. *Bioresour. Technol.* 216, 511–519. https://doi.org/10.1016/j.biortech.2016.05.070

Nawaz, A., Singh, B., Mishra, R.K., Kumar, P., 2023. Pyrolysis of low-value waste Trapa natans peels: An exploration of thermal decomposition characteristics, kinetic behaviour, and pyrolytic liquid product. *Sustain. Energy Technol. Assessments* 56, 103128. https://doi.org/10.1016/j.seta.2023.103128

Neethu, B., Bhowmick, G.D., Ghangrekar, M.M., 2019. A novel proton exchange membrane developed from clay and activated carbon derived from coconut shell for application in microbial fuel cell. *Biochem. Eng. J.* 148, 170–177. https://doi.org/10.1016/j.bej.2019.05.011

Neumann, A., Dörsam, S., Oswald, F., Ochsenreither, K., 2015. Microbial Production of Value-Added Chemicals from Pyrolysis Oil and Syngas BT - Sustainable Production of Bulk Chemicals: Integration of Bio-Chemo- Resources and Processes, in: Xian, M. (Ed.), Springer Netherlands, Dordrecht, pp. 69–105. https://doi.org/10.1007/978-94-017-7475-8_4

Oasmaa, A., Peacocke, C., 2001. A guide to physical property characterisation of biomass-derived fast pyrolysis liquids. VTT Publ. 2–65.

Özsin, G., Pütün, A.E., 2017. Kinetics and evolved gas analysis for pyrolysis of food processing wastes using TGA/MS/FT-IR. *Waste Manag*. 64, 315–326. https://doi.org/10.1016/j.wasman.2017.03.020

Park, Y.-K., Yoo, M.L., Heo, H.S., Lee, H.W., Park, S.H., Jung, S.-C., Park, S.-S., Seo, S.-G., 2012. Wild reed of Suncheon Bay: Potential bio-energy source. *Renew. Energy* 42, 168–172. https://doi.org/10.1016/j.renene.2011.08.025

Phukan, M.M., Chutia, R.S., Konwar, B.K., Kataki, R., 2011. Microalgae Chlorella as a potential bio-energy feedstock. *Appl. Energy* 88, 3307–3312. https://doi.org/10.1016/j.apenergy.2010.11.026

Pinheiro Pires, A.P., Arauzo, J., Fonts, I., Domine, M.E., Fernández Arroyo, A., Garcia-Perez, M.E., Montoya, J., Chejne, F., Pfromm, P., Garcia-Perez, M., 2019. Challenges and opportunities for bio-oil refining: A Review. *Energy & Fuels* 33, 4683–4720. https://doi.org/10.1021/acs.energyfuels.9b00039

Piskorz, J., Majerski, P., Radlein, D., Vladars-Usas, A., Scott, D.S., 2000. Flash pyrolysis of cellulose for production of anhydro-oligomers. *J. Anal. Appl. Pyrolysis* 56, 145–166. https://doi.org/10.1016/S0165-2370(00)00089-9

Prosen, E.M., Radlein, D., Piskorz, J., Scott, D.S., Legge, R.L., 1993. Microbial utilization of levoglucosan in wood pyrolysate as a carbon and energy source. *Biotechnol. Bioeng*. 42, 538–541. https://doi.org/10.1002/bit.260420419

Qian, K., Kumar, A., Zhang, H., Bellmer, D., Huhnke, R., 2015. Recent advances in utilization of biochar. *Renew. Sustain. Energy Rev*. 42, 1055–1064. https://doi.org/10.1016/j.rser.2014.10.074

Quan, C., Gao, N., Song, Q., 2016. Pyrolysis of biomass components in a TGA and a fixed-bed reactor: Thermochemical behaviors, kinetics, and product characterization. *J. Anal. Appl. Pyrolysis* 121, 84–92. https://doi.org/10.1016/j.jaap.2016.07.005

Ravi, R.K., Neeraj, A., Yadav, R.H., 2021. 12 - Assessment of microbial biomass for production of ecofriendly single-cell protein, bioenergy, and other useful products, in: Singh, J.S., Tiwari, S., Singh, C., Singh, A.K.B.T.-M. in L.U.C.M. (Eds.), Elsevier, pp. 267–284. https://doi.org/10.1016/B978-0-12-824448-7.00015-2

Ringer, M., Putsche, V., Scahill, J., 2006. Large-Scale Pyrolysis Oil Production and Economic Analysis. Tech. Rep. NREL/TP-510–37779 1–93.

Senatham, S., Chamduang, T., Kaewchingduang, Y., Thammasittirong, A., Srisodsuk, M., Elliston, A., Roberts, I.N., Waldron, K.W., Thammasittirong, S.N.-R., 2016. Enhanced xylose fermentation and hydrolysate inhibitor tolerance of Scheffersomyces shehatae for efficient ethanol production from non-detoxified lignocellulosic hydrolysate. Springerplus 5, 1040. https://doi.org/10.1186/s40064-016-2713-4

Shafizadeh, F., Stevenson, T.T., 1982. Saccharification of douglas-fir wood by a combination of prehydrolysis and pyrolysis. *J. Appl. Polym. Sci*. 27, 4577–4585. https://doi.org/10.1002/app.1982.070271205

Shahbeik, H., Shafizadeh, A., Gupta, V.K., Lam, S.S., Rastegari, H., Peng, W., Pan, J., Tabatabaei, M., Aghbashlo, M., 2023. Using nanocatalysts to upgrade pyrolysis bio-oil: A critical review. *J. Clean. Prod*. 413, 137473. https://doi.org/10.1016/j.jclepro.2023.137473

Shemfe, M.B., Gu, S., Ranganathan, P., 2015. Techno-economic performance analysis of biofuel production and miniature electric power generation from biomass fast pyrolysis and bio-oil upgrading. *Fuel* 143, 361–372. https://doi.org/10.1016/j.fuel.2014.11.078

Singh, S., Pant, K.K., Krishania, M., 2023. Current perspective for bio-oil production from agricultural residues in commercialization aspect: A review. *J. Anal. Appl. Pyrolysis* 175, 106160. https://doi.org/10.1016/j.jaap.2023.106160

So, K.S., Brown, R.C., 1999. Economic analysis of selected lignocellulose-to-ethanol conversion technologies. *Appl. Biochem. Biotechnol.* 79, 633–640. https://doi.org/10.1385/ABAB:79:1-3:633

Solantausta, Y., Nylund, N.-O., Westerholm, M., Koljonen, T., Oasmaa, A., 1993. Wood-pyrolysis oil as fuel in a diesel-power plant. *Bioresour. Technol.* 46, 177–188. https://doi.org/10.1016/0960-8524(93)90071-I

Stegen, S., Kaparaju, P., 2020. Effect of temperature on oil quality obtained through pyrolysis of sugarcane bagasse. *Fuel* 276, 118112. https://doi.org/10.1016/j.fuel.2020.118112

Tahir, M.H., Çakman, G., Goldfarb, J.L., Topcu, Y., Naqvi, S.R., Ceylan, S., 2019. Demonstrating the suitability of canola residue biomass to biofuel conversion via pyrolysis through reaction kinetics, thermodynamics and evolved gas analyses. *Bioresour. Technol.* 279, 67–73. https://doi.org/10.1016/j.biortech.2019.01.106

Wang, C., Thygesen, A., Liu, Y., Li, Q., Yang, M., Dang, D., Wang, Z., Wan, Y., Lin, W., Xing, J., 2013. Bio-oil based biorefinery strategy for the production of succinic acid. *Biotechnol. Biofuels* 6, 74. https://doi.org/10.1186/1754-6834-6-74

Wang, H., Livingston, D., Srinivasan, R., Li, Q., Steele, P., Yu, F., 2012. Detoxification and Fermentation of Pyrolytic Sugar for Ethanol Production. *Appl. Biochem. Biotechnol.* 168, 1568–1583. https://doi.org/10.1007/s12010-012-9879-1

Welker, C.M., Balasubramanian, V.K., Petti, C., Rai, K.M., DeBolt, S., Mendu, V., 2015. Engineering plant biomass lignin content and composition for biofuels and bioproducts. *Energies*. https://doi.org/10.3390/en8087654

Wright, M.M., Daugaard, D.E., Satrio, J.A., Brown, R.C., 2010. Techno-economic analysis of biomass fast pyrolysis to transportation fuels. *Fuel* 89, S2–S10. https://doi.org/10.1016/j.fuel.2010.07.029

Xiao, R., Yang, W., 2013. Influence of temperature on organic structure of biomass pyrolysis products. *Renew. Energy* 50, 136–141. https://doi.org/10.1016/j.renene.2012.06.028

Xu, J., Liu, N., Qiao, K., Vogg, S., Stephanopoulos, G., 2017. Application of metabolic controls for the maximization of lipid production in semicontinuous fermentation. *Proc. Natl. Acad. Sci.* 114, E5308–E5316. https://doi.org/10.1073/pnas.1703321114

Xu, P., 2018. Production of chemicals using dynamic control of metabolic fluxes. *Curr. Opin. Biotechnol.* 53, 12–19. https://doi.org/10.1016/j.copbio.2017.10.009

Xu, P., Li, L., Zhang, F., Stephanopoulos, G., Koffas, M., 2014. Improving fatty acids production by engineering dynamic pathway regulation and metabolic control. *Proc. Natl. Acad. Sci.* 111, 11299–11304. https://doi.org/10.1073/pnas.1406401111

Yaqoob, A.A., Ibrahim, M.N.M., Guerrero-Barajas, C., 2021. Modern trend of anodes in microbial fuel cells (MFCs): An overview. *Environ. Technol. Innov.* 23, 101579. https://doi.org/10.1016/j.eti.2021.101579

You, K., Zhou, Z., Gao, C., Yang, Q., 2021. One-step preparation of biochar electrodes and their applications in sediment microbial electrochemical systems. *Catalysts*. https://doi.org/10.3390/catal11040508

Yu, Z., Zhang, H., 2003. Ethanol fermentation of acid-hydrolyzed cellulosic pyrolysate with Saccharomyces cerevisiae. *Bioresour. Technol.* 90, 95–100. https://doi.org/10.1016/S0960-8524(03)00093-2

Zhang, F., Cheng, S., Pant, D., Bogaert, G. Van, Logan, B.E., 2009. Power generation using an activated carbon and metal mesh cathode in a microbial fuel cell. *Electrochem. Commun.* 11, 2177–2179. https://doi.org/10.1016/j.elecom.2009.09.024

Zhou, M., Bodenmuller, N., Hedlund, J., 2023. Enhanced bio-oil upgrading by sub-microscale dispersed silanol-free ZSM-5 nanosheets and evidence for revealing an unconventional mechanism. *Chem. Eng. J.* 478, 147457. https://doi.org/10.1016/j.cej.2023.147457

Zhuang, X.L., Zhang, H.X., Yang, J.Z., Qi, H.Y., 2001. Preparation of levoglucosan by pyrolysis of cellulose and its citric acid fermentation. *Bioresour. Technol.* 79, 63–66. https://doi.org/10.1016/S0960-8524(01)00023-2

Index

Pages in *italics* refer to figures and pages in **bold** refer to tables.

U

V

W

X

Y

Z